Organic Chemistry

Organic Chemistry

Melody Frost

Larsen & Keller
www.larsen-keller.com

Organic Chemistry
Melody Frost
ISBN: 978-1-64172-150-9 (Hardback)

Larsen & Keller
Published by Larsen and Keller Education,
5 Penn Plaza,
19th Floor,
New York, NY 10001, USA

Cataloging-in-Publication Data

Organic chemistry / Melody Frost.
p. cm.
Includes bibliographical references and index.
ISBN 978-1-64172-150-9
1. Chemistry, Organic. 2. Chemistry. I. Frost, Melody.
QD251.3 .O74 2019
547--dc23

Printed and bound in China.

For more information regarding Larsen and Keller Education and its products, please visit the publisher's website www.larsen-keller.com

Table of Contents

Preface

Compounds whose molecules contain carbon and hydrogen are organic compounds. They can be classified into functional groups, aliphatic compounds, aromatic compounds, heterocyclic compounds, polymers, etc. These compounds are the basis of all life on earth and constitute the majority of known chemicals, such as agrichemicals, petrochemicals and pharmaceuticals. Many materials such as solvents, plastics, lubricants, fuels and explosives are made from organic compounds. The science concerned with the study of the properties, structures and reactions of organic compounds is called organic chemistry. It studies the properties such as qualitative and quantitative characteristics, including melting and boiling points, conductivity, solubility, refractive index, odor, color, etc. Various techniques such as gas chromatography and HPLC as well as separation methods of crystallization, distillation and solvent extraction are commonly used in organic chemistry for purity analyses and separation. This textbook is a compilation of chapters that discuss the most vital aspects in the field of organic chemistry. It presents this complex subject in the most comprehensible and easy to understand language. As this field is emerging at a rapid pace, the contents of this book will help the readers understand the modern concepts and applications of the subject.

A foreword of all Chapters of the book is provided below:

Chapter 1, The study of the structure, properties and reactions of organic compounds and organic materials is under the domain of organic chemistry. This is an introductory chapter, which will introduce briefly all the significant aspects of organic chemistry, such as organic compounds, organic matter, etc.; **Chapter 2**, Any chemical compound, which contains carbon is called an organic compound. The aim of this chapter is to explore the fundamental aspects of organic compounds. Varied topics, such as carbine, methylcyclohexane, organic peroxide, organic superconductor, alicyclic compound, volatile organic compound, etc. which are crucial for a complete understanding of the subject, have been elaborated in this chapter; **Chapter 3**, Compounds that contain one or more chemical bonds between a metal, such as alkaline, transition and alkaline earth metals, and a carbon atom of an organic molecule are called organometallic compounds. The topics elaborated in this chapter on s- and p-Block organometallic compounds, organoboron compounds, metal hydrides, organoaluminium compounds, etc. will help in providing a better perspective about organometallic compounds; **Chapter 4**, Chemical reactions involving organic compounds are called organic reactions. All the diverse types of organic reactions such as addition reactions, elimination reactions, redox reactions and substitution reactions have been carefully analyzed in this chapter; **Chapter 5**, The synthesis of organic compounds is under the domain of organic synthesis. In order to completely understand organic synthesis, it is necessary to understand the processes related to it, such as Achmatowicz reaction, carbohydrate synthesis, Wöhler synthesis and retrosynthesis. The following chapter elucidates these varied processes and mechanisms associated with this area of study; **Chapter 6**, A complete understanding of organic chemistry requires a comprehensive study of aromaticity, covalent bonding, functional groups, heterocyclic compounds, the IUPAC nomenclature of organic chemistry, etc. which have been elaborately discussed in this chapter; **Chapter 7**, Organic chemistry is a vast subject that branches out into a number of significant sub-disciplines such as organometallic chemistry, biomolecular chemistry, heterocyclic chemistry, organosulfur chemistry and organophosphorus chemistry, which have been thoroughly discussed in this chapter; **Chapter 8**, An acid is an ion or a molecule, which donates a hydrogen ion or a proton or alternatively forms a covalent bond with a pair of electrons. Bases are substances that release hydroxide ions in aqueous solution. This chapter discusses in extensive detail the fundamental concepts of acids and bases and includes vital topics such as Lewis acids and bases and the Brønsted–Lowry acid–base theory.

I would like to thank the entire editorial team who made sincere efforts for this book and my family who supported me in my efforts of working on this book. I take this opportunity to thank all those who have been a guiding force throughout my life.

Melody Frost

1

Introduction to Organic Chemistry

The study of the structure, properties and reactions of organic compounds and organic materials is under the domain of organic chemistry. This is an introductory chapter, which will introduce briefly all the significant aspects of organic chemistry, such as organic compounds, organic matter, etc.

Chemistry

Chemistry is frequently defined as the study of matter and the reactions that matter undergoes. Chemistry is part of everything in our lives. Every material in existence is made up of matter even our own bodies. Chemistry is involved in everything we do, from growing and cooking food to cleaning our homes and bodies to launching a space shuttle. Chemistry is one of the physical sciences that help us to describe and explain our world.

Dynamics

Refers to the details of that rearrangements of atoms that occur during chemical change, and that strongly affect the rate at which change occurs.

Energetics

Refers to the thermodynamics of chemical change, relating to the uptake or release of heat. More importantly, this aspect of chemistry controls the direction in which change occurs, and the mixture of substances that results.

Composition and Structure

Define the substances that are results of chemical change. Structure refers specifically to the relative arrangements of the atoms in space. The extent to which a given structure can persist is itself determined by energetics and dynamics.

Synthesis

Strictly speaking, refers to formation of new (and usually more complex) substances from simpler ones, but in the present context we use it in the more general sense to denote the operations required to bring about chemical change and to isolate the desired products.

Five Branches

There are five main branches of chemistry, each of which has many areas of study:

1. Analytical chemistry uses qualitative and quantitative observation to identify and measure the physical and chemical properties of substances. In a sense, all chemistry is analytical.
2. Physical chemistry combines chemistry with physics. Physical chemists study how matter and energy interact. Thermodynamics and quantum mechanics are two of the important branches of physical chemistry.
3. Organic chemistry specifically studies compounds that contain the element carbon. Carbon has many unique properties that allow it to form complex chemical bonds and very large molecules. Organic chemistry is known as the "Chemistry of Life" because all of the molecules that make up living tissue have carbon as part of their makeup.
4. Inorganic chemistry studies materials such as metals and gases that do not have carbon as part of their makeup.
5. Biochemistry is the study of chemical processes that occur within living organisms.

Environmental Chemistry

Environmental chemists study how chemicals interact with the natural environment. Environmental chemistry is an interdisciplinary study that involves both analytical chemistry and an understanding of environmental science. Environmental chemists must first understand the chemicals and chemical reactions present in natural processes in the soil water and air. Sampling and analysis can then determine if human activities have contaminated the environment or caused harmful reactions to affect it.

Water quality is an important area of environmental chemistry. "Pure" water does not exist in nature; it always has some minerals or other substance dissolved in it. Water quality chemists test rivers, lakes and ocean water for characteristics such as dissolved oxygen, salinity, turbidity, suspended sediments, and pH. Water destined for human consumption must be free of harmful contaminants and may be treated with additives like fluoride and chlorine to increase its safety.

Agricultural Chemistry

Agricultural chemistry is concerned with the substances and chemical reactions that are involved with the production, protection and use of crops and livestock. It is a highly interdisciplinary field that relies on ties to many other sciences.

Agricultural chemists develop fertilizers, insecticides and herbicides necessary for large-scale crop production. They must also monitor how these products are used and their impacts on the environment. Nutritional supplements are developed to increase the productivity of meat and dairy herds.

Agricultural biotechnology is a fast-growing focus for many agricultural chemists. Genetically manipulating crops to be resistant to the herbicides used to control weeds in the fields requires detailed understanding of both the plants and the chemicals at the molecular level. Biochemists must understand genetics, chemistry and business needs to develop crops that are easier to transport or that have a longer shelf life.

Geochemistry

Geochemists combine chemistry and geology to study the makeup and interaction between substances found in the Earth.

Forensic Chemistry

Forensic chemists capture and analyze the physical evidence left behind at a crime scene to help determine the identities of the people involved as well as to answer other vital questions regarding how and why the crime was carried out. Forensic chemists use a wide variety of analyzation methods, such as chromatography, spectrometry and spectroscopy.

Modern Principles

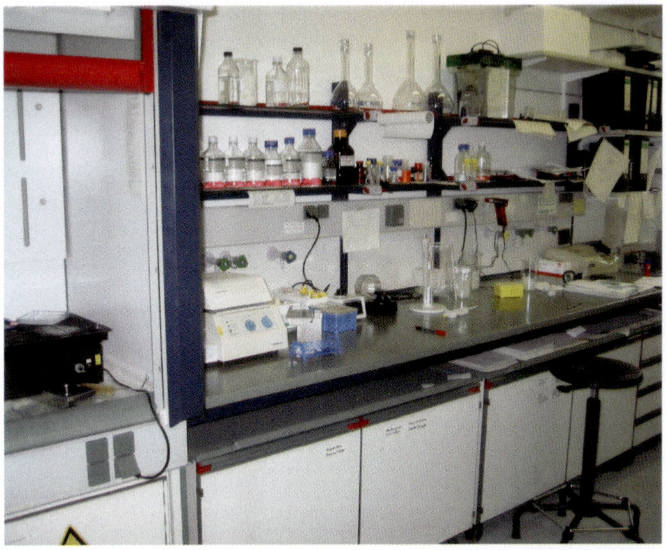

Laboratory, Institute of Biochemistry, University of Cologne in Germany.

The current model of atomic structure is the quantum mechanical model. Traditional chemistry starts with the study of elementary particles, atoms, molecules, substances, metals, crystals and other aggregates of matter. This matter can be studied in solid, liquid, or gas states, in isolation or in combination. The interactions, reactions and transformations that are studied in chemistry are usually the result of interactions between atoms, leading to rearrangements of the chemical bonds which hold atoms together. Such behaviors are studied in a chemistry laboratory.

The chemistry laboratory stereotypically uses various forms of laboratory glassware. However glassware is not central to chemistry, and a great deal of experimental (as well as applied/industrial) chemistry is done without it.

Solutions of substances in reagent bottles, including ammonium hydroxide and nitric acid, illuminated in different colors

A chemical reaction is a transformation of some substances into one or more different substances. The basis of such a chemical transformation is the rearrangement of electrons in the chemical bonds between atoms. It can be symbolically depicted through a chemical equation, which usually involves atoms as subjects. The number of atoms on the left and the right in the equation for a chemical transformation is equal. (When the number of atoms on either side is unequal, the transformation is referred to as a nuclear reaction or radioactive decay.) The type of chemical reactions a substance may undergo and the energy changes that may accompany it are constrained by certain basic rules, known as chemical laws.

Energy and entropy considerations are invariably important in almost all chemical studies. Chemical substances are classified in terms of their structure, phase, as well as their chemical compositions. They can be analyzed using the tools of chemical analysis, e.g. spectroscopy and chromatography. Scientists engaged in chemical research are known as chemists. Most chemists specialize in one or more sub-disciplines. Several concepts are essential for the study of chemistry; some of them are:

Matter

In chemistry, matter is defined as anything that has rest mass and volume (it takes up space) and is made up of particles. The particles that make up matter have rest mass as well – not all particles have rest mass, such as the photon. Matter can be a pure chemical substance or a mixture of substances.

Atom

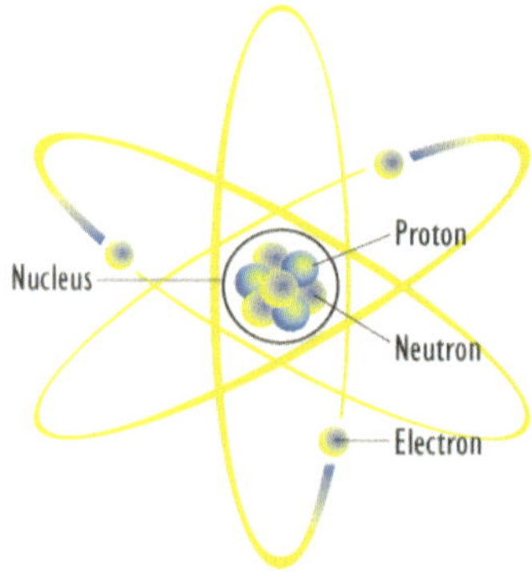

A diagram of an atom based on the Bohr model

The atom is the basic unit of chemistry. It consists of a dense core called the atomic nucleus surrounded by a space occupied by an electron cloud. The nucleus is made up of positively charged protons and uncharged neutrons (together called nucleons), while the electron cloud consists of negatively charged electrons which orbit the nucleus. In a neutral atom, the negatively charged electrons balance out the positive charge of the protons. The nucleus is dense; the mass of a nucleon is appromixately 1,836 times that of an electron, yet the radius of an atom is about 10,000 times that of its nucleus.

The atom is also the smallest entity that can be envisaged to retain the chemical properties of the element, such as electronegativity, ionization potential, preferred oxidation state(s), coordination number, and preferred types of bonds to form (e.g., metallic, ionic, covalent).

Element

A chemical element is a pure substance which is composed of a single type of atom, characterized by its particular number of protons in the nuclei of its atoms, known as the atomic number and represented by the symbol Z. The mass number is the sum of the number of protons and neutrons in a nucleus. Although all the nuclei of all atoms belonging to one element will have the same atomic number, they may not necessarily have the same mass number; atoms of an element which have different mass numbers are known as isotopes. For example, all atoms with 6 protons in their nuclei are atoms of the chemical element carbon, but atoms of carbon may have mass numbers of 12 or 13.

Group→ ↓Period	1	2		3	4	5	6	7	8	9	10	11	12	13	14	15	16	17	18
1	1 H																		2 He
2	3 Li	4 Be												5 B	6 C	7 N	8 O	9 F	10 Ne
3	11 Na	12 Mg												13 Al	14 Si	15 P	16 S	17 Cl	18 Ar
4	19 K	20 Ca		21 Sc	22 Ti	23 V	24 Cr	25 Mn	26 Fe	27 Co	28 Ni	29 Cu	30 Zn	31 Ga	32 Ge	33 As	34 Se	35 Br	36 Kr
5	37 Rb	38 Sr		39 Y	40 Zr	41 Nb	42 Mo	43 Tc	44 Ru	45 Rh	46 Pd	47 Ag	48 Cd	49 In	50 Sn	51 Sb	52 Te	53 I	54 Xe
6	55 Cs	56 Ba	*	71 Lu	72 Hf	73 Ta	74 W	75 Re	76 Os	77 Ir	78 Pt	79 Au	80 Hg	81 Tl	82 Pb	83 Bi	84 Po	85 At	86 Rn
7	87 Fr	88 Ra	**	103 Lr	104 Rf	105 Db	106 Sg	107 Bh	108 Hs	109 Mt	110 Ds	111 Rg	112 Cn	113 Nh	114 Fl	115 Mc	116 Lv	117 Ts	118 Og

*	57 La	58 Ce	59 Pr	60 Nd	61 Pm	62 Sm	63 Eu	64 Gd	65 Tb	66 Dy	67 Ho	68 Er	69 Tm	70 Yb
**	89 Ac	90 Th	91 Pa	92 U	93 Np	94 Pu	95 Am	96 Cm	97 Bk	98 Cf	99 Es	100 Fm	101 Md	102 No

Standard form of the periodic table of chemical elements.
The colors represent different categories of elements.

The standard presentation of the chemical elements is in the periodic table, which orders elements by atomic number. The periodic table is arranged in groups, or columns, and periods, or rows. The periodic table is useful in identifying periodic trends.

Compound

Carbon dioxide (CO_2), an example of a chemical compound.

A compound is a pure chemical substance composed of more than one element. The properties of a compound bear little similarity to those of its elements. The standard nomenclature of compounds is set by the International Union of Pure and Applied Chemistry (IUPAC). Organic compounds are named according to the organic nomenclature system. The names for Inorganic compounds are created according to the inorganic nomenclature system. When a compound has more than one component, then they are divided into two classes, the electropositive and the electronegative components. In addition the Chemical Abstracts Service has devised a method to index chemical substances. In this scheme each chemical substance is identifiable by a number known as its CAS registry number.

Molecule

A ball-and-stick representation of the caffeine molecule ($C_8H_{10}N_4O_2$).

A molecule is the smallest indivisible portion of a pure chemical substance that has its unique set of chemical properties, that is, its potential to undergo a certain set of chemical reactions with other substances. However, this definition only works well for substances that are composed of molecules, which is not true of many substances. Molecules are typically a set of atoms bound together by covalent bonds, such that the structure is electrically neutral and all valence electrons are paired with other electrons either in bonds or in lone pairs.

Thus, molecules exist as electrically neutral units, unlike ions. When this rule is broken, giving the "molecule" a charge, the result is sometimes named a molecular ion or a polyatomic ion. However, the discrete and separate nature of the molecular concept usually requires that molecular ions be present only in well-separated form, such as a directed beam in a vacuum in a mass spectrometer. Charged polyatomic collections residing in solids (for example, common sulfate or nitrate ions) are generally not considered "molecules" in chemistry. Some molecules contain one or more unpaired electrons, creating radicals. Most radicals are comparatively reactive, but some, such as nitric oxide (NO) can be stable.

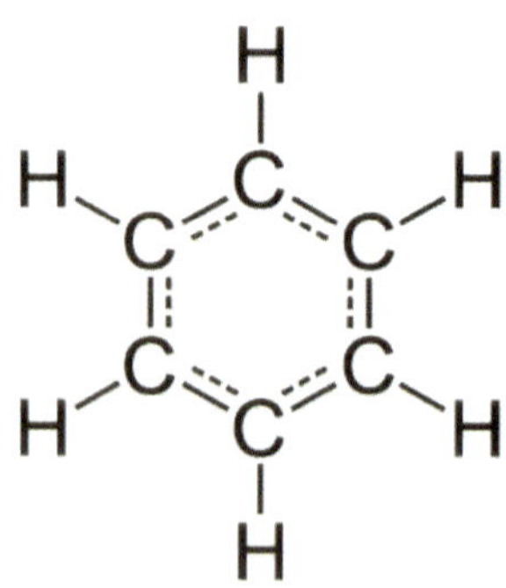

A 2-D skeletal model of a benzene molecule (C_6H_6)

The "inert" or noble gas elements (helium, neon, argon, krypton, xenon and radon) are composed of lone atoms as their smallest discrete unit, but the other isolated chemical elements consist of either molecules or networks of atoms bonded to each other in some way. Identifiable molecules compose familiar substances such as water, air, and many organic compounds like alcohol, sugar, gasoline, and the various pharmaceuticals.

However, not all substances or chemical compounds consist of discrete molecules, and indeed most of the solid substances that make up the solid crust, mantle, and core of the Earth are chemical compounds without molecules. These other types of substances, such as ionic compounds and network solids, are organized in such a way as to lack the existence of identifiable molecules per se. Instead, these substances are discussed in terms of formula units or unit cells as the smallest repeating structure within the substance. Examples of such substances are mineral salts (such as table salt), solids like carbon and diamond, metals, and familiar silica and silicate minerals such as quartz and granite.

One of the main characteristics of a molecule is its geometry often called its structure. While the structure of diatomic, triatomic or tetra atomic molecules may be trivial, (linear, angular pyramidal etc.) the structure of polyatomic molecules, that are constituted of more than six atoms (of several elements) can be crucial for its chemical nature.

Substance and Mixture

Examples of pure chemical substances. From left to right: the elements tin (Sn) and sulfur (S), diamond (an allotrope of carbon), sucrose (pure sugar), and sodium chloride (salt) and sodium bicarbonate (baking soda), which are both ionic compounds.

A chemical substance is a kind of matter with a definite composition and set of properties. A collection of substances is called a mixture. Examples of mixtures are air and alloys.

Mole and Amount of Substance

The mole is a unit of measurement that denotes an amount of substance (also called chemical amount). The mole is defined as the number of atoms found in exactly 0.012 kilogram (or 12 grams) of carbon-12, where the carbon-12 atoms are unbound, at rest and in their ground state. The number of entities per mole is known as the Avogadro constant, and is determined empirically to be approximately 6.022×1023 mol^{-1}. Molar concentration is the amount of a particular substance per volume of solution, and is commonly reported in mol/dm^3.

Phase

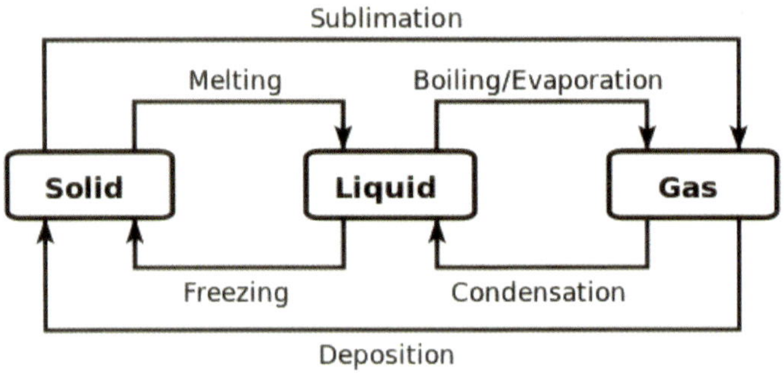

Diagram showing relationships among the phases and the terms used to describe phase changes.

In addition to the specific chemical properties that distinguish different chemical classifications, chemicals can exist in several phases. For the most part, the chemical classifications are independent of these bulk phase classifications; however, some more exotic phases are incompatible with certain chemical properties. A phase is a set of states of a chemical system that have similar bulk structural properties, over a range of conditions, such as pressure or temperature.

Physical properties, such as density and refractive index tend to fall within values characteristic of the phase. The phase of matter is defined by the phase transition, which is when energy put into or taken out of the system goes into rearranging the structure of the system, instead of changing the bulk conditions.

Sometimes the distinction between phases can be continuous instead of having a discrete boundary, in this case the matter is considered to be in a supercritical state. When three states meet based on the conditions, it is known as a triple point and since this is invariant, it is a convenient way to define a set of conditions.

The most familiar examples of phases are solids, liquids, and gases. Many substances exhibit multiple solid phases. For example, there are three phases of solid iron (alpha, gamma, and delta) that vary based on temperature and pressure. A principal difference between solid phases is the crystal structure, or arrangement, of the atoms. Another phase commonly encountered in the study of chemistry is the aqueous phase, which is the state of substances dissolved in aqueous solution (that is, in water).

Less familiar phases include plasmas, Bose–Einstein condensates and fermionic condensates and the paramagnetic and ferromagnetic phases of magnetic materials. While most familiar phases deal with three-dimensional systems, it is also possible to define analogs in two-dimensional systems, which has received attention for its relevance to systems in biology.

Bonding

Atoms sticking together in molecules or crystals are said to be bonded with one another. A chemical bond may be visualized as the multipole balance between the positive charges in the nuclei and the negative charges oscillating about them. More than simple attraction and repulsion, the energies and distributions characterize the availability of an electron to bond to another atom.

A chemical bond can be a covalent bond, an ionic bond, a hydrogen bond or just because of Van der Waals force. Each of these kinds of bonds is ascribed to some potential. These potentials create the interactions which hold atoms together in molecules or crystals. In many simple compounds, valence bond theory, the Valence Shell Electron Pair Repulsion model (VSEPR), and the concept of oxidation number can be used to explain molecular structure and composition.

An ionic bond is formed when a metal loses one or more of its electrons, becoming a positively charged cation, and the electrons are then gained by the non-metal atom, becoming a negatively charged anion. The two oppositely charged ions attract one another, and the ionic bond is the electrostatic force of attraction between them. For example, sodium (Na), a metal, loses one electron to become an Na^+ cation while chlorine (Cl), a non-metal, gains this electron to become Cl^-. The ions are held together due to electrostatic attraction, and that compound sodium chloride (NaCl), or common table salt, is formed.

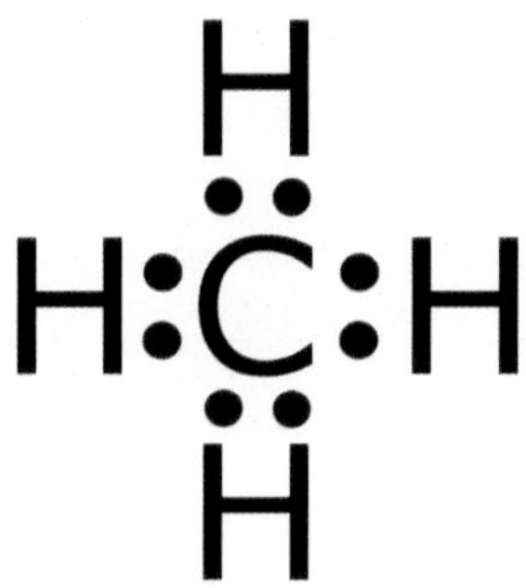

In the methane molecule (CH_4), the carbon atom shares a pair of valence electrons with each of the four hydrogen atoms. Thus, the octet rule is satisfied for C-atom (it has eight electrons in its valence shell) and the duet rule is satisfied for the H-atoms (they have two electrons in their valence shells).

In a covalent bond, one or more pairs of valence electrons are shared by two atoms: the resulting electrically neutral group of bonded atoms is termed a molecule. Atoms will share valence electrons in such a way as to create a noble gas electron configuration (eight electrons in their outermost shell) for each atom. Atoms that tend to combine in such a way that they each have eight electrons in their valence shell are said to follow the octet rule. However, some elements like hydrogen and lithium need only two electrons in their outermost shell to attain this stable configuration; these atoms are said to follow the duet rule, and in this way they are reaching the electron configuration of the noble gas helium, which has two electrons in its outer shell.

Similarly, theories from classical physics can be used to predict many ionic structures. With more complicated compounds, such as metal complexes, valence bond theory is less applicable and alternative approaches, such as the molecular orbital theory, are generally used.

Energy

In the context of chemistry, energy is an attribute of a substance as a consequence of its atomic, molecular or aggregate structure. Since a chemical transformation is accompanied by a change in one or more of these kinds of structures, it is invariably accompanied by an increase or decrease of energy of the substances involved. Some energy is transferred between the surroundings and the reactants of the reaction in the form of heat or light; thus the products of a reaction may have more or less energy than the reactants.

A reaction is said to be exergonic if the final state is lower on the energy scale than the initial state; in the case of endergonic reactions the situation is the reverse. A reaction is said to be exothermic if the reaction releases heat to the surroundings; in the case of endothermic reactions, the reaction absorbs heat from the surroundings.

Chemical reactions are invariably not possible unless the reactants surmount an energy barrier known as the activation energy. The speed of a chemical reaction (at given temperature T) is related to the activation energy E, by the Boltzmann's population factor $e^{-E/kT}$ – that is the probability of a molecule to have energy greater than or equal to E at the given temperature T. This exponential dependence of a reaction rate on temperature is known as the Arrhenius equation. The activation energy necessary for a chemical reaction to occur can be in the form of heat, light, electricity or mechanical force in the form of ultrasound.

A related concept free energy, which also incorporates entropy considerations, is a very useful means for predicting the feasibility of a reaction and determining the state of equilibrium of a chemical reaction, in chemical thermodynamics. A reaction is feasible only if the total change in the Gibbs free energy is negative, $\Delta G \leq 0$; if it is equal to zero the chemical reaction is said to be at equilibrium.

There exist only limited possible states of energy for electrons, atoms and molecules. These are determined by the rules of quantum mechanics, which require quantization of energy of a bound system. The atoms/molecules in a higher energy state are said to be excited. The molecules/atoms of substance in an excited energy state are often much more reactive; that is, more amenable to chemical reactions.

The phase of a substance is invariably determined by its energy and the energy of its surroundings. When the intermolecular forces of a substance are such that the energy of the surroundings is not sufficient to overcome them, it occurs in a more ordered phase like liquid or solid as is the case with water (H_2O); a liquid at room temperature because its molecules are bound by hydrogen bonds. Whereas hydrogen sulfide (H_2S) is a gas at room temperature and standard pressure, as its molecules are bound by weaker dipole-dipole interactions.

The transfer of energy from one chemical substance to another depends on the size of energy quanta emitted from one substance. However, heat energy is often transferred more easily from almost any substance to another because the phonons responsible for vibrational and rotational energy levels in a substance have much less energy than photons invoked for the electronic energy transfer. Thus, because vibrational and rotational energy levels are more closely spaced than electronic energy levels, heat is more easily transferred between substances relative to light or other forms of electronic energy. For example, ultraviolet electromagnetic radiation is not transferred with as much efficacy from one substance to another as thermal or electrical energy.

The existence of characteristic energy levels for different chemical substances is useful for their identification by the analysis of spectral lines. Different kinds of spectra are often used in chemical spectroscopy, e.g. IR, microwave, NMR, ESR, etc. Spectroscopy is also used to identify the composition of remote objects – like stars and distant galaxies – by analyzing their radiation spectra.

Emission spectrum of iron

The term chemical energy is often used to indicate the potential of a chemical substance to undergo a transformation through a chemical reaction or to transform other chemical substances.

Reaction

During chemical reactions, bonds between atoms break and form, resulting in different substances with different properties. In a blast furnace, iron oxide, a compound, reacts with carbon monoxide to form iron, one of the chemical elements, and carbon dioxide.

When a chemical substance is transformed as a result of its interaction with another substance or with energy, a chemical reaction is said to have occurred. A chemical reaction is therefore a concept related to the "reaction" of a substance when it comes in close contact with another, whether as a mixture or a solution; exposure to some form of energy, or both. It results in some energy exchange between the constituents of the reaction as well as with the system environment, which may be designed vessels—often laboratory glassware.

Chemical reactions can result in the formation or dissociation of molecules, that is, molecules breaking apart to form two or more smaller molecules, or rearrangement of atoms within or across molecules. Chemical reactions usually involve the making or breaking of chemical bonds. Oxidation, reduction, dissociation, acid-base neutralization and molecular rearrangement are some of the commonly used kinds of chemical reactions.

A chemical reaction can be symbolically depicted through a chemical equation. While in a non-nuclear chemical reaction the number and kind of atoms on both sides of the equation are equal, for a nuclear reaction this holds true only for the nuclear particles. protons and neutrons.

The sequence of steps in which the reorganization of chemical bonds may be taking place in the course of a chemical reaction is called its mechanism. A chemical reaction can be envisioned to take place in a number of steps, each of which may have a different speed. Many reaction intermediates with variable stability can thus be envisaged during the course of a reaction. Reaction mechanisms are proposed to explain the kinetics and the relative product mix of a reaction. Many physical chemists specialize in exploring and proposing the mechanisms of various chemical reactions. Several empirical rules, like the Woodward–Hoffmann rules often come in handy while proposing a mechanism for a chemical reaction.

According to the IUPAC gold book, a chemical reaction is "a process that results in the interconversion of chemical species." Accordingly, a chemical reaction may be an elementary reaction or a stepwise reaction. An additional caveat is made, in that this definition includes cases where the interconversion of conformers is experimentally observable. Such detectable chemical reactions normally involve sets of molecular entities as indicated by this definition, but it is often conceptually convenient to use the term also for changes involving single molecular entities (i.e. 'microscopic chemical events').

Ions and Salts

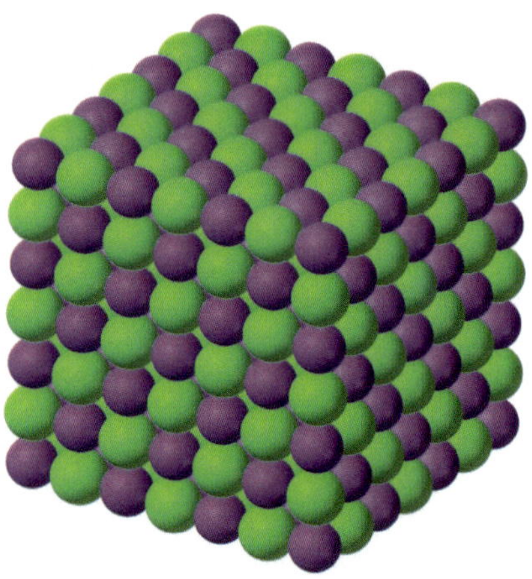

The crystal lattice structure of potassium chloride (KCl), a salt which is formed due to the attraction of K^+ cations and Cl^- anions. Note how the overall charge of the ionic compound is zero.

An ion is a charged species, an atom or a molecule, that has lost or gained one or more electrons. When an atom loses an electron and thus has more protons than electrons, the atom is a positively charged ion or cation. When an atom gains an electron and thus has more electrons than protons, the atom is a negatively charged ion or anion. Cations and anions can form a crystalline lattice of neutral salts, such as the Na^+ and Cl^- ions forming sodium chloride, or NaCl. Examples of polyatomic ions that do not split up during acid-base reactions are hydroxide (OH^-) and phosphate (PO_4^{3-}).

Plasma is composed of gaseous matter that has been completely ionized, usually through high temperature.

Acidity and Basicity

Hydrogen bromide (HBr), pictured, is dissolved in water, it forms the strong acid hydrobromic acid

A substance can often be classified as an acid or a base. There are several different theories which explain acid-base behavior. The simplest is Arrhenius theory, which states than an acid is a substance that produces hydronium ions when it is dissolved in water, and a base is one that produces hydroxide ions when dissolved in water. According to Brønsted–Lowry acid–base theory, acids are substances that donate a positive hydrogen ion to another substance in a chemical reaction; by extension, a base is the substance which receives that hydrogen ion. A third common theory is Lewis acid-base theory, which is based on the formation of new chemical bonds. Lewis theory explains that an acid is a substance which is capable of accepting a pair of electrons from another substance during the process of bond formation, while a base is a substance which can provide a pair of electrons to form a new bond. According to this theory, the crucial things being exchanged are charges. There are several other ways in which a substance may be classified as an acid or a base, as is evident in the history of this concept.

Acid strength is commonly measured by two methods. One measurement, based on the Arrhenius definition of acidity, is pH, which is a measurement of the hydronium ion concentration in a solution, as expressed on a negative logarithmic scale. Thus, solutions that have a low pH have a high hydronium ion concentration, and can be said to be more acidic. The other measurement, based on the Brønsted–Lowry definition, is the acid dissociation constant (K_a), which measures the relative ability of a substance to act as an acid under the Brønsted–Lowry definition of an acid. That is, substances with a higher K_a are more likely to donate hydrogen ions in chemical reactions than those with lower K_a values.

Redox

Redox (reduction-oxidation) reactions include all chemical reactions in which atoms have their oxidation state changed by either gaining electrons (reduction) or losing electrons (oxidation). Substances that have the ability to oxidize other substances are said to be oxidative and are known as oxidizing agents, oxidants or oxidizers. An oxidant removes electrons from another substance. Similarly, substances that have the ability to reduce other substances are said to be reductive and are known as reducing agents, reductants, or reducers.

A reductant transfers electrons to another substance, and is thus oxidized itself. And because it "donates" electrons it is also called an electron donor. Oxidation and reduction properly refer to a change in oxidation number—the actual transfer of electrons may never occur. Thus, oxidation is better defined as an increase in oxidation number, and reduction as a decrease in oxidation number.

Equilibrium

Although the concept of equilibrium is widely used across sciences, in the context of chemistry, it arises whenever a number of different states of the chemical composition are possible, as for example, in a mixture of several chemical compounds that can react with one another, or when a substance can be present in more than one kind of phase.

A system of chemical substances at equilibrium, even though having an unchanging composition, is most often not static; molecules of the substances continue to react with one another thus giving rise to a dynamic equilibrium. Thus the concept describes the state in which the parameters such as chemical composition remain unchanged over time.

Organic Chemistry

Organic chemistry is the study of the structure, properties, composition, reactions, and preparation of carbon-containing compounds, which include not only hydrocarbons but also compounds with any number of other elements, including hydrogen (most compounds contain at least one carbon–hydrogen bond), nitrogen, oxygen, halogens, phosphorus, silicon, and sulfur. This branch of chemistry was originally limited to compounds produced by living organisms but has been broadened to include human-made substances such as plastics. The range of application of organic compounds is enormous and also includes, but is not limited to, pharmaceuticals, petrochemicals, food, explosives, paints, and cosmetics.

The field of organic chemistry includes more than twenty million compounds for which properties have been determined and recorded in the literature. Many hundreds of new compounds are added every day. Much more than half of the world's chemists are organic chemists. Some new organic compounds are simply isolated from plants or animals; some are made by modifying naturally occurring chemicals, but most new organic compounds are actually synthesized in the laboratory from other (usually smaller) organic molecules. Over the years organic chemists have developed a broad array of reactions that allow them to make all kinds of complex products from simpler starting materials.

Singular Attributes of Carbon

When one considers the millions of chemical compounds that are known and notes that more than 95 percent of them are compounds of carbon, one realizes that carbon is unique. Why are there so many carbon compounds? It turns out that atoms of carbon are quite remarkable in a number of ways.

Carbon atoms form very strong bonds with other carbon atoms. The bonds are so strong that carbon can form long chains, some containing thousands of carbon atoms. (Carbon is the only element that can do this).

A carbon atom forms four bonds, therefore carbon not only can form long chains, but it also forms chains that have branches. It is a major reason why carbon compounds exhibit so much isomerism. The simple compound decane ($C_{10}H_{22}$), for example, has 75 different isomers.

Carbon atoms can be bonded by double or triple bonds as well as single bonds. This multiple bonding is much more prevalent with carbon than with any other element.

Carbon atoms can form rings of various sizes. The rings may be saturated or unsaturated. The unsaturated 6-membered ring known as the benzene ring is the basis for an entire subfield of "aromatic" organic chemistry.

Carbon atoms form strong bonds not only with other carbon atoms but also with atoms of other elements. In addition to hydrogen, many carbon compounds also contain oxygen. Nitrogen, sulfur, phosphorus, and the halogens also frequently occur in carbon compounds.

Various kinds of functional groups occur widely among carbon compounds, and many different kinds of isomers are possible.

Hydrocarbons

Compounds of carbon and hydrogen only are called hydrocarbons. These are the simplest compounds of organic chemistry. The most basic group of hydrocarbons are the alkanes, which contain only single bonds. The simplest member of the alkane series is methane, CH_4, the main component of natural gas. The names of some alkanes are listed in table.

ALKANES			
Formula		**Name**	
CH_4	CH_4	methane	gases
C_2H_6	CH_3CH_3	ethane	
C_3H_8	$CH_3CH_2CH_3$	propane	
C_4H_{10}	$CH_3CH_2CH_2CH_3$	butane	
C_5H_{12}	$CH_3(CH_2)_3CH_3$	pentane	liquids
C_6H_{14}	$CH_3(CH_2)_4CH_3$	hexane	
C_7H_{16}	$CH_3(CH_2)_5CH_3$	heptane	
C_8H_{18}	$CH_3(CH_2)_6CH_3$	octane	
C_9H_{20}	$CH_3(CH_2)_7CH_3$	nonane	
$C_{10}H_{22}$	$CH_3(CH_2)_8CH_3$	decane	
$C_{11}H_{24}$	$CH_3(CH_2)_9CH_3$	undecane	
$C_{12}H_{26}$	$CH_3(CH_2)_{10}CH_3$	dodecane	
$C_{13}H_{28}$	$CH_3(CH_2)_{11}CH_3$	tridecane	
$C_{14}H_{30}$	$CH_3(CH_2)_{12}CH_3$	tetradecane	
$C_{15}H_{32}$	$CH_3(CH_2)_{13}CH_3$	pentadecane	
$C_{16}H_{34}$	$CH_3(CH_2)_{14}CH_3$	hexadecane	
$C_{17}H_{36}$	$CH_3(CH_2)_{15}CH_3$	heptadecane	
$C_{18}H_{38}$	$CH_3(CH_2)_{16}CH_3$	octadecane	solids
$C_{19}H_{40}$	$CH_3(CH_2)_{17}CH_3$	nonadecane	
$C_{20}H_{42}$	$CH_3(CH_2)_{18}CH_3$	eicosane	

have ring structures. Since a 4-carbon chain of the alkane series is called butane, a ring of 4 carbon atoms is called.

$CH_3CH_2CH_2CH_3$

butane (C_4H_{10})

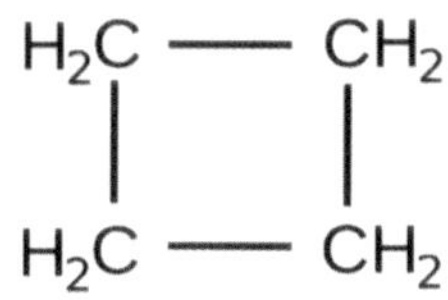

cyclobutane (C_4H_8)

Simple hydrocarbons that contain one or more double bonds are called alkenes. They are named like alkanes, but their names end in " –ene." The simplest alkene has two carbon atoms and is called ethene. A 3-carbon chain that has a double bond is called propene.

$H_2C{=}CH_2$ ethene (C_2H_4) $H_2C{=}CHCH_3$ propene (C_3H_6) $CH_3CH{=}CHCH_2CH_3$ 2–pentene (C_5H_{10})

A 5-carbon hydrocarbon chain with a double bond is called pentene, and if the double bond links the second and third carbons, it is 2- pentene. Like cycloalkanes, alkenes have the general formula C_nH_{2n}. Alkenes having ring structures are called cycloalkenes. A 5-carbon ring with a double bond is called cyclopentene.

Hydrocarbons that contain one or more triple bonds are called alkynes, and is the name ending is yne. 2-carbon alkyne is therefore named ethyne. (However, the compound is often referred to by its common name, which is acetylene).

Compounds that contain double or triple bonds are said to be "unsaturated"—because they are not "saturated" with hydrogen atoms. Unsaturated compounds are reactive materials that readily add hydrogen when heated over a catalyst such as nickel. The reverse reaction also occurs. Heating ethane with steam is an important commercial process for making ethene (or ethylene). This is an important commercial process called "steam cracking."

When a 6-carbon ring contains 2 double bonds, it is called cyclohexadiene, but when it has 3 double bonds, it is not called cyclohexatriene; this is because a 6-carbon ring with three double bonds takes on a special kind of stability. The double bonds become completely conjugated and no longer behave as double bonds. The ring, known as a "benzene ring," is said to be aromatic.

The removal of a hydrogen atom from a hydrocarbon molecule leaves an alkyl group that readily attaches to a functional group, or forms a branch on a hydrocarbon chain. The groups are named after the corresponding hydrocarbons. For example, CH_3^- is named methyl; $CH_3CH_2^-$, ethyl; $CH_2 = CH^-$, ethenyl; $CH_3\,CH_2\,CH_2^-$, propyl; and so on. A benzene ring from which a hydrogen atom has been removed is often referred to as a phenyl. The branched molecules shown here would be given names as follows:

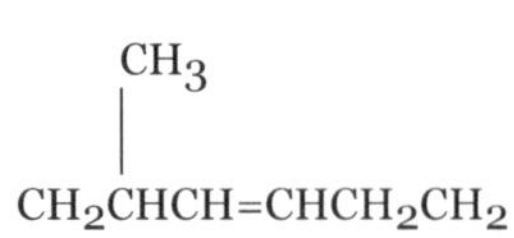

2-methyl-3-hexene

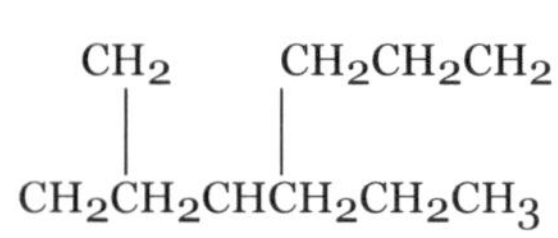

3-methyl-5-propyloctane

Theoretically there is no limit to the length of hydrocarbon chains. Very large hydrocarbon molecules (polymers) have been made containing as many as 100,000 carbon atoms. However, such molecules are hard to make and very difficult to melt and to shape into useful products.

Hydrocarbons are obtained primarily from fossil fuels—especially petroleum and natural gas. Natural gas is a mixture that is largely methane mixed with varying amounts of ethane and other light hydrocarbons, while petroleum is a complex mixture of many different hydrocarbons. Coal, the other fossil fuel, is a much more complicated material from which many kinds of organic compounds, some of them hydrocarbons, can be obtained.

Functional Groups

Alkane molecules are rather unreactive (except for being very flammable), but alkenes react with many other substances. When a drop of bromine is added to an alkene, for example, the deep orange color of the bromine immediately disappears as the bromine adds across the double bond to form a dibromo derivative. The double bond is called a "functional group" because its presence in a molecule causes reactivity at that particular site. There are a dozen or so functional groups that appear frequently in organic compounds. Some of the most common ones are listed in table. The same molecule may contain several functional groups. Aspirin, for example, is both a carboxylic acid and an ester, and cholesterol is an alkene as well as an alcohol.

Isomerism

Isomers are molecules with the same molecular formula but different structures. There is only one structure for methane, ethane, or propane; but butane, C_4H_{10}, can have either of two different structure:

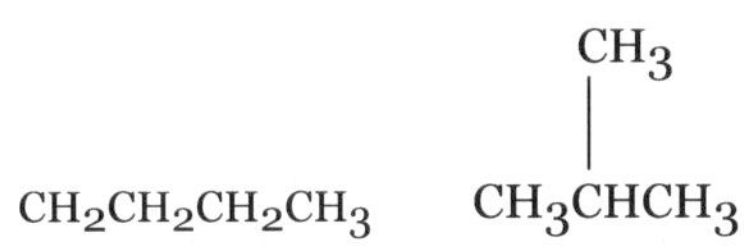

(a) (b)

The linear molecule (a) is called butane, or normal butane (n-butane), whereas the branched molecule (b) is methylpropane (rather than 2-methylpropane, as the methyl group has to be in a 2-position). If the methyl group of (b) were attached to a terminal carbon, the resultant molecule would be the same as (a). Methylpropane (b) is also called iso butane.

In a conjugated system, there are alternating double and single bonds, allowing electrons to flow back and forth. Molecules that contain such conjugated systems are said to be stabilized by "resonance." In the benzene ring every other bond is a double bond, all the way around the ring. This results in a special kind of stabilization called "aromaticity," in which the electrons are delocalized and free to travel all around the ring. Certain ring compounds, like benzene, that contain such a conjugated system of double and single bonds are described as "aromatic."

Pentane has 3 isomers: pentane (or n-pentane), methylbutane (or iso-pentane), and dimethyl propane (or neo pentane). Hexane has 5 isomers: hexane, 2-methylpentane, 3-methylpentane, 2, 2-dimethylbutane, and 2, 3-dimethylbutane. Heptane has 9 different isomers, octane has

18, nonane has 35, and decane has 75. An increase in the number of carbon atoms greatly increases the possibilities for isomerism. There are more than 4,000 isomers of $C_{15}H_{32}$ and more than 366,000 isomers of $C_{20}H_{42}$. The formula $C_{30}H_{62}$ has more than 4 billion. Of course, most of them have never been isolated as pure compounds (but could be, if there were any point in doing it).

For molecules other than hydrocarbons, still other kinds of isomers are possible. The simple formula C_2H_6O can represent ethyl alcohol or dimethyl ether; and C_3H_6O could stand for an alcohol, an ether, an aldehyde, or a ketone (among other things). The larger a molecule is, and the greater the variety of atoms and functional groups it contains, the more numerous its isomers.

There is still another kind of isomerism that stems from the existence of "right-handed" and "left-handed" molecules. It is sometimes referred to as optical isomerism because the molecules that make up a pair of these isomers usually differ only in the way they rotate plane polarized light.

Nomenclature

There are so many millions of organic compounds that simply finding names for them all is a major challenge. It was not until the late nineteenth century that chemists developed a logical system for naming organic compounds. Compounds had often been named according to their sources. The 1-carbon carboxylic acid, for example, was first obtained from ants, and so it was called formic acid, from the Latin word for ants (formicae). The 2-carbon acid was obtained from vinegar, and was called acetic acid.

To bring some order to the naming process an international meeting was held in 1892 at Geneva, Switzerland. The group later became known as the International Union of Pure and Applied Chemistry (IUPAC). Its objective was to establish a naming process that would provide each compound with a unique and systematic name. An initial set of rules was adopted at that first meeting in Geneva, and IUPAC has continued that work. Its systematic naming rules are used by organic chemists all over the world. The names of the alkanes form the basis for the system, with functional groups usually being indicated with appropriate suffixes.

Organic Reactions

Organic chemistry is concerned with the many compounds of carbon, their names, their isomers, and their properties, but it is mostly concerned with their reactions. Organic chemists have developed a huge array of chemical reactions that can convert one organic compound to another. Some reactions involve addition of one molecule to another; some involve decomposition of molecules; some involve substitution of one atom or group by another; and some even involve the rearrangement of molecules, with some atoms moving into new positions. Some reactions require energy in the form of heat or radiation; and some require a special kind of catalyst or some sort of solvent. Of course, not all organic reactions are highly successful. One reaction might be a very simple one giving essentially 100 percent of the desired product; but another might be a complex multistep process yielding less than 5 percent overall of the wanted product.

Naming Organic Compounds					
Functional Group		**Type Compound**	**Example**	**IUPAC Name**	**Common Name**
C=C	double bond	alkene	$H_2C=CH_2$	ethene	ethylene
C≡C	triple bond	alkyne	HC≡CH	ethyne	acetylene
–OH	hydroxyl	alcohol	CH_3OH	methanol	methyl alcohol
–O–	oxy	ether	H_3COCH_3	methoxymethane	methyl ether
	carbonyl	aldehyde	$H_2C=o$	methanal	formaldehyde
	carbonyl	ketone	CH_3COCH_3	propanone	acetone
	carboxyl	carboxylic acid	HCOOH	methanoic acid	formic acid
	carboxyl	ester	$HCOOCH_2CH_3$	ethyl methanoate	ethyl formate
$-NH_2$	amino	amine	CH_3NH_2	aminomethane	methylamine
–CN	cyano	nitrile	CH_3CN	ethanenitrile	acetonitrile
–X	halogen	haloalkane	CH_3Cl	chloromethane	methyl chloride

Organic reactions can often give remarkable control as to what products should be formed. Adding water to propene for example, produces 2-propanol in the presence of acid, but it yields 1-propanol if treated first with B_2H_6 and then H_2O_2 in the presence of base.

Future Sources of Organic Chemicals

Fossil fuels have been our primary natural source for many organic chemicals for more than a century, but our fossil fuel resources are finite, and they are being rapidly depleted (especially oil and gas). What will be our sources of organic materials in the future? Since fossil fuels are non-renewable resources, it is believed that the twenty-first century will see a shift toward greater dependence on renewable raw materials. The largest U.S. chemical company has a goal of becoming 25 percent based on renewable resources by 2010. It is already producing 1,3-propanediol from cornstarch using a gene-tailored E. coli bacterium. This diol is used in Du Pont's fiber Sorona, which is said to combine the best features of both polyester and nylon fibers. Succinic acid and polyhydroxybutyrate are also obtainable from renewable crops, and the list of such renewable raw materials is destined to grow. For example, ethylene (or ethene), $CH_2=CH_2$, which is a highly important commercial chemical used in making many industrial chemicals and polymers, is presently made by steam cracking of ethane obtained from oil or natural gas; but ethylene can also be made by dehydration of ethyl alcohol made by fermentation of sugar. Efforts are even being made to use bio waste materials, such as corn husks, nutshells, and wood chips as industrial raw materials.

Analytical Tools

Organic chemists often need to examine products for identification, purity analysis, or structure determination. There are some marvelous tools available to help them do these things. Chromatography, spectroscopy, and crystallography are especially widely used in organic chemistry.

Column chromatography, gas chromatography, and liquid chromatography are all important methods for separating mixtures of organic compounds. Spectroscopic tools include ultraviolet (UV), infrared (IR), nuclear magnetic resonance (NMR), and mass spectroscopy (MS), each capable of providing a different kind of information about an organic compound. Although it is limited to substances that can be prepared as pure crystals, x-ray crystallography is probably the ultimate tool for determining molecular structure.

Applications of Organic Chemistry

With applications in the field of medicine, textile, pesticides, petroleum and more, organic chemistry is quite substantial for our undertakings. organic chemistry is applied in:

- Analysis: Organic chemistry is what enables us to analyze a substance to its elemental stage. To test a compound for impurities and processes like chromatography and more help us to completely analyze a substance.
- Synthesis: To synthesize naturally occurring substances in large quantities, organic chemistry enables us to recreate the required substance in quantities we need with various processes.
- For better molecules: With a proper understanding of the structure of a molecule, we can improve its performance in the way we require by fabricating a new molecule by altering its functional groups and chains.

Organic Compounds

Organic compounds are quite simply any molecule that contains carbon atoms. Some definitions also include the need to contain hydrogen atoms, as carbon and hydrogen are both components necessary for living things. This definition holds for the molecules, regardless of what state of matter they are in. Historically there are a few carbon-containing molecules such as the carbides that were not considered organic.

An entire field of science, also known as organic chemistry, evolved to study the carbon-containing substances. Organic chemistry was actually born out of a very old, first-century understanding of the root causes and requirements for life called "vitalism."

Examples of Organic Compounds

1. Natural vs Synthetic

Some organic compounds occur naturally in plants and animals, while others are created. What scientists have further classified is the fact that some synthetic organic compounds are just what you'd think the word means: created artificially, such as in laboratory settings. But researchers have also discovered that some synthetic organic compounds are made as a byproduct inside plants or animals. They're still organic compounds because they contain carbon, but they're also still synthetic since they are "made" by the organism. One example could be methane, as it's a byproduct of digestion.

2. Saccharides

There are several key types of saccharides-monosaccharide, disaccharides, and polysaccharides, to name a few-and they are organic compounds made up of carbohydrates and chains of simple sugars. Some of these include cellulose, carboxymethylcellulose, hemicellulose, arabinoxylan, sucrose, maltose, lactose, fructose, galactose, glucose, and ribose.

3. Lipids

Lipids are classified as organic compounds due to the presence of carbon molecules in their make-up. Lipids are fatty or waxy substances that are produced naturally in the cells of plants and animals, and have multiple functions. These functions include creating a protective layer over sensitive areas, such as the coating outside leaves or the ear canals of humans, and in the transport of substances throughout the cell. Some of the most well-known lipids are the unsaturated and saturated fats, trans fats, and glyceride, although there are many more.

4. Other Organic Compounds

There are many other categories of organic compounds, including proteins, hormones like estrogen, cellular components like T-cell receptors, various alcohols and sugar alcohols, and many more.

Organic Matter

Organic matter is matter that contains a large amount of carbon-based compounds. The largest component of organic matter is dead matter. In soils, dead matter makes up roughly 85% of the organic matter. Organic matter includes dead matter, living microbes, and living parts of plants (e.g., roots). Organic matter can be exceedingly beneficial for its nutrient value when incorporated into the soil used by plants.

Maximum Yield Explains Organic Matter

One prime example of organic matter is compost. Organic matter is created through the process of decomposition. For example, when a plant dies, its compounds are broken down by fungi and other organisms in the soil. This process will eventually change the chemical composition of the soil. Decomposition has several different stages, and can take a varied amount of time. However, decomposition will only occur if the conditions are suitable.

Compounds like sugars, starches, and proteins are relatively easy to break down, and do not take as much time. Fungi and bacteria in the soil can break down these compounds making the energy they provide readily available. However, some compounds like cellulose can take much longer to be broken down, on the order of 10 to 1,000 years.

Type of Organic Matter

The types of organic matter in supplies are very diverse and may be present in suspension, or in colloidal or true solution. It is largely decaying vegetable matter but there are many other possible sources including run-off from fields and domestic and industrial wastes:

- Biochemical Oxygen Demand

  Biochemical Oxygen Demand (BOD) is an indirect measure of biodegradable organic compounds in water, and is determined by measuring the dissolved oxygen decrease in a controlled water sample over a five-day period. During this five-day period, aerobic (oxygen-consuming) bacteria decompose organic matter in the sample and consume dissolved oxygen in proportion to the amount of organic material that is present.

- Nutrients

  Nutrients are chemical elements or compounds essential for plant and animal growth. Nutrient parameters include ammonia, organic nitrogen, Kjeldahl nitrogen, nitrate nitrogen (for water only) and total phosphorus. High amounts of nutrients have been associated with eutrophication, or over fertilization of a water body, while low levels of nutrients can reduce plant growth and (for example) starve higher level organisms that consume phytoplankton.

- Organic Carbon

  Most organic carbon in water occurs as partly degraded plant and animal materials, some of which are resistant to microbial degradation. Organic carbon is important in the estuarine food web and is incorporated into the ecosystem by photosynthesis of green plants, then consumed as carbohydrates and other organic compounds by higher animals. In another process, formerly living tissue containing carbon is decomposed as detritus by bacteria and other microbes.

- Oil and Grease constitutes one of the most common parameters for quantifying organics from human sources and, to a lesser extent, biogenic sources (e.g., algae and fish). Some examples of oil and grease loadings are leaks from automobile crankcases, illegal dumping into storm sewers, motorboats, oil spills, and discharge from oil production platforms in the bay.

References

- Carsten Reinhardt. Chemical Sciences in the 20th Century: Bridging Boundaries. Wiley-VCH, 2001. ISBN 3-527-30271-9. pp. 1–2
- Derewenda, Zygmunt S.; Derewenda, ZS (2007). “On wine, chirality and crystallography”. Acta Crystallographica Section A. 64 (Pt 1): 246–258 [247]. Bibcode:2008AcCrA..64..246D. doi:10.1107/S0108767307054293. PMID 18156689
- What-is-chemistry-45986: livescience.com, Retrieved 27 May 2018
- Chemistry 412 course notes. “A Brief History of the Development of Periodic Table”. Western Oregon University. Retrieved July 20, 2015
- “International Year of Chemistry – The History of Chemistry”. G.I.T. Laboratory Journal Europe. Feb 25, 2011. Retrieved March 12, 2013
- Organic-chemistry, areas-of-chemistry, college-to-career: acs.org, Retrieved 21 April 2018

- Theodore L. Brown, H. Eugene Lemay, Bruce Edward Bursten, H. Lemay. Chemistry: The Central Science. Prentice Hall; 8 edition (1999). ISBN 0-13-010310-1. pp. 3–4
- Eagle, Cassandra T.; Jennifer Sloan (1998). "Marie Anne Paulze Lavoisier: The Mother of Modern Chemistry". The Chemical Educator. 3 (5): 1–18. doi:10.1007/s00897980249a
- Organic-Chemistry: chemistryexplained.com, Retrieved 17 April 2018
- Winter, Mark. "WebElements: the periodic table on the web". The University of Sheffield. Archived from the original on January 4, 2014. Retrieved January 27, 2014
- Herbst, Eric (May 12, 2005). "Chemistry of Star-Forming Regions". Journal of Physical Chemistry A. 109 (18): 4017–4029. Bibcode:2005JPCA..109.4017H. doi:10.1021/jp050461c. PMID 16833724
- Organic-matter-584: maximumyield.com, Retrieved 18 July 2018

2

Organic Compounds

Any chemical compound, which contains carbon is called an organic compound. The aim of this chapter is to explore the fundamental aspects of organic compounds. Varied topics, such as carbine, methylcyclohexane, organic peroxide, organic superconductor, alicyclic compound, volatile organic compound, etc. which are crucial for a complete understanding of the subject, have been elaborated in this chapter.

The chemical compounds of living things are known as *organic compounds* because of their association with organisms and because they are carbon-containing compounds. Organic compounds, which are the compounds associated with life processes, are the subject matter of organic chemistry. Among the numerous types of organic compounds, four major categories are found in all living things: carbohydrates, lipids, proteins, and nucleic acids.

Characteristics of Organic Compounds

The general characteristics of Organic Compounds include:

- Can be isolated as well as prepared in laboratory;
- Comprise almost 90% of all known compounds;
- Mostly built up of only three elements- carbon, hydrogen and oxygen. Other elements like halogen, nitrogen as well as phosphorous are also present but to a lesser extent;
- Possess complex structures and high molecular weights;
- Their properties are decided by certain active atom or group of atoms known as the functional group;
- They are mostly insoluble in water but soluble in organic solvents;
- They are combustible in nature;
- Chemical reactions involving organic compounds proceed at slower rates.

Characteristics Due to Presence of Covalent Bonds

A covalent bond is a chemical *bond* that involves the sharing of electron pairs between atoms that result in a stable balance of attractive and repulsive forces between the atoms. The presence of a covalent bond renders certain characteristics to the organic compounds. These include:

- Low melting points and boiling points in comparison to the inorganic compounds.
- Organic acids and bases are less stronger and thus they have a limited dissociation in an aqueous medium.

- They exhibit the phenomenon of isomerism in which a single molecular formula represents several organic compounds differing in physical and chemical properties.
- They are volatile in nature.

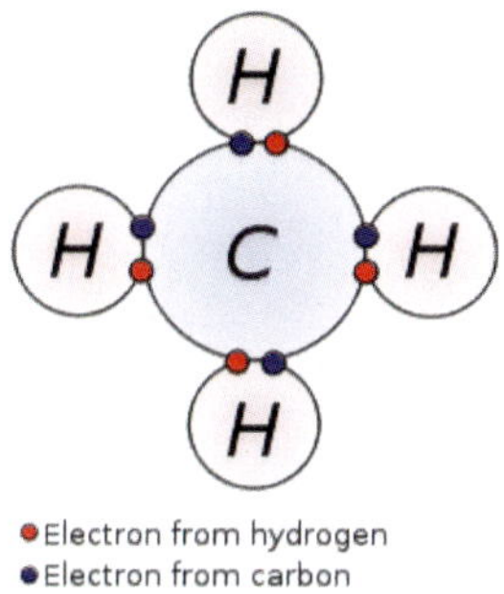

General Characteristics of Members of Homologous Series

A Homologous Series is a group of organic chemical compounds, usually listed in order of increasing size, that have a similar structure (and hence, also similar properties) and whose structures differ only by the number of CH_2–CH_2 units in the main carbon chain. They possess the following general characteristics:

- A general formula describes the members of the homologous series.
- Successive members differ from each other by CH_2CH_2.
- Physical properties change regularly with increasing number of carbon atoms.
- Members have similar chemical properties because they have same functional group.
- Members of the homologous series can be prepared using the same method.

Classification of Organic Compounds

Depending upon the arrangement of carbon atoms in their structure, organic compounds are broadly categorized into:

- Acyclic or Open Chain compounds.
- Cyclic or Closed Chain compounds.

The following diagram will give you a clear idea about the classification of organic compounds:

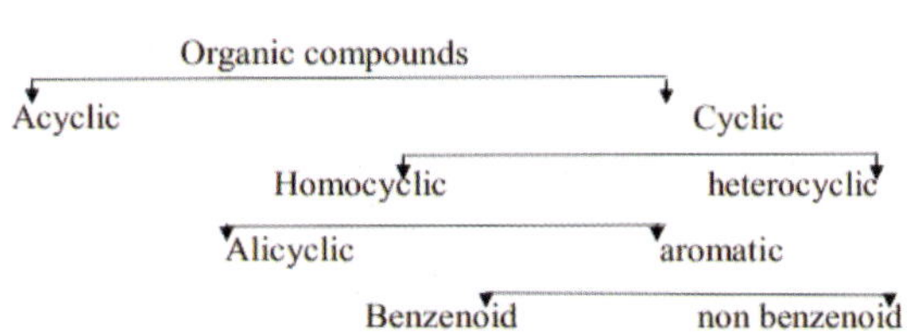

Acyclic or Open Chain Compounds

The carbon atoms are present in the form of an open chain. This chain may either be a straight chain or a branched chain. These were initially known as aliphatic compounds because the compounds of this class were derived from either animal or vegetable fats:

- Straight Chain Compounds: The carbon skeleton is in the form of a straight chain.

  Examples:

  n-Propane CH_3-CH_2-CH_3

  Propene CH_2=CH-CH_3

- Branched Chain Compounds: The carbon skeleton is in the form of a branched chain.

  Examples: Isobutylene

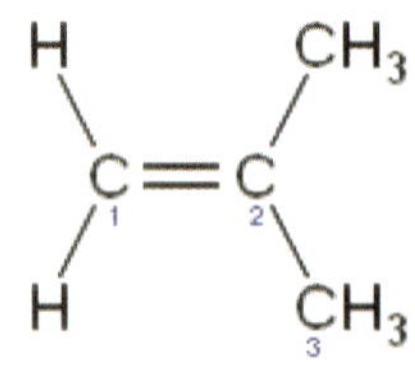

Cyclic or Closed Chain Compounds

They are marked by the presence of one or more closed chains or ring of atoms in their structure. Depending on whether there is a presence of any other atom apart from carbon in the constitution of the ring, they are further classified as:

- Homocyclic or Carbocyclic Compounds;
- Heterocyclic Compounds.

Homocyclic or Carbocyclic Compounds

The rings in these compounds are entirely made up of carbon atoms. No other atom is present in the ring skeleton. These can be further divided into two sub-classes:

- Alicyclic Compounds;
- Aromatic Compounds.

Nomenclature

Scientists introduced a proper method in order to name the organic compounds. This uniform system for naming the compounds is the IUPAC system, which is the International Union of Pure and Applied Chemistry.

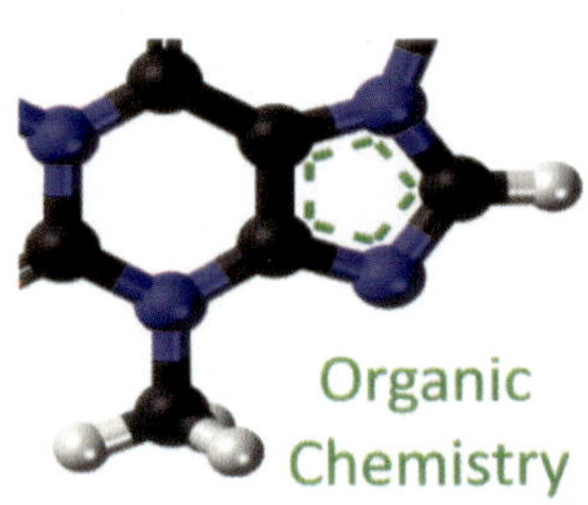

Features of the Trivial System

The name of an organic compound, when in a non-systematic manner or vernacular name is what is known as a trivial system. There are no particular set of rules for the trivial name of the compound. In this system, names are usually simple like acetic acid, toluene, and phenol etc. For example, tartaric acid is a carbolic acid that we usually find in tamarind. But in IUPAC, it is 2,3-dihydroxy-1,4-Butanedioic acid.

Drawbacks of this System

- Many trivial names are present for a single compound. For example, Phenol has different names like hydroxybenzene, carbolic acid, and phenol.
- This system is limited to few compounds in each group. For example, the first two members of the carbolic acid family have trivial names, formic acid, and acetic acid respectively but carbolic acid with more atoms does not have any trivial names.
- There are no particular guidelines for naming complex compounds.

Chemical Nomenclature: IUPAC Rules

According to the IUPAC system, the nomenclature of organic compounds consists of the following parts:

- Steps Involved
    - Longest Chain Rule: In this step, all we have to do is identify the parent hydrocarbon and give the name to it. The parent chain of the compound is usually the longest chain of carbon atoms. This chain could be straight or of a different shape.
    - Lowest number of Locants: We start the numbering of the carbon atoms in the longest chain from the end that gives the lowest number to the carbon atoms carrying the substituents.
    - Multiple Presence of the same substituent: Prefixes such as di, tri, etc. are added to the substituents that are present twice, thrice respectively in the parent chain.
    - Naming the various substituents: If more than one substituent is present, then we need to arrange the substituents in an alphabetical order of their names.
    - Naming different substituents at equivalent positions: If we find the presence of two different substituents on the same position from the two ends, what do we do? In such cases, the substituent first in the alphabetical order gets the lowest number.
    - The Naming of Complex Substituents: We name the complex substituent when the substituent on the parent chain has a branched structure (i.e complex structure). We name these substituents as a substituted alkyl group. It is also important to note that the carbon atom of this substituent gets the first step. We write the name of these type of substituents in brackets.

The final name will be in format: *Locant + Prefix + Root + Locant + Suffix*. Now, we will look at some more details of nomenclature of organic compounds.

- Word root

It indicates the number of carbon atoms in the longest selected carbon chain. For example, C_1 is 'Meth' and C_5 is 'Pent'.

- Suffix

A suffix is generally a functional group in the molecule which follows the word root. We can divide it into:

  - Primary suffix: We write it immediately after the word root. For example, in alkanes the suffix is ane.
  - Secondary suffix: We write it after the primary suffix. For instance, if a compound has alkane and alcohol group attached to it, the naming will be alkanol, -ol being the suffix for alcohol.

- Prefix

We add the prefix to the word root while naming the compound. It indicates the presence of substituent groups or side chains in the organic molecule. It reveals the cyclic and acyclic nature of the compound.

  - Primary prefix: Indicates whether the molecule is cyclic or not. For example, for cyclic compounds the prefix used is cyclo.
  - Secondary prefix: Indicates the presence of substituent groups or any side chain. For example $-CH_3$ is known as Methyl and –Br is Bromo.

Types of Chemical Nomenclature

1) Compositional Nomenclature of Organic Compounds

This term denotes the named constructions based on the composition of species or substances being named, against the systems that involve structural composition or information. One among them is the generalised stoichiometric name. Substances or the elements are named with multiple prefixes in order to give the overall stoichiometry of an element or a compound.

When there are more components, then we divide them into 2 classes namely, electropositive and electronegative components. These names will sound like salt names and this does not imply the chemical nature and behaviour of those species. Examples: Sodium Chloride – NaCl, Trioxygen – O_3, Phosphorous trichloride – PCl_3.

2) Substitutive Nomenclature of Organic Compounds

It is based on the approach where parent hydride is changed by replacement of hydrogen atoms with other atoms or a group of atoms. It is a system where we name the organic compounds using functional groups as the suffix or prefix to the name of the parent compound. We use this system in naming compounds derived from hydrides of specific group elements in the periodic table.

Similar to that of carbon, these elements may form rings and chains that will have many derivatives. Rules come in handy in naming the parent or main compounds and their substituents. Hydrides belonging to group 13-17 of the periodic table get the suffix – ane. For example – Borane, Phosphane, and oxidane etc.

Structural Representation of Organic Compounds

The Complete Structural Formulae

The Lewis structures can be simplified easily by representing the two – electron covalent bond by a single dash. Such a type of structural formula focuses on the electrons involved in the bond formation.

A single dash represents a single bond, a double dash represents a double bond and a triple dash represents a triple bond respectively. Lone pairs of electrons on heteroatoms such as oxygen, halogens and more, may or may not be shown.

Condensed Structural Formulae

We can further abbreviate complete structural formulae by omitting some or all the dashes that represent covalent bonds. Identical repetitive units are put in parenthesis and subscripts are used to indicate their repetition. For example, we can further condense $CH_3CH_2CH_2CH_2CH_2CH_3$ to $CH_3(CH_2)_4CH_3$.

The role of parentheses is very important in condensed structural formulae. We can further condense these structural formulae by enclosing the repetitive structural unit within a bracket and placing an integer as a subscript that indicates the number times the structural unit gets repeated. Further examples of this phenomenon are:

- $CH_3CH_2CH_2CH_2CH_2CH_2CH_3$ changes to CH_3 $(CH_2)_5CH_3$
- $CH_3CH_2CH_2CH_2CH_2CH_2CH_2COOH$ changes to $CH_3(CH_2)_6COOH$

Bond Line Structural Representation

In a Bond line structural representation, carbon and hydrogen atoms are not shown and the lines representing carbon-carbon bonds are drawn in a zig-zag fashion. Only heteroatoms are written in bond line representation. The terminals in the bond – line structures denote methyl (CH_3) groups unless indicated otherwise by a functional group, while the line junctions denote carbon atoms bonded to an appropriate number of hydrogen molecules required to satisfy the valency of the carbon atoms.

Cyclic compounds are usually represented by bond line formulae as well. It is a simple, short and convenient method of representing organic molecule. Each carbon on the line end or intersection is attached to the required number of hydrogen atoms. Thus, the terminal denotes CH_3 group and an unsubstituted intersection denotes a CH_2 group.

Polygon Formulae

There are many organic compounds, in which carbon atoms are not joined in a chain but they are actually joined in a ring. These cyclic compounds are usually represented by polygon without showing carbon and hydrogen atoms.

The corners of a polygon represent a carbon atom and the sides of a polygon denote a carbon-carbon bond. Similarly, if an atom or a group of atoms other than hydrogen is attached to carbon, then that atom or a group of atoms manifest in this structure.

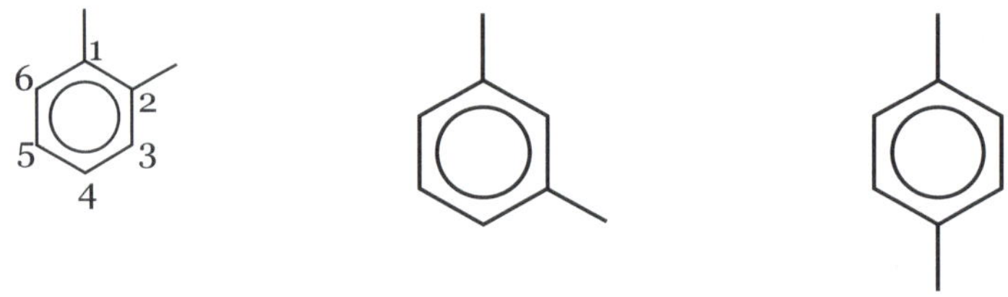

1, 2-dimethylbenzene (ortho-xylene) 1, 3-dimethylbenzene (meta-xylene) 1, 4-dimethylbenzene (para-xylene)

Importance of Organic Compounds

- Organic compounds are important because all living organisms contain carbon.
- While carbohydrates, proteins and fats, the basic structures of life, are organic compounds.
- They are the basic components of many of the cycles that drive the earth. For example, the carbon cycle that includes the exchange of carbon between plants and animals in photosynthesis and cellular respiration.
- Organic compounds combine with metals to form organometallic compounds. These compounds are important industrially. They are used as catalysts, promoters, analysers as well as stabilizers.

Alicyclic Compound

Alicyclic compound is a large class of organic compounds in which three or more atoms of the element carbon are linked together in a ring. The bonds between pairs of adjacent atoms may all be of the type designated single bonds (involving two electrons), or some of them may be double or triple bonds (with four or six electrons, respectively); 6-membered rings for which a system of alternating single and double bonds may be envisioned, however, belong to another important class (aromatic compounds), distinguished from the alicyclic compounds by a characteristically different pattern of chemical reactivity. Those alicyclic compounds in which the ring contains three or four carbon atoms are less stable than the compounds having larger rings, because the angles formed by adjacent covalent bonds are smaller than is necessary for maximum effectiveness. In the larger rings all the bond angles have the preferred value (about 109.5°); consequently, the atoms in the ring do not lie in one plane. Similar restrictions on the angles in double and triple bonds affect the stability of alicyclic compounds containing such bonds.

Nomenclature of Alicyclic Compounds

The following rules are generally used to name alicyclic compounds:

- Cyclo: Prefix 'cyclo' is added to the word root of the alicyclic compounds.

Cyclobutane

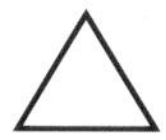

Cyclopropane

- Lowest sum rule: When two or more substituents are present in the alicyclic ring, the numbering is done in such a way that the sum of the positional numbers (locants) of the substituents is the lowest.

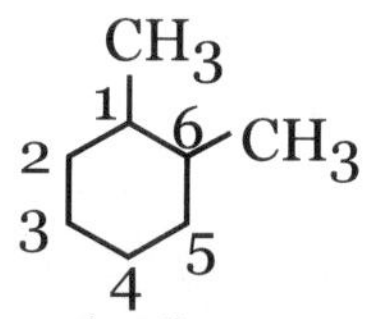

sum (1+6)=7
wrong way

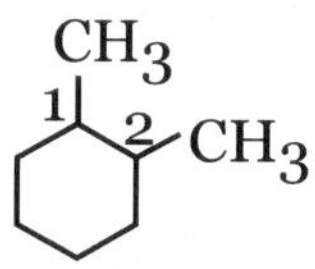

sum (1+2)=3
correct way

- Number alphabetically: If two or more substituents are present, then the numbering is done from the substituent which comes first in the alphabetical order provided it satisfies the lowest sum rule.

Examples:

1-Ethyl-2-methylcyclohexane

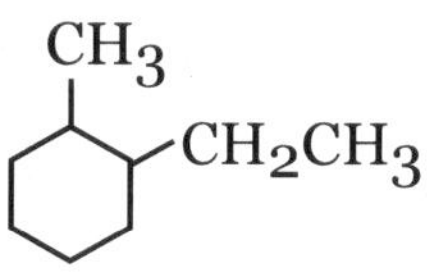

2-Bromo-1-chloro-3-methylcyclohexane

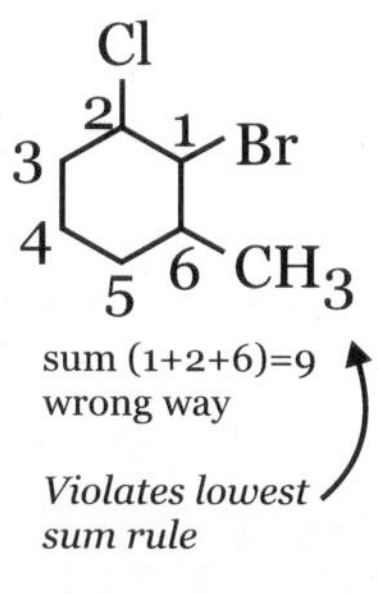

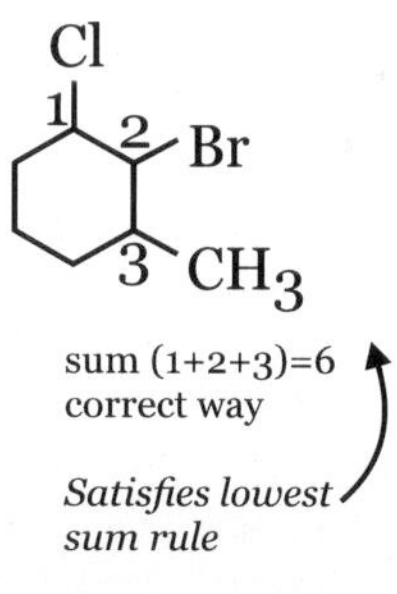

- Priority: According to IUPAC 2013 recommendation, if both alicyclic ring and side chain are present in the compound, the compound is named as a derivative of ring regardless of the length. Prior to 2013 rules, priority was given to the part having greater number of carbon atoms, but not anymore.

Example: Since the priority of ring is higher than that of side chain, the following compound is a derivative of ring.

Hexylcyclohexane

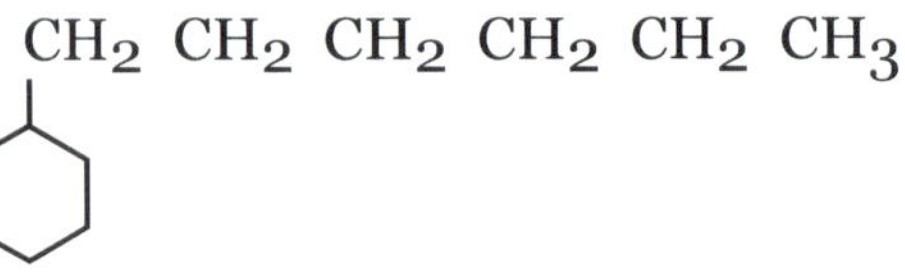

Example: Although the side chain has more carbon atoms, the following compound is a derivative of alicyclic ring. Notice that earlier it was named cyclopropylbutane as length of the carbon chain used to be given priority.

Butylcyclopropane (according to iupac 2013 rules)

Cyclopropylbutane (according to earlier rules)

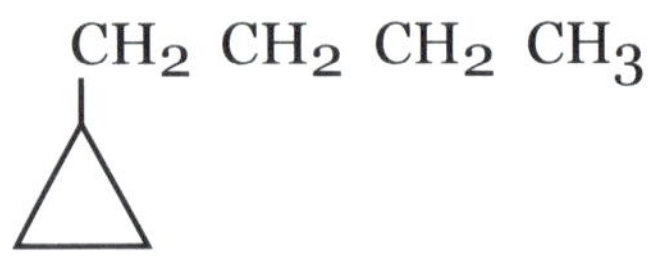

Example: Look at the way the following compound is named.

(1-Methylethyl)cyclohexane.

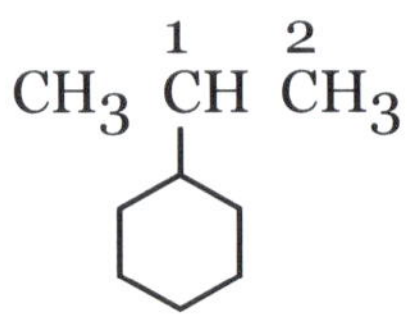

Carbene

In chemistry, a carbene is a highly reactive organic compound with the general molecular formula "R^1R^2C:." This formula indicates that each molecule has a carbon atom (C) attached to two substituents (R^1 and R^2), and this carbon atom has two additional (valence) electrons in its outermost shell that account for the molecule's high reactivity. Most carbenes are very short lived, but some persistent carbenes are also known. They can be stabilized in the form of organometallic complexes.

The prototypical carbene is H_2C:, also called methylene. One well-studied carbene is Cl_2C:, or dichlorocarbene, which can be generated in situ by the reaction of chloroform with a strong base.

Structure

The orbital distribution of electrons in singlet and triplet carbenes.

Generally, there are two types of carbenes, known as *singlet* and *triplet* carbenes.They differ in structure based on the distribution of electrons in orbitals of the reactive carbon atom.

- In a singlet carbene, the reactive carbon atom has three sp^2 hybrid orbitals, with a pair of electrons occupying one of these orbitals. In addition, it has one empty p orbital crossing the plane containing R^1, R^2, and the free electron pair.
- In a triplet carbene, the reactive carbon has two unpaired electrons distributed in one of two possible orbital configurations:
    - The reactive carbon has three sp^2 hybrid orbitals and one unhybridized p orbital. One unpaired electron occupies an sp^2 hybrid orbital and the other occupies a p orbital.
    - The reactive carbon atom has two sp hybrid orbitals (in a linear structure) and two unhybridized p orbitals. The two unpaired electrons occupy the latter two p orbitals.

Most carbenes have a nonlinear triplet ground state, except for those with nitrogen, oxygen, or sulfur atoms, and dihalocarbenes. Carbenes are called singlet or triplet depending on the electronic spins they possess. Triplet carbenes are paramagnetic and may be observed by electron paramagnetic resonance spectroscopy (EPR) if they persist long enough. The total spin of singlet carbenes is zero while that of triplet carbenes is one (in units of $\hbar$). Bond angles are 125°-140° for triplet methylene and 102° for singlet methylene (as determined by EPR). Triplet carbenes are generally stable in the gaseous state, while singlet carbenes occur more often in aqueous media.

For simple hydrocarbons, triplet carbenes usually have energies 8 kcal/mol (33 kJ/mol) lower than singlet carbenes. Thus, in general, triplet is the more stable state (the ground state) and singlet is the excited state species. Substituents that can donate electron pairs may stabilize the singlet state by delocalizing the pair into an empty p-orbital. If the energy of the singlet state is sufficiently reduced, it will actually become the ground state.

No viable strategies exist for triplet stabilization. The carbene called 9-fluorenylidene has been shown to be a rapidly equilibrating mixture of singlet and triplet states with an approximately 1.1 kcal/mol (4.6 kJ/mol) energy difference. It is, however, debatable whether diaryl carbenes such as the fluorene carbene are true carbenes because the electrons can delocalize to such an extent that they become in fact biradicals. In silico experiments suggest that triplet carbenes can be stabilized with electropositive groups such as trifluorosilyl groups.

Reactivity

Singlet and triplet carbenes do not demonstrate the same reactivity. Singlet carbenes generally participate in cheletropic reactions as either electrophiles or nucleophiles. Singlet carbene with its unfilled p-orbital should be electrophilic. Triplet carbenes should be considered to be diradicals, and participate in stepwise radical additions. Triplet carbenes have to go through an intermediate with two unpaired electrons whereas singlet carbene can react in a single concerted step. Addition of singlet carbenes to olefinic double bonds is more stereoselective than that of triplet carbenes. Addition reactions with alkenes can be used to determine whether the singlet or triplet carbene is involved.

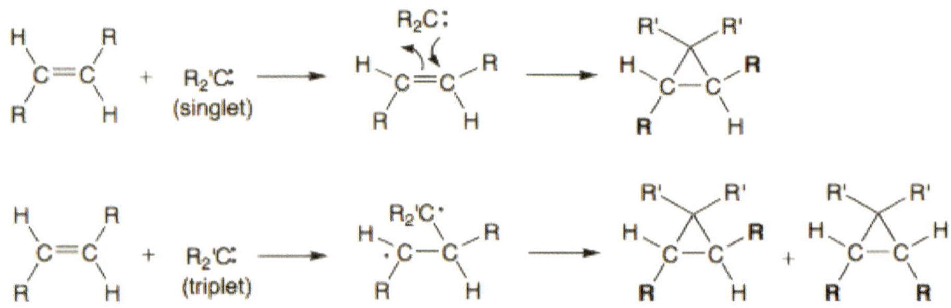

Addition of singlet and triplet carbenes to alkenes.

Reactions of singlet methylene are stereospecific while those of triplet methylene are not. For instance the reaction of methylene generated from photolysis of diazomethane with cis-2-butene and trans-2-butene is stereospecific which proves that in this reaction methylene is a singlet.

Reactivity of a particular carbene depends on the substituent groups, preparation method, reaction conditions such as presence or absence of metals. Some of the reactions carbenes can do are insertions into C-H bonds, skeletal rearrangements, and additions to double bonds. Carbenes can be classified as nucleophilic, electrophilic, or ambiphilic. Reactivity is especially strongly influenced by substituents. For example, if a substituent is able to donate a pair of electrons, most likely carbene will not be electrophilic. Alkyl carbenes insert much more selectively than methylene, which does not differentiate between primary, secondary, and tertiary C-H bonds.

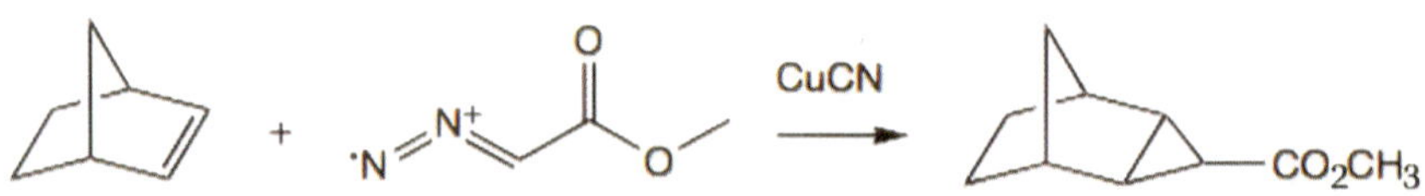

Carbene cyclopropanation.

Carbenes add to double bonds to form cyclopropanes. A concerted mechanism is available for singlet carbenes. Triplet carbenes do not retain stereochemistry in the product molecule. Addition reactions are commonly very fast and exothermic. The slow step in most instances is generation of carbene. A well-known reagent employed for alkene-to-cyclopropane reactions is Simmons-Smith reagent. It is a system that includes copper, zinc, and iodine, where the active reagent is believed to be iodomethylzinc iodide.

Carbenes are also involved in insertion reactions, in which the carbene interposes itself into an existing bond. The order of preference is commonly:

- X-H bonds, where X is not carbon;
- C-H bond,
- C-C bond. Insertions may or may not occur in single step.

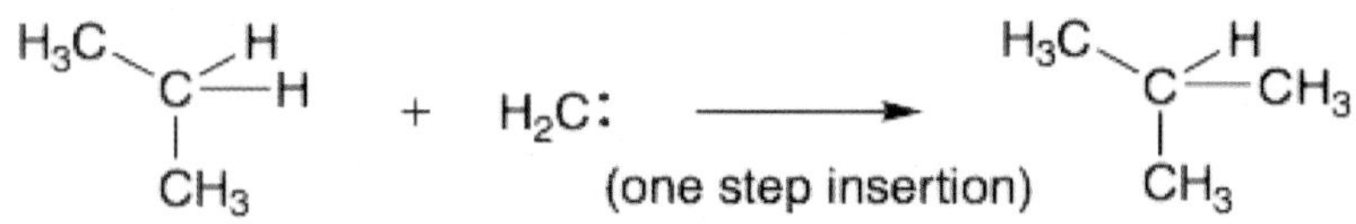

Carbene insertion.

Intramolecular insertion reactions present new synthetic solutions. Generally, rigid structures favor such insertions to happen. When an intramolecular insertion is possible, no intermolecular insertions are seen. In flexible structures, five-membered ring formation is preferred to six-membered ring formation. Both inter and intramolecular insertions are amendable to asymmetric induction by choosing chiral ligands on metal centers.

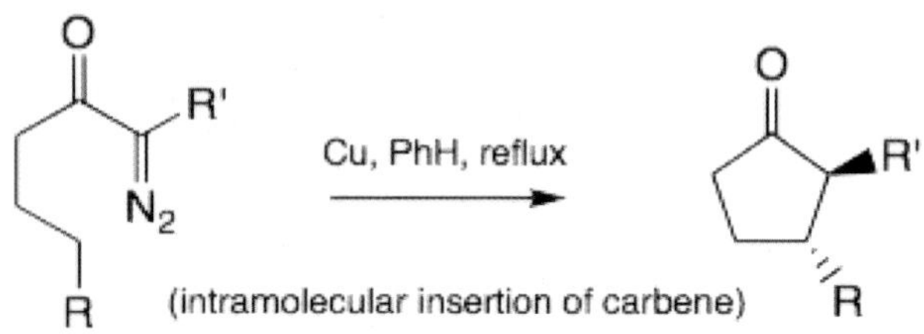

Carbene intramolecular reaction

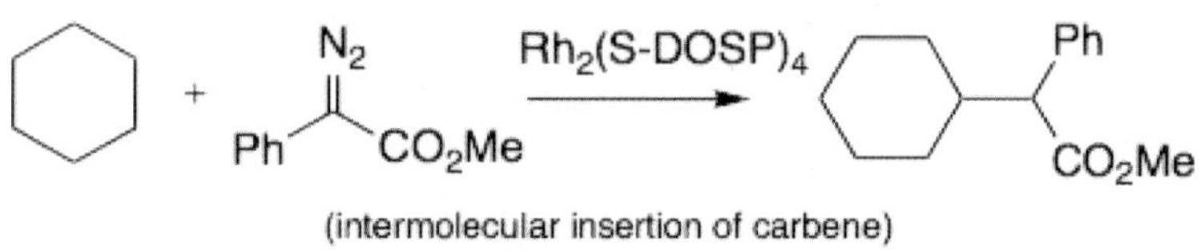

Carbene intermolecular reaction

Alkylidene carbenes are alluring in that they offer formation of cyclopentene moieties. To generate an alkylidene carbene a ketone can be exposed to trimethylsilyl diazomethane.

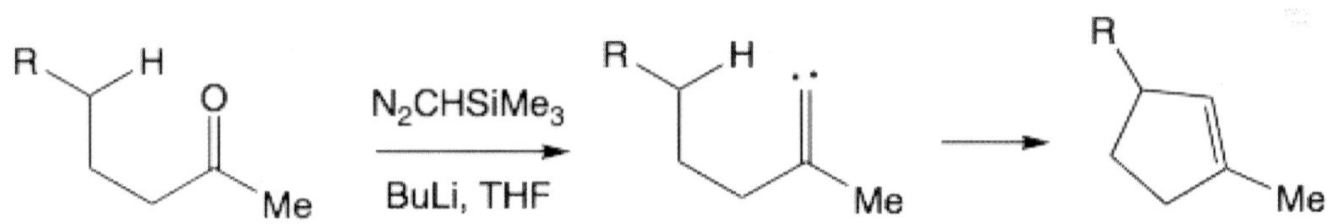

Alkylidene carbene

Generation of Carbenes

Carbenes may be produced by a number of different reactions, some of which are noted below:

- Most commonly, photolytic, thermal, or transition metal catalyzed decomposition of diazoalkanes is used to create carbene molecules. A variation on catalyzed decomposition of diazoalkanes is the Bamford-Stevens reaction, which gives carbenes in aprotic solvents and carbenium ions in protic solvents.
- Another method is induced elimination of halogen from gem-dihalides or HX from CHX_3 moiety, employing organolithium reagents (or another strong base). It is not certain that in these reactions actual free carbenes are formed. In some cases there is evidence that completely free carbene is never present. It is likely that instead a metal-carbene complex forms. Nevertheless, these metallocarbenes (or carbenoids) give the expected products.

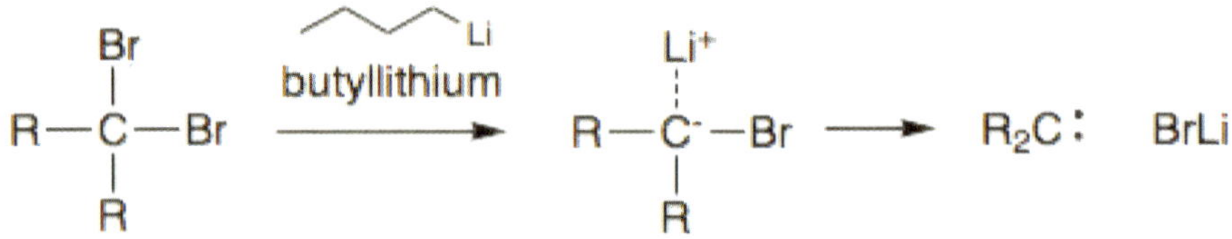

Carbene preparation.

- Photolysis of diazirines and epoxides can also be employed. Diazirines contain 3-membered rings and are cyclic forms of diazoalkanes. The strain of the small ring makes photoexcitation easy. Photolysis of epoxides gives carbonyl compounds as side products. With asymmetric epoxides, two different carbonyl compounds can potentially form. The nature of substituents usually favors formation of one over the other. One of the C-O bonds will have a greater double bond character and thus will be stronger and less likely to break. Resonance structures can be drawn to determine which part will contribute more to the formation of carbonyl. When one substituent is alkyl and another aryl, the aryl-substituted carbon is usually released as a carbene fragment.
- Thermolysis of alpha-halomercury compounds is another method to generate carbenes.
- Rhodium and copper complexes promote carbene formation.
- Carbenes are intermediates in the Wolff rearrangement.

Stabilization of Carbenes and Carbene Ligands

Carbenes can be stabilized as organometallic species. These transition metal carbene complexes fall into the following three categories, of which the first two are the most clearly defined:

- Fischer carbenes, in which the carbene is tethered to a metal that bears an electron-withdrawing group (usually a carbonyl).
- Schrock carbenes, in which the carbene is tethered to a metal that bears an electron-donating group.
- Persistent carbenes, also known as stable carbenes or Arduengo carbenes. They include the class of *N*-heterocyclic carbenes (NHCs) and are often are used as ancillary ligands in organometallic chemistry.

An additional group of carbenes, known as foiled carbenes, derive their stability from the proximity of a double bond—that is, their ability to form conjugated systems.

Covalent Organic Framework

Covalent organic frameworks (COFs) are an emerging class of porous covalent organic structures whose backbones were composed of light elements (B, C, N, O, Si) and linked by robust covalent bonds to endow such material with desirable properties, *i.e.*, inherent porosity, well-defined pore aperture, ordered channel structure, large surface area, high stability, and multi-dimension.

Porous organic frameworks (POFs) constructed from light elements and linked by covalent bonds are a new generation of porous materials, which have been of particular interest due to their desirable properties of high stability, versatile functionality, large surface area. To date, several types of POFs have been prepared including porous aromatic frameworks (PAFs) polymers of intrinsic microporosity (PIMs) conjugated microporous polymers (CMPs), hyper-cross linked polymers (HCPs), and covalent organic frameworks (COFs).

Synthesis of COFs by Linking Organic Building Blocks

B—O Linkages

Boronic acids were first used to synthesize COFs *via* producing the linkages of boroxine anhydride in the form of B_3O_3 rings. COF-1 was synthesized by the self-condensation of 1,4-phenylenediboronic acid (BDBA) to give an extended layers with hexagonal pores of 15 Å diameter and BET surface area of 711 m²/g. Besides, COFs can also be synthesized by the reaction of boronic acids with catechols to produce boronateesters-based linkages as in case of COF-5 and COF-105. Additionally, boronic acids can react with silanols to produce borosilicate bonds and then to form structures of COFs such as COF-202. Undoubtedly, B—O linkages are remarkably general in the synthesis of COFs and COFs can be tailored by selecting diverse building blocks and reaction strategies, *i.e.*, tunable topologies and multi-dimension of COFs can be obtained.

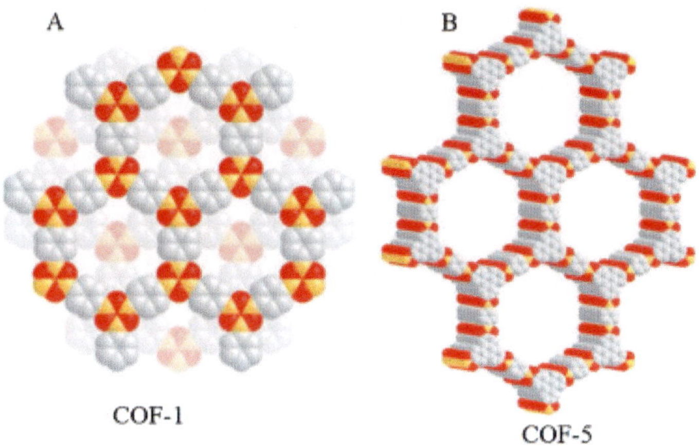

C—N Linkages

In addition to the B—O bonds, COFs can also be prepared through the formation of C—N bonds. For example, COF-300 with a surface area of 1360 m²/g and a 5-fold interpenetrated dia topology was obtained by the imine condensation of aldehyde and amine linkers. The formation of imine bonds is a versatile strategy for the preparation of COFs and the polymerization reaction of Schiff bases is a similar method used to enhance the stability of imine-based COFs. For instance, Wan's group synthesized a bilayer COF by imine reaction between tetrathiafulvalene tetraaldehyde and *p*-phenylenediamine. Zhao's group prepared two kinds of 2D COFs which are constructed from truxene-based building block and 1,4-diaminobenzene or benzidine. More importantly, they studied the influence of substituents on interlayers stacking of 2D COFs, which may open a new path for the design of COF-based functional materials. To obtain higher stability, COFs have been synthesized through the formation of hydrazone linkages, another formation of C—N bonds with the condensation of aldehydes and hydrazide linkers. As in the case of COF-42 and COF-43 , these two examples of hydrazone COFs possessed enhanced chemical stabilities than imine-based COFs. In addition to the abovementioned C—N bonds, other azine linkages and triazine rings have also been applied for the synthesis of COFs. Remarkably, Wang's group indicated a facile strategy for the synthesis of —C=N— linked COFs, *i.e.*, imine-linked LZU-20, hydrazone-linked LZU-21, and azine-linked LZU-22 *via* the condensation of dimethyl acetals and amines with good thermal stability.

Other Linkages

Many other linkages have also been explored for the synthesis of COFs in addition to the above-mentioned general strategies because of the diversity of reversible reactions, which is critical for the preparation of COFs with perfect order. Squaraine-linked COFs, which is composed of squaric acids and amines, allowed for the integration of zwitterion structures in their structures. Subsequently, another ionic COF was synthesized based on azodioxy linkages *via* the self-condensation of hydroxylamine with tetrabutylammonium fluoride as catalyst. Afterwards, spiroborate-linked ionic COFs was prepared *via* the condensation of diols and trialkyl borate in the presence of basic catalysts. In addition, double-stage strategies for the synthesis of COFs with multiple topologies also received wide attention, such as the combination of an imine and boronate ester or an imine and boroxine.

Secondary Building Units

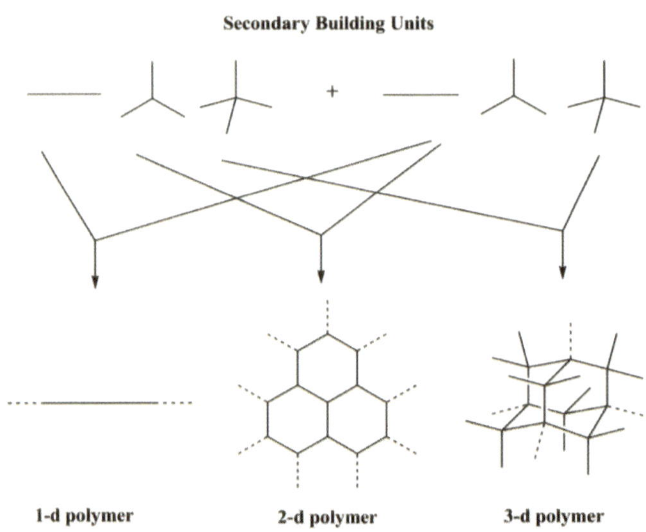

Schematic Figure of Reticular Chemistry

The term 'secondary building unit' has been used for some time to describe conceptual fragments which can be compared as bricks used to build a house of zeolites; in the context of this page it refers to the geometry of the units defined by the points of extension. Recently, 279 of new secondary building units have been found on the crystal structure database.

Reticular Synthesis

Although the synthesis of new materials has long been recognized as the most essential element in advancing technology, it generally remains more of an art than a science—in that the discovery of new compounds has mostly been serendipitous, using methods referred to by critics as 'shake and bake', 'mix and wait', 'mash and smash' and 'heat and beat'. The reason is that the starting entities do not maintain their structure during the reaction, leading to poor correlation between reactants and products. However, the design of an extended network that will maintain its structural integrity throughout the construction process can be realized by starting with well-defined and rigid molecular building blocks.

In essence, reticular synthesis can be described as the process of assembling judiciously designed rigid secondary building units into predetermined ordered structures (networks), which are held together by strong bonding. It is different from retrosynthesis of organic compounds, because the structural integrity and rigidity of the building blocks in reticular synthesis remain unaltered throughout the construction process—an important aspect that could help to fully realize the benefits of design in crystalline solid-state frameworks. Similarly, reticular synthesis should be distinguished from supramolecular assembly, because in the former, building blocks are linked by strong bonds throughout the crystal.

Applications

Hydrogen Storage

Omar M. Yaghi and William A. Goddard III reported COFs as exceptional hydrogen storage materials. They predicted the highest excess H_2 uptakes at 77 K are 10.0 wt % at 80 bar for COF-105, and 10.0 wt % at 100 bar for COF-108, which have higher surface area and free volume, by grand canonical Monte Carlo (GCMC) simulations as a function of temperature and pressure. This is the highest value reported for associative H_2 storage of any material. Thus 3-D COFs are most promising new candidates in the quest for practical H_2 storage materials. In 2012, the lab of William A. Goddard III reported the uptake for COF102, COF103, and COF202 at 298 K and they also proposed new strategies to obtain higher interaction with H_2. Such strategy consist on metalating the COF with alkaline metals such as Li. These complexes composed of Li, Na and K with benzene ligands (such as 1,3,5-benzenetribenzoate, the ligand used in MOF-177) have been synthesized by Krieck et al. and Goddard showed that the THF is important of their stability. If the metalation with alkaline is performed in the COFs, Goddard et al. calculated that some COFs can reach 2010 DOE gravimetric target in delivery units at 298 K of 4.5 wt %: COF102-Li (5.16 wt %), COF103-Li (4.75 wt %), COF102-Na (4.75 wt %) and COF103-Na (4.72 wt %). COFs also perform better in delivery units than MOFs because the best volumetric performance is for COF102-Na (24.9), COF102-Li (23.8), COF103-Na (22.8), and COF103-Li (21.7), all using delivery g H_2/L units for 1–100 bar. These are the highest gravimetric molecular hydrogen uptakes for a porous material under these thermodynamic conditions. Other strategies to increase the interaction of COFs with molecular hydrogen have been reviewed recently. In 2012, the new COF-301-PdCl2 is predicted to reach 60 g total H_2/L at 100 bar, which is 1.5 times the DOE 2015 target of 40 g/L and close to the ultimate (2050) target of 70 g/L.

Methane Storage

Omar M. Yaghi and William A. Goddard III also reported COFs as exceptional methane storage materials. The best COF in terms of total volume of CH_4 per unit volume COF absorbent is COF-1, which can store 195 v/v at 298 K and 30 bar, exceeding the U.S. Department of Energy target for CH_4 storage of 180 v/v at 298 K and 35 bar. The best COFs on a delivery amount basis (volume adsorbed from 5 to 100 bar) are COF-102 and COF-103 with values of 230 and 234 v(STP: 298 K, 1.01 bar)/v, respectively, making these promising materials for practical methane storage. More recently, new COFs with better delivery amount have been designed in the lab of William A. Goddard III, and they have been shown to be stable and overcome the DOE target in delivery basis. COF-103-Eth-trans and COF-102-Ant, are found to exceed the DOE target of 180 v(STP)/v at 35 bar for methane storage. They reported that using thin vinyl bridging groups aid performance by minimizing the interaction methane-COF at low pressure. This is a new feature that can be used to enhance loading in addition to the common practice of adding extra fused benzene rings.

Optical Properties

A highly ordered π-conjugation TP-COF, consisting of pyrene and triphenylene functionalities alternately linked in a mesoporous hexagonal skeleton, is highly luminescent, harvests a wide wavelength range of photons, and allows energy transfer and migration. Furthermore, TP-COF is electrically conductive and capable of repetitive on–off current switching at room temperature.

Porosity/Surface-area Effects

Most studies to date have focused on the development of synthetic methodologies with the aim of maximizing pore size and surface area for gas storage. That means the functions of COFs have not yet been well explored, but COFs can be used as catalyst, or gas separation etc.

Carbon Capture

In 2015 the use of highly porous, catalyst-decorated COFs for converting carbon dioxide into carbon monoxide.

Diazonium Compound

Diazonium compound is also called Diazonium Salt. It is a class of organic compounds. that have the molecular structure:

$$R\text{-}\overset{+}{N}\equiv N,$$

in which R is an atomic grouping formed by removal of a hydrogen atom from an organic compound. Diazonium salts are usually prepared by the reaction (diazotization) of primary amines with nitrous acid; their most striking property is their instability. The aliphatic diazonium salts exist only as transient intermediates, quickly decomposing into a nitrogen molecule and a carbonium ion; certain aromatic diazonium salts are stable enough to be isolated but react readily either by loss of nitrogen or by formation of azo compounds.

Diazonium salts were first obtained from aromatic amines in 1858, and their utility in the preparation of azo compounds was soon recognized by the dye industry. By varying the chemical structures of the amines that are diazotized (the diazo components) and of the compounds with which they react (the coupling components), colors throughout the visible spectrum are imparted to dyes applicable to many types of fibres by several techniques.

The diazonium group can be replaced by numerous atoms or groups of atoms, often with the aid of copper or a copper salt; these reactions make possible the preparation of a wide variety of aromatic derivatives. Chemical reduction of aromatic diazonium salts leads to formation of hydrazine derivatives.

Properties of Diazonium Salts

- They are ionic in nature.
- They are water soluble.
- Aryl diazonium salts are colorless crystalline solids.
- Benzenediazonium chloride is soluble in water. But, it reacts with it only when warmed.
- Benzenediazonium fluoroborate is not soluble in water. It is pretty stable at room temperature.

Method of Preparation of Diazonium Salt

The process of converting an organic compound, generally primary aromatic amines, into diazonium salts is referred to as diazotization or dissociation. It has been seen that diazonium group is highly unstable under normal conditions and hence it is not generally stored, most of the times it is used just after its preparation. One of the most common methods of preparation of diazonium salt is by reacting nitrous acid with aromatic amines. The reaction of aniline (aromatic amine) with nitrous acid results in the diazonium salt formation that is benzene diazonium chloride. Nitrous acid is a highly toxic gas and hence it is generally prepared during the reaction itself (in situ) by reacting $NaNO_2$ with a mineral acid. Another important parameter that governs the product formation during preparation of these salts is temperature; most diazonium salts are stable below the temperature of 5° C. It is, therefore, important to keep the reaction conditions below 5°C or else the diazonium group will decompose to give N_2 as soon as it is formed. The reaction of aniline with $NaNO_2$ and hydrochloric acid is given below, the product of the reaction is benzene diazonium chloride.

$$C_6H_5NH_2 + NaNO_2 + 2HCl \xrightarrow{Temperature\ 0-5^{\circ}C} C_6H_5N + -Cl + NaCl + 2H_2O$$

The diazonium group finds application in many ways. From dyes and pigment industry to synthesis of various organic compounds, this combination of organic and inorganic components has been a boon for the scientists.

Diazo Coupling Reactions

The most widely practiced reaction of diazonium salts is azo coupling. In this process, the diazonium compound is attacked by, i.e., coupled to, electron-rich substrates. When the coupling partners are arenes such as anilines and phenols, the process is an example of electrophilic aromatic substitution:

$$ArN_2^+ + Ar'H \rightarrow ArN_2Ar' + H^+$$

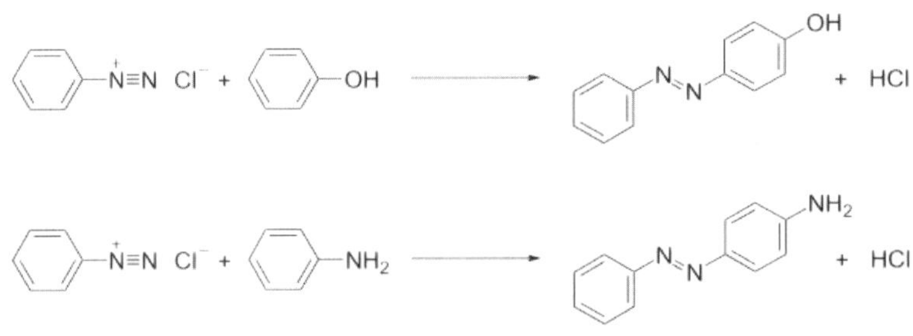

Another commercially important class of coupling partners are acetoacetic amides, as illustrated by the preparation of Pigment Yellow 12, a diarylide pigment.

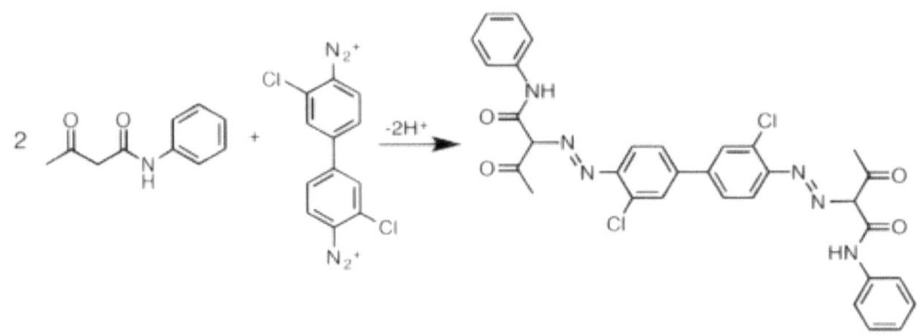

The resulting azo compounds are often useful dyes and in fact are called azo dyes. The deep colors of the dyes reflects their extended conjugation. For example, the dye called aniline yellow is produced by mixing aniline and cold solution of diazonium salt and then shaking it vigorously. Aniline yellow is obtained as an yellow solid. Similarly, a cold basic solution of Naphthalen-2-ol (B-naphthol) give the intensely orange-red precipitate. Methyl orange is an example of an azo dye that is used in the laboratory as a pH indicator.

Displacement of the N_2 Group

Arenediazonium cations undergo several reactions in which the N_2 group is replaced by another group or ion. Some of the major ones are the following:

Replacement by Halogens

Sandmeyer Reaction

Benzenediazonium chloride heated with cuprous chloride or cuprous bromide respectively dissolved in HCl or HBr yield chlorobenzene or bromobenzene, respectively.

$$C_6H_5N_2^+ + CuCl \rightarrow C_6H_5Cl + N_2 + Cu^+$$

Gatterman Reaction

In the Gatterman reaction, benzenediazonium chloride is warmed with copper powder and HCl or HBr to produce chlorobenzene and bromobenzene respectively. It is named after the German chemist Ludwig Gattermann.

$$C_6H_5N_2^+ + CuX \rightarrow C_6H_5X + N_2 + Cu^+$$

Replacement by Iodide

Iodine is not easily introduced into the benzene ring directly. However it can be introduced by treating aryldiazonium cations with potassium iodide:

$$C_6H_5N_2^+ + KI \rightarrow C_6H_5I + K^+ + N_2$$

Replacement by Fluoride

Fluorobenzene is produced by thermal decomposition of benzenediazonium fluoroborate. The conversion is called the Balz-Schiemann reaction.

$$[C_6H_5N_2^+]BF_4^- \rightarrow C_6H_5F + BF_3 + N_2$$

Miscellaneous Replacements

Replacement by Hydrogen

Arenediazonium cations are reduced by hypophosphorous acid or sodium stannite gives benzene:

$$[C_6H_5N_2^+]Cl^- + H_3PO_2 + H_2O \rightarrow C_6H_6 + N_2 + H_3PO_3 + HCl$$

Replacement by a Hydroxyl Group

Phenols are produced by heating aqueous solutions of aryldiazonium salts to 100 °C.

$$C_6H_5N_2^+ + H_2O \rightarrow C_6H_5OH + N_2 + H^+$$

This reaction goes by the German name *Phenolverkochung* ("cooking down to yield phenols"). The phenol formed may react with the diazonium salt and hence the reaction is carried in the presence of an acid which helps in suppressing this further reaction.

Replacement by a Nitro Group

Nitrobenzene can be obtained by treating benzenediazonium fluoroborate with sodium nitrite in presence of copper. Alternatively, the diazotisation of the aniline can be conducted in presence of cuprous oxide, which generates cuprous nitrite in situ:

$$C_6H_5N_2^+ + CuNO_2 \rightarrow C_6H_5NO_2 + N_2 + Cu^+$$

Replacement by a Cyano Group

The cyano group usually cannot be introduced by nucleophilic substitution of haloarenes, but such compounds can be easily prepared from diazonium salts. Illustrative is the preparation of benzonitrile using the reagent cuprous cyanide:

$$C_6H_5N_2^+ + CuCN \rightarrow C_6H_5CN + Cu^+ + N_2$$

This reaction is a special type of Sandmeyer reaction.

Replacement by a Thiol Group

Diazonium salts can be converted to thiols in a two-step procedure. Treatment of benzenediazonium chloride with potassium ethylxanthate followed by hydrolysis of the intermediate xanthate ester gives thiophenol:

$$C_6H_5N_2^+ + C_2H_5OCS^-{}_2 \rightarrow C_6H_5SC(S)OC_2H_5$$

$$C_6H_5SC(S)OC_2H_5 + H_2O \rightarrow C_6H_5SH + HOC(S)OC_2H_5$$

Replacement by an Aryl Group

The aryl group can be coupled to another using aryldiazonium salts. For example, treatment of benzenediazonium chloride with benzene (an aromatic compound) in the presence of sodium hydroxide gives diphenyl:

$$[C_6H_5N_2^+]Cl^- + C_6H_6 \rightarrow C_6H_5{-}C_6H_5 + N_2 + HCl$$

This reaction is known as the Gomberg–Bachmann reaction. A similar conversion is also achieved by treating benzenediazonium chloride with ethanol and copper powder.

Replacement by a Carboxyl Group

Diazonium fluoroborates react with an aliphatic carboxylic acid yield the corresponding benzoic acid. This reaction provides a method to prepare aromatic carboxylic acids from aliphatic carboxylic acids:

$$[C_6H_5N_2^+]BF_4^- + RCO_2H \rightarrow C_6H_5CO_2H + BF_3 + N_2 + RF$$

Meerwein Reaction

Benzenediazonium chloride reacts with compounds containing activated double bonds to produces phenylated products. The reaction is called the Meerwein arylation:

$$[C_6H_5N_2^+]Cl^- + ArCH{=}CHCO_2H \rightarrow ArC{=}C{-}C_6H_5 + N_2 + CO_2 + HCl$$

Metal Complexes

In their reactions with metal complexes, diazonium cations behave similarly to NO^+. For example, low-valent metal complexes add with diazonium salts. Illustrative complexes are $[Fe(CO)_2(PPh_3)_2(N_2Ph)]^+$ and the chiral-at-metal complex $Fe(CO)(NO)(PPh_3)(N_2Ph)$.

Other Methods for Dediazotization

- By organic reduction at an electrode;
- By mild reducing agents such as ascorbic acid (vitamin C);
- By gamma radiation from solvated electrons generated in water;
- Photoinduced electron transfer;

- Reduction by metal cations, most commonly a cuprous salt;
- Anion-induced dediazoniation: a counterion such as iodine gives electron transfer to the diazonium cation forming the aryl radical and an iodine radical;
- Solvent-induced dediazoniation with solvent serving as electron donor.

Grafting Reactions

In a potential application in nanotechnology, the diazonium salts 4-chlorobenzenediazonium tetrafluoroborate very efficiently functionalizes single wall nanotubes. In order to exfoliate the nanotubes, they are mixed with an ionic liquid in a mortar and pestle. The diazonium salt is added together with potassium carbonate, and after grinding the mixture at room temperature the surface of the nanotubes are covered with chlorophenyl groups with an efficiency of 1 in 44 carbon atoms. These added subsituents prevent the tubes from forming intimate bundles due to large cohesive forces between them, which is a recurring problem in nanotube technology.

It is also possible to functionalize silicon wafers with diazonium salts forming an aryl monolayer. In one study, the silicon surface is washed with ammonium hydrogen fluoride leaving it covered with silicon–hydrogen bonds (hydride passivation). The reaction of the surface with a solution of diazonium salt in acetonitrile for 2 hours in the dark is a spontaneous process through a free radical mechanism:

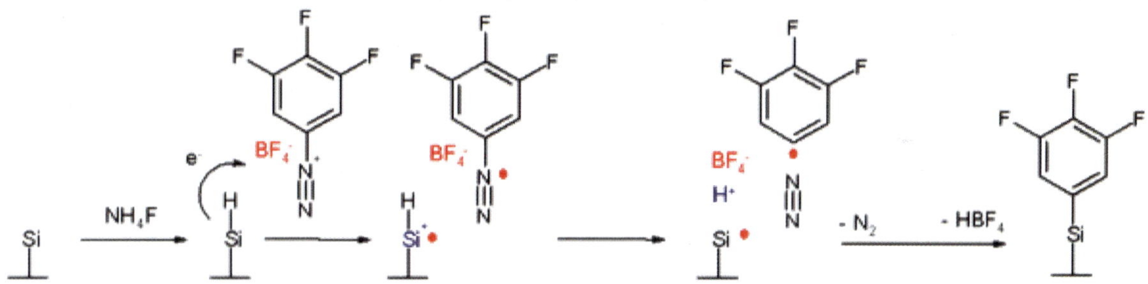

So far grafting of diazonium salts on metals has been accomplished on iron, cobalt, nickel, platinum, palladium, zinc, copper and gold surfaces. Also grafting to diamond surfaces has been reported. One interesting question raised is the actual positioning on the aryl group on the surface. An in silico study demonstrates that in the period 4 elements from titanium to copper the binding energy decreases from left to right because the number of d-electrons increases. The metals to the left of iron are positioned tilted towards or flat on the surface favoring metal to carbon pi bond formation and those on the right of iron are positioned in an upright position, favoring metal to carbon sigma bond formation. This also explains why diazonium salt grafting thus far has been possible with those metals to right of iron in the periodic table.

Reduction to a Hydrazine Group

Diazonium salts can be reduced with stannous chloride ($SnCl_2$) to the corresponding hydrazine derivatives. This reaction is particularly useful in the Fischer indole synthesis of triptan compounds and indometacin. The use of sodium dithionite is an improvement over stannous chloride since it is a cheaper reducing agent with fewer environmental problems.

Applications

The first use of diazonium salts was to produce water-fast dyed fabrics by immersing the fabric in an aqueous solution of the diazonium compound, followed by immersion in a solution of the coupler (the electron-rich ring that undergoes electrophilic substitution). The major applications of diazonium compounds remains in the dye and pigment industry.

Other Uses

Diazonium compounds are standard reagents used in synthesis of organic compounds, especially aryl derivatives.

Diazonium salts are light sensitive and break down under near UV or violet light. This property has led to their use in document reproduction. In this process, paper or film is coated with a diazonium salt. After contact exposure under light, the residual diazo is converted to a stable azo dye with an aqueous solution of coupler. A more common process uses a paper coated with diazo, coupler and an acid to inhibit coupling; after exposure the image is developed by a vapor mixture of ammonia and water which forces coupling.

Safety

Solid diazonium halides are often dangerously explosive, and fatalities and injuries have been reported.

The nature of the anions affects stability of the salt. Aryl diazonium perchlorates, such as nitrobenzenediazonium perchlorate, have been used to initiate explosives.

Methylcyclohexane

Methylcyclohexane is a colorless liquid with a faint, Benzene-like odor. It is used as a solvent and to manufacture organic chemicals.

A substituent group in a substituted cyclohexane, such as the methyl group in methyl cyclohexane, can be in either an equatorial or an axial position.

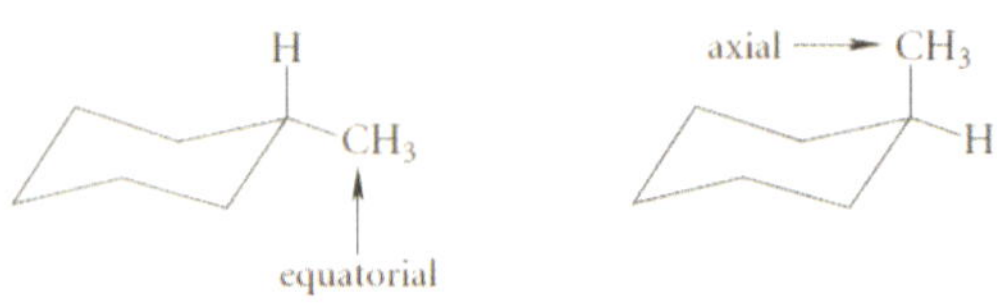

These two compounds are not identical, yet they have the same connectivity, so they are stereoisomers. Because they are not enantiomers, they must be diastereomers. Like cyclohexane itself, substituted cyclohexanes such as methylcyclohexane also undergo the chair interconversion. As figure above shows, axial methylcyclohexane and equatorial methylcyclohexane are interconverted by this process. Note in this interconversion that a down methyl remains down and an up methyl remains up. Because this process is rapid at room temperature, methylcyclohexane is a mixture of two conformational diastereomers. Because diastereomers have different energies, one form is more stable than the other.

Equatorial methylcyclohexane is more stable than axial methylcyclohexane. In fact, it is usually the case that the equatorial conformation of a substituted cyclohexane is more stable than the axial conformation. Why should this be so?

Examination of a space-filling model of axial methylcyclohexane shows that van der Waals repulsions occur between one of the methyl hydrogens and the two axial hydrogens on the same face of the ring. Such unfavorable interactions between axial groups are called 1,3-diaxial interactions. These vander Waals repulsions destabilize the axial conformation relative to the equatorial conformation, in which such vander Waals repulsions are absent.

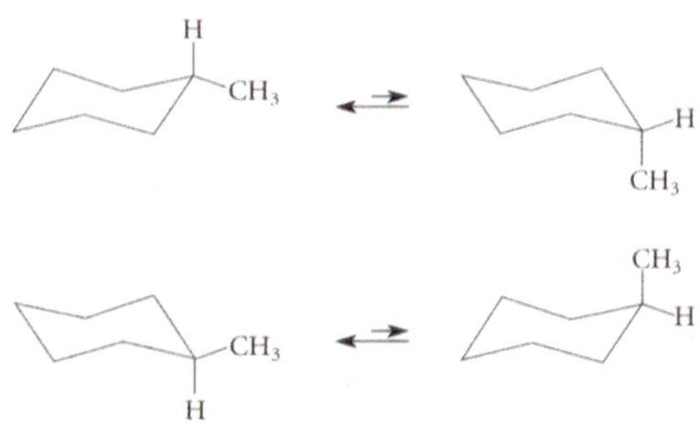

Figure: The chair interconversion results in an equilibrium between equatorial (left) and axial (right) conformations of methylcyclohexane.The conversion is shown with two different ring perspectives. Notice in this interconversion that a down methyl remains down and an up methyl remains up.

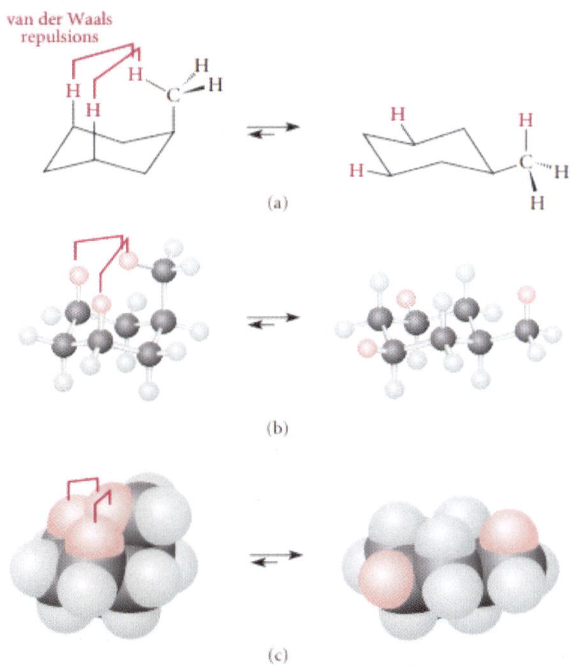

Figure, the equilibrium between axial and equatorial conformations of methylcyclohexane is shown with (a) Lewis structures, (b) ball-and-stick models, and (c) space-filling models. The hydrogens involved in 1,3-diaxial interactions in the axial conformation are shown in color, and the interactions themselves are indicated with red brackets.

As figure shows, the energy (enthalpy) difference between axial and equatorial conformations of methylcyclohexanes is 7.4 kJ mol^{-1} (1.8 kcal mol^{-1}). Because there are two 1,3-diaxial interactions in methylcyclohexane, each interaction is responsible for one-half of the enthalpy difference, or 3.7 kJ mol^{-1} (0.9 kcal mol^{-1}). We'll find that we can use this value in predicting the relative energies of other methyl-substituted cyclohexanes. In other words, each methyl–hydrogen 1,3-diaxial interaction in a cyclohexane derivative raises the enthalpy by 3.7 kJ mol^{-1} (0.9 kcal mol^{-1}).

As shown in figure, the 1,3-diaxial interaction of a methyl group and a hydrogen in axialmethylcyclohexane looks a lot like the van der Waals interaction between methyl hydrogens in gauche-butane. The energy cost of this interaction in gauche-butane is 2.8 kJ mol^{-1}. Because there are two such 1,3-diaxial interactions in axial-methylcyclohexane, the gauche-butane analogy would predict an energy cost of 2 3 2.8 = 5.6 kJ mol^{-1}. The actual value, 7.4 kJ mol^{-1}, is in fair agreement with this prediction. For this reason, 1,3-diaxial methyl–hydrogen interactions in cyclohexane derivatives are sometimes called gauche-butane interactions.

The energy cost of placing a methyl group in the axial position of a cyclohexane ring is reflected in the relative amounts of axial and equatorial methylcyclohexanes present at equilibrium.

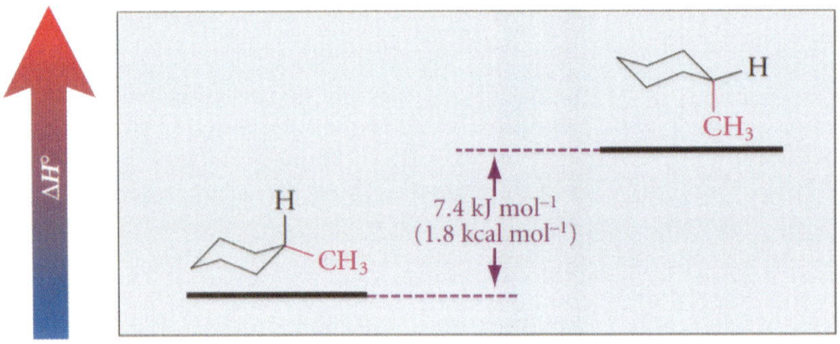

Figure: Relative enthalpies of axial and equatorial methylcyclohexane.

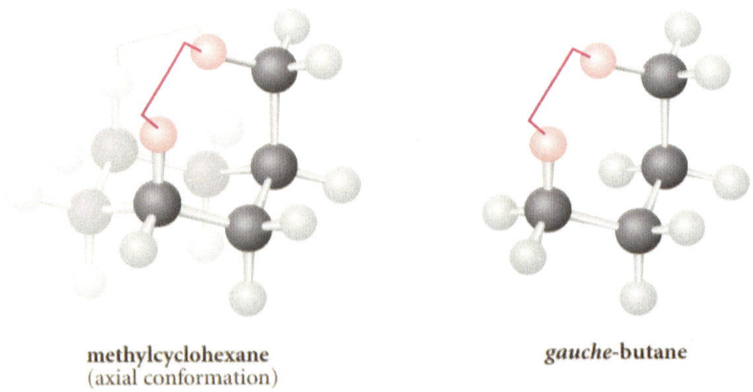

Figure: The relationship between the axial conformation of methylcyclohexane and gauche-butane.

One gauche-butane part of methylcyclohexane is highlighted, and the corresponding van der Waals repulsion is shown with a colored bracket. The second gauche-butane interaction in methylcyclohexane is shown with the gray bracket.

The investigation of molecular conformations and their relative energies is called conformational analysis. We have just carried out a conformational analysis of methylcyclohexane. The conformational analyses of many different substituted cyclohexanes have been performed. As might be expected, the 1,3-diaxial interactions of large substituent groups are greater than the interactions in methylcyclohexane. For example, the equatorial conformation of tert-butylcyclohexane is favored over the axial conformation by about 20 kJ mol^{-1} (about 5 kcal mol^{-1}).

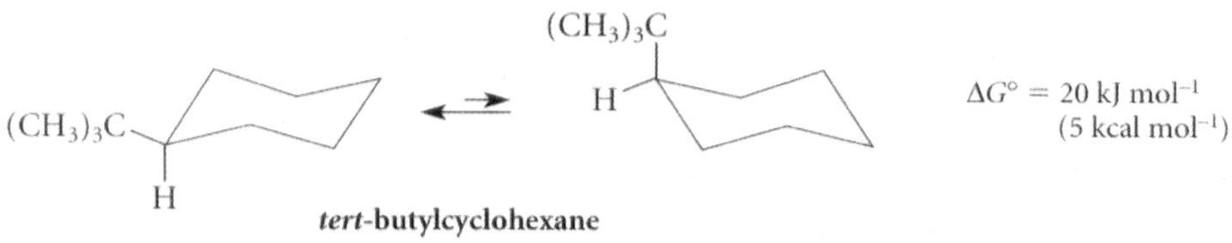

(7.7)

This means that a sample of tert-butylcyclohexane contains a truly minuscule amount of the axial conformation.

Production and Use

It can be also produced by hydrogenation of toluene:

$$CH_3C_6H_5 + 3\ H_2 \rightarrow CH_3C_6H_{11}$$

Methylcyclohexane, as a component of a mixture, is usually dehydrogenated to toluene, which increases the octane rating of gasoline.

It is also one of a host substances in jet fuel surrogate blends, e.g., for Jet A fuel.

Solvent

Methylcyclohexane has no particular applications, although it used as an organic solvent, with properties similar to related saturated hydrocarbons such as heptane.

Structure

Methylcyclohexane is a monosubstituted cyclohexane because it has one branching via the attachment of one methyl group on one carbon of the cyclohexane ring. Like all cyclohexanes, it can interconvert rapidly between two chair conformers. The lowest energy form of this monosubstituted methylcyclohexane occurs when the methyl group occupies an equatorial rather than an axial position. This equilibrium is embodied in the concept of A value. In the axial position, the methyl group experiences steric crowding (steric strain) because of the presence of axial hydrogen atoms on the same side of the ring (known as the 1,3-diaxial interactions). There are two such interactions, with each pairwise methyl/hydrogen combination contributing approximately 7.61 kJ/mol of strain energy. The equatorial conformation experiences no such interaction, and so it is the energetically favored conformation.

Flammability and Toxicity

Methylcyclohexane is flammable.

Furthermore, it is considered “very toxic to aquatic life”. Note, while methylcyclohexane is a substructure of 4-methylcyclohexanemethanol (MCHM), it is distinct in its physical, chemical, and biological (ecologic, metabolic, and toxicologic) properties.

Organic Peroxide

An organic peroxide is any organic (carbon-containing) compound having two oxygen atoms joined together (-O-O-). This chemical group is called a "peroxy" group. Organic peroxides can be severe fire and explosion hazards.

The plastics and rubber industries are the heaviest users of organic peroxides. Organic peroxides and mixtures containing an organic peroxide are used as accelerators, activators, catalysts, cross-linking agents, curing agents, hardeners, initiators and promoters. Organic peroxides and mixtures containing an organic peroxide are often referred to by these terms. However, using terms like accelerator, activator, etc. to mean "organic peroxide" can be misleading since they can also refer to materials that do not contain organic peroxides. This can cause confusion and a serious accident could result if these substances were mixed with organic peroxides.

Organic peroxides are available as solids (usually fine powders), liquids or pastes. Some materials, such as water, odourless mineral spirits, and some phthalate esters do not react with organic peroxides and are often used to dilute them. The diluted mixtures or formulations are less likely to explode when exposed to heat or physical shock than the undiluted organic peroxide. Dilution makes the unstable peroxides safer to produce, handle, and use. We use the term "organic peroxide" to refer to both undiluted and diluted organic peroxides, unless otherwise specified.

Hazardousness Associated with Organic Peroxides

The main hazard related to organic peroxides are their fire and explosion hazards. Organic peroxides may also be toxic or corrosive. Depending on the material, route of exposure (inhalation, eye or skin contact, or swallowing) and dose or amount of exposure, they could harm the body. Corrosive organic peroxides can also attack and destroy metals.

It is the double oxygen of the "peroxy" group that makes organic peroxides both useful and hazardous. The peroxy group is chemically unstable. It can easily decompose, giving off heat at a rate that increases as the temperature rises. Many organic peroxides give off flammable vapours when they decompose. These vapours can easily catch fire.

Most undiluted organic peroxides can catch fire easily and burn very rapidly and intensely. This is because they combine both fuel (carbon) and oxygen in the same compound. Some organic peroxides are dangerously reactive. They can decompose very rapidly or explosively if they are exposed to only slight heat, friction, mechanical shock or contamination with incompatible materials.

Organic peroxides can also be strong oxidizing agents. Combustible materials contaminated with most organic peroxides can catch fire very easily and burn very intensely (i.e., deflagrate). This means that the burn rate is very fast: it can vary from 1 m/sec to hundreds of metres per second. Also the combustion rate increases as the pressure increases and the combustion (or reaction) zone can travel through air or a gaseous medium faster than the speed of sound. However, the speed of combustion in a solid medium does not exceed the speed of sound.

This is one characteristic that distinguishes deflagration from detonation. We mention these two terms because they are used in classifying organic peroxide formulations. Deflagrations and detonations are similar chemical reactions except that in detonations the burn rate in a solid medium is faster than the speed of sound. This supersonic speed results in a shock wave being produced. They can transmit the shock wave at speeds of about 2,000 to 9,000 m/sec and is not dependent on the surrounding pressure. This is another difference between detonation and deflagration: deflagration rates increase as the pressure becomes greater.

Explosive decomposition is a rapid chemical reaction resulting in almost instantaneous release of energy. This term includes both deflagration and detonation.

Properties

The O–O bond length in peroxides is about 1.45 Å, and the R–O–O angles (R = H, C) are about 110° (water-like). Characteristically, the C–O–O–R (R = H, C) dihedral angles are about 120°. The O–O bond is relatively weak, with a bond dissociation energy of 45–50 kcal/mol (190–210 kJ/mol), less than half the strengths of C–C, C–H, and C–O bonds.

Major classes of organic peroxides include:

- Hydroperoxides, compounds with the functionality ROOH (R = alkyl).
- Peroxy acids and esters, compounds with the functionality RC(O)OOH and RC(O)OOR' (R,R' = alkyl, aryl).
- Diacyl peroxides, compounds with the functionality RC(O)OOC(O)R (R = alkyl, aryl).

- Dialkylperoxides, compounds with the functionality ROOR (R = alkyl).

These compounds occur in nature or are useful in commercial settings. Still other more specialized peroxy compounds are known.

Biology

Peroxides play important roles in biology. Hundreds of peroxides and hydroperoxides are known, being derived from fatty acids, steroids, and terpenes. Derived from fatty acids are a number of 1,2-dioxenes. The biosynthesis prostaglandins proceeds via an endoperoxide, a class of bicyclic peroxides.

Illustrative biosynthetic transformation involving a hydroperoxide. Here cis-3-hexenal is generated by conversion of linolenic acid to the hydroperoxide by the action of a lipoxygenase followed by the lyase-induced formation of the hemiacetal.

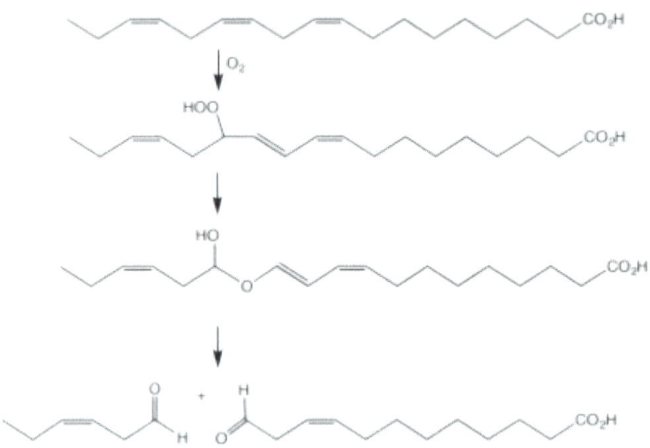

In firefly, oxidation of luciferins, which is catalyzed by luciferases, yields a peroxy compound 1,2-dioxetane. The dioxetane is unstable and decays spontaneously to carbon dioxide and excited ketones, which release excess energy by emitting light (bioluminescence).

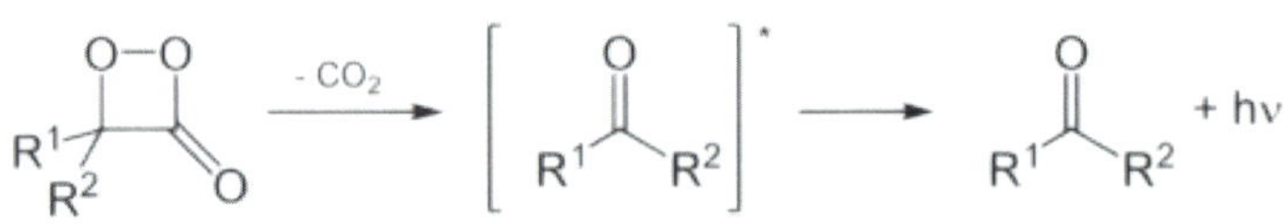

Loss of CO_2 of a dioxetane, giving rise to an excited ketone, which relaxes by emitting light.

Firefly Lampyris noctiluca

Industrial Uses

In polymer Chemistry

Dibenzoyl peroxide is used as a radical initiator, assist polymerization of acrylates. Industrial resins based on acrylic and/or methacrylic acid esters are invariably produced by radical polymerization with organic peroxides at elevated temperatures. The polymerization rate is adjusted by suitable choice of temperature and type of peroxide.

Drying oils, as found in many paints and varnishes function via the formation of hydroperoxides.

Methyl ethyl ketone peroxide, benzoyl peroxide and to a smaller degree acetone peroxide are used as initiators for radical polymerisation of some resins, e.g. polyester and silicone, often encountered when making fiberglass. Pinane hydroperoxide is used in production of styrene-butadiene (synthetic rubber).

Bleaching and Disinfecting Agents

Benzoyl peroxide and hydrogen peroxide are used as bleaching and "maturing" agents for treating flour to make its grain release gluten more easily; the alternative is letting the flour slowly oxidize by air, which is too slow for the industrialized era. Benzoyl peroxide is an effective topical medication for treating most forms of acne.

In the Synthesis of Organic Compounds

Hydroperoxides are intermediates in the production of many organic compounds in industry. For example, the cobalt catalyzed oxidation of cyclohexane to cyclohexanone:

$$C_6H_{12} + O_2 \rightarrow (CH_2)_5CO + H_2O$$

Acetone and phenol are produced by the so-called cumene process, which proceeds via cumene hydroperoxide.

Many epoxides are prepared using peroxides as reagents, such as the Halcon process that gives propylene oxide. The Sharpless epoxidation is a related reaction conducted on laboratory scale. Tert-butyl hydroperoxide (TBHP) is an organic-soluble oxidant employed in these operations.

Preparation

From Hydrogen Peroxide

Dialkyl sulfates react with alkaline hydrogen peroxide. In this method, the alkyl sulfate donates the alkyl group and the sulfate ion forms the leaving group:

$$R_2SO_4 + H_2O_2 \rightarrow R-O-O-R + H_2SO_4$$

This method can also yield cyclic peroxides. The four-membered dioxetanes can be obtained by 2+2 cycloaddition of oxygen to alkenes.

From O_2

The formation of hydroperoxides is effected extensively by enzymes.

This process is used industrially on a very large scale for the production of phenol from benzene and is called the Cumene process or Hock process for its cumene and cumene hydroperoxide intermediates. Here the active site formed by a radical initiator reacts with oxygen to form a hydroperoxyl. The addition of oxygen results in a more active radical which can further extract hydrogen atoms and release the hydroperoxide, leaving a new radical.

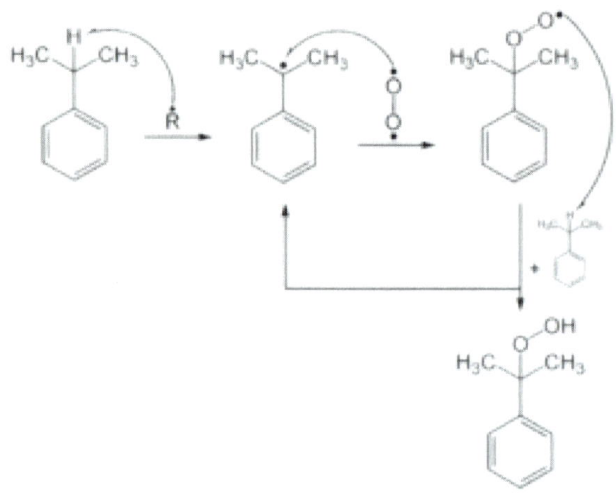

Synthesis of cumene hydroperoxide

In the laboratory, the production of peroxides from O_2 is not practiced often. Compounds with allylic and benzylic C–H bonds are especially susceptible to oxygenation.

Auto-oxidation reaction is observed with common ethers, such as diethyl ether, diisopropyl ether, tetrahydrofuran or 1,4-dioxane. An illustrative product is diethyl ether peroxide. Such compounds can result in a serious explosion.

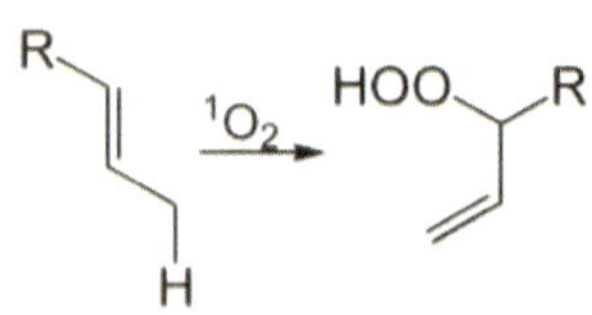

Synthesis of hydroperoxides of alkene and singlet oxygen in an ene reaction

Reactions

Hydroperoxides and diorganoperoxides can be reduced to alcohols with lithium aluminium hydride, as described in this idealized equation:

$$4ROOH + LiAlH_4 \rightarrow LiAlO_2 + 2H_2O + 4ROH$$

The phosphite esters and tertiary phosphines also effect reduction:

$$ROOH + PR_3 \rightarrow OPR_3 + ROH$$

Cleavage to ketones and alcohols in the base catalyzed Kornblum–DeLaMare rearrangement

Some peroxides are drugs, whose action is based on the formation of radicals at desired locations in the organism. For example, artemisinin and its derivatives, such as such artesunate, possess the most rapid action of all current drugs against falciparum malaria. Artesunate is also efficient in reducing egg production in *Schistosoma haematobium* infection.

Identifying Reactions

Several analytical methods are used for qualitative and quantitative determination of peroxides. A simple qualitative detection of peroxides is carried out with the iodine-starch reaction. Here peroxides, hydroperoxides or peracids oxidize the added potassium iodide into iodine, which reacts with starch producing a deep-blue color. Commercial paper indicators using this reaction are available. This method is also suitable for quantitative evaluation, but it can not distinguish between different types of peroxide compounds. Discoloration of various indigo dyes in presence of peroxides is used instead for this purpose. For example, the loss of blue color in leuco-methylene blue is selective for hydrogen peroxide.

Iodine-starch test. Note the blackening (left) of initially yellowish (right) starch.

Quantitative analysis of hydroperoxides can be performed using potentiometric titration with lithium aluminium hydride. Another way to evaluate the content of peracids and peroxides is the volumetric titration with alkoxides such as sodium ethoxide.

Active Oxygen in Peroxides

Each peroxy group is considered to contain one active oxygen atom. The concept of active oxygen content is useful for comparing the relative concentration of peroxy groups in formulations, which is related to the energy content. In general, energy content increases with active oxygen content, and thus the higher the molecular weight of the organic groups, the lower the energy content and, usually, the lower the hazard.

The term active oxygen is used to specify the amount of peroxide present in any organic peroxide formulation. One of the oxygen atoms in each peroxide group is considered "active". The theoretical amount of active oxygen can be described by the following equation:

$$A[O]_{\text{theoretical}}\,(\%) = 16p/m \times 100,$$

where, p is the number of peroxide groups in the molecule, and m is the molecular mass of the pure peroxide.

Organic peroxides are often sold as formulations that include one or more phlegmatizing agents. That is, for safety sake or performance benefits the properties of an organic peroxide formulation are commonly modified by the use of additives to phlegmatize (desensitize), stabilize, or otherwise enhance the organic peroxide for commercial use. Commercial formulations occasionally consist of mixtures of organic peroxides, which may or may not be phlegmatized.

Thermal Decomposition of Organic Peroxides

Organic peroxides are useful in chemical synthesis due to their propensity to decompose. In doing so they generate useful radicals that can initiate polymerization to create polymers, modify polymers by grafting or visbreaking, or cross-link polymers to create a thermoset. When used for these purposes, the peroxide is highly diluted, so the heat generated by the exothermic decomposition is safely absorbed by the surrounding medium (e.g. polymer compound or emulsion). But when a peroxide is in a more pure form, the heat evolved by its decomposition may not dissipate as quickly as it is generated, which can result in increasing temperature, which further intensifies the rate of exothermic decomposition. This can create a dangerous situation known as a *self-accelerating decomposition*.

A self-accelerating decomposition occurs when the rate of peroxide decomposition is sufficient to generate heat at a faster rate than it can be dissipated to the environment. Temperature is the main factor in the rate of decomposition. The lowest temperature at which a packaged organic peroxide will undergo a self-accelerating decomposition within a week is defined as the *self-accelerating decomposition temperature* (SADT).

Safety

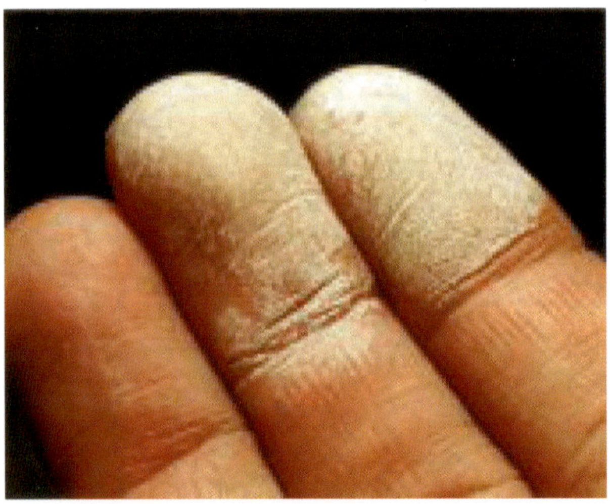

Skin shortly after exposure to 35% H_2O_2

Peroxides are also strong oxidizers and easily react with skin, cotton and wood pulp. For safety reasons, peroxidic compounds are stored in a cool, opaque container, as heating and illumination accelerate their chemical reactions. Small amounts of peroxides, which emerge from storage or reaction vessels are neutralized using reducing agents such as iron(II) sulfate. Safety measures in industrial plants producing large amounts of peroxides include the following:

- The equipment is located within reinforced concrete structures with foil windows, which would relieve pressure and not shatter in case of explosion.
- The products are bottled in small containers and are moved to a cold place promptly after the synthesis.
- The containers are made of non-reactive materials such as stainless steel, some aluminium alloys or dark glass.

For safe handling of concentrated organic peroxides, an important parameter is temperature of the sample, which should be maintained below the self accelerating decomposition temperature of the compound.

Organic Superconductor

Organic superconductivity refers to the regular phenomenon of superconductivity as it is observed in some metals and metallic inorganic compounds. However, what makes organic superconductivity so distinctive is that conduction in organic molecular conductors is linked to the transport of free charges (electrons or holes) between π-like molecular orbitals of neighbouring open shell molecules. In addition, these new materials reveal quasi one dimensional features of their electronic transport properties due to their peculiar crystal structure. They are considered as textbook examples for low dimensional physics. They have also been stimulating for the research on high *Tc* superconductors and pnictides.

Superconductivity

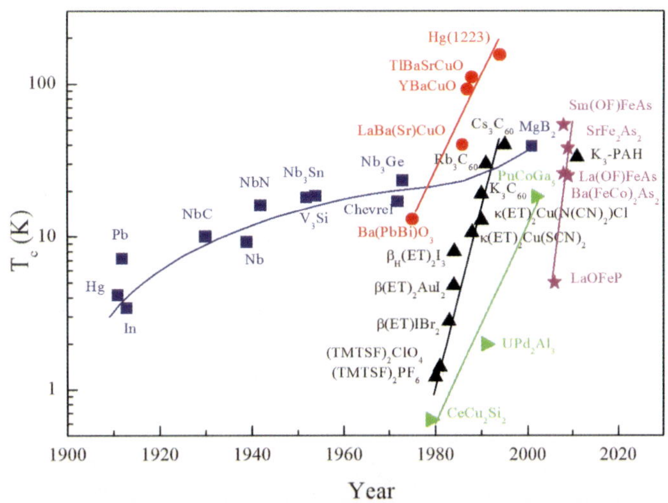

Figure: The figure displays the evolution of *Tc* in materials according to the date of the discovery of their superconducting properties.

It clearly shows a renewed effort in superconductivity at the beginning of the eighties which triggered the research of novel materials with the discovery of organic superconductors in 1980 and high *Tc* cuprates in 1986.

Superconductivity (a new state of matter) is one of the major discovery in physics for the twentieth century given its experimental and theoretical impact on science. From an experimental point of view the finding by G. Holst in the laboratory of H. Kamerlingh Onnes in 1911 of a current travelling without resistance through a metal cooled at very low temperature paved the way to a very large number of industrial applications. This experimental discovery happened long before the proposal of a satisfactory theoretical framework by Bardeen, Cooper and Schrieffer (BCS) in 1957 explaining the two basic properties of the superconducting ground state of a metal namely, the zero resistance state and the magnetic flux expulsion.

Superconductivity requires low temperature conditions to be observed in metals and intermetallic compounds, typically the helium temperature range. Until the years in the sixties, the critical temperature for superconductivity had been steadily rising as new intermetallic compounds were synthesized according to the paradigm of the BCS theory predicting an isotope effect namely, an electron-electron pairing in the superconducting ground state mediated by the electron-phonon interaction with the ionic mass dependence of the critical temperature, $Tc\ aM^{-1/2}$.

Searching for new materials exhibiting the highest possible values of superconducting *Tc* was already a strong motivation in materials sciences in the early 70's, and the term high temperature superconductor was already commonly used referring to the intermetallic compounds of the A15 structure, namely, (Nb_3Sn or V_3Si).

Expending the successful idea of the isotope effect of the BCS theory other models were proposed in which excitations of the lattice responsible for the electron pairing had been replaced by higher energy excitations namely, electronic excitations, with the hope of new materials with *Tc* higher than those explained by the BCS theory. Consequently, the small electronic mass *me* of the polarizable medium would lead to an enhancement of *Tc* of the order of $(M/me)1/2$ times the value which is observed in a conventional superconductor, admittedly a huge factor. V.L. Ginzburg considered in 1964 the possibility for the pairing of electrons in metal layers sandwiched between polarizable dielectrics through virtual excitations at high energy.

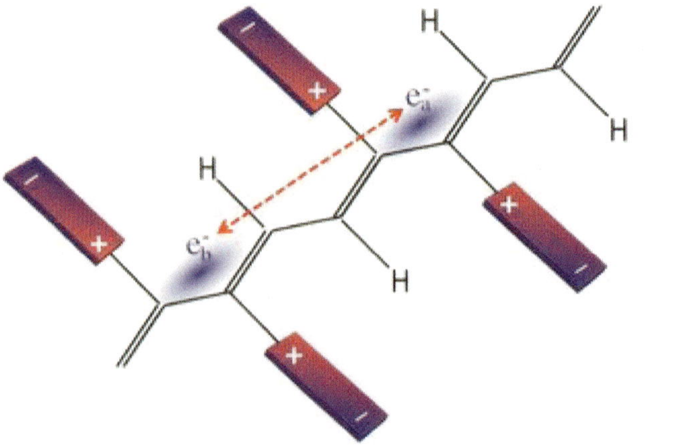

Figure: Schematic picture for the Little's suggestion. Charge carriers a and b forming a Cooper pair in the conducting spine are bound via a virtual electronic excitation of the polarizable side groups.

An even more provocative suggestion came from W.A. Little namely, a new mechanism for superconductivity which could lead to a drastic enhancement of the superconducting *Tc* to be observed in especially designed macromolecules. The idea of Little was still rooted in the extension of the isotope effect proposed by BCS replacing the mediating phonon by an electronic excitation which especially designed macromolecules could reveal. However a prerequisite to the model of Little was the achievement of metallic conduction in organic molecular crystals, not a trivial problem in the sixties.

Organic Conductors

Electronic conduction proceeds in organic molecular conductors through the transport of free charges (electrons or holes) between π -like molecular orbitals of neighbouring open shell molecules rather than between atomic orbitals in regular metallic conductors. One cannot find organic conductors in nature, unlike the metallic ones. The former must always be synthesized in a chemistry laboratory.

The first successful attempt to promote metal-like conduction between open shell molecular species came out in 1954 with the synthesis of the molecular salt of perylene oxidized with bromine although this salt was unstable in air.

Organic Superconductors Part of a Generic Phase Diagram

The Copenhagen group led by Klaus Bechgaard and experienced with the chemistry of selenium succeeded in the synthesis of a new series of conducting salts all based on the TMTSF molecule namely, $(TMTSF)_2X$ where X is an inorganic mono-anion with various possible symmetry, spherical (PF_6, AsF_6, SbF_6, TaF_6), tetrahedral (BF_4, ClO_4, ReO_4) or triangular (NO_3). All these compounds but the one with X=ClO_4 did reveal an insulating ground state under ambient pressure.

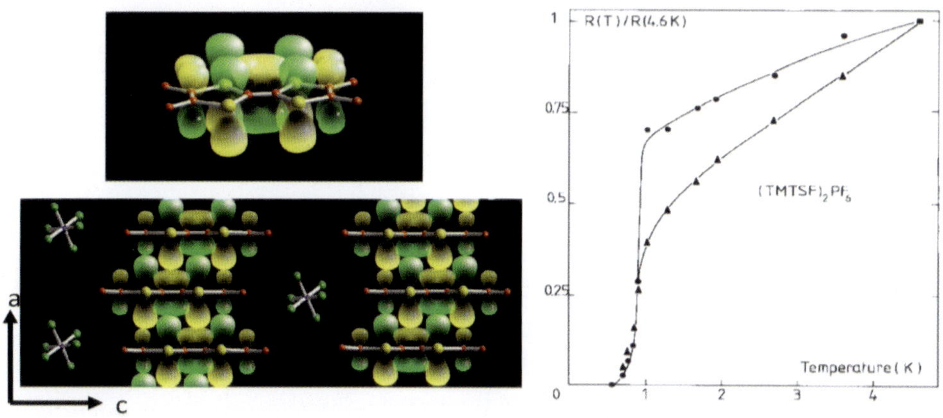

Side view of the TMTSF molecule (yellow and red dots are selenium and carbon atoms respectively, hydrogens not shown) and $(TMTSF)_2PF_6$ quasi 1D structure along the *b* axis, *courtesy* of J.Ch. Ricquier, IMN, Nantes (left). First observation of superconductivity in $(TMTSF)_2PF_6$ under a pressure of 9 kbar. The resistance of two samples is normalized to the 4.5 K value (right).

What is so special with $(TMTSF)_2PF_6$, the prototype of the so-called Bechgaard salts? Unlike previously investigated TTF-TCNQ, it is the magnetic origin of the ambient pressure insulating state contrasting with the Peierls-like ground states discovered previously in charge transfer compounds. The ground state of $(TMTSF)_2PF_6$ turned out to be a spin density wave state similar to the predictions made by Lomer in 1962 and by Overhauser for metals. It is the onset of itinerant antiferromagnetism which opens a gap at Fermi level. Since the Fermi surface is nearly planar, the exchange gap develops over the whole surface. The magnetic origin of the insulating ground state of $(TMTSF)_2PF_6$ was thus the first experimental hint for the prominent role played by correlations in these organic conductors.

Superconductivity of $(TMTSF)_2PF_6$ occurred at 1K as shown by transport, once the *SDW* insulating ground state could be suppressed increasing the transverse overlap between molecular stacks under a pressure of about 9 kbar. AC susceptibility measurements performed on the same material detected an anomaly at T_c indicative of a transition into a diamagnetic state. A confirmation of the bulk nature of organic superconductivity has been given subsequently by the observation of shielding and Meissner signals in $(TMTSF)_2PF_6$.

The synthesis of the superconducting compound $(TMTSF)_2PF_6$ allowed to establish a link between this compound and the isostructural family comprising the sulfur molecule with the same series of monoanions discovered earlier. Thanks to an intensive study under pressure, it was realized that $(TMTTF)_2X$ and $(TMTSF)_2X$ salts both belong to a common class of materials forming the generic $(TM)_2X$ phase diagram.

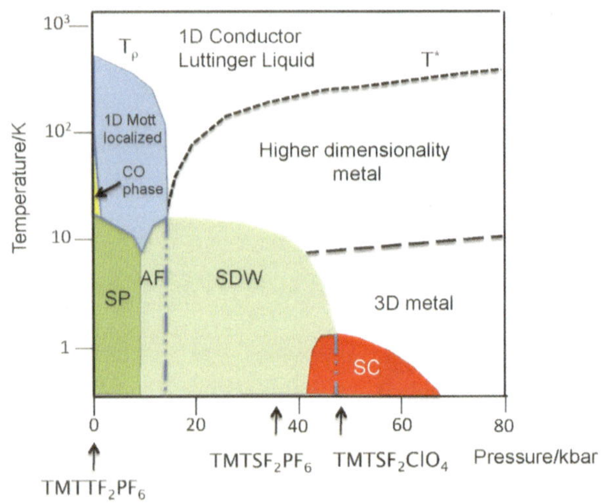

Generic phase diagram for the $(TM)_2X$ family based on the sulfur compound $(TMTTF)_2PF_6$ under ambient pressure taken as the origin for the pressure scale. Phases in yellow, green, light green and red colors are long range ordered, Charge Order (CO), Spin Peierls (SP), Magnetic phases Neel antiferromagnet and Spin Density Wave phase and Superconductivity respectively. There exists a small pressure window around 45 kbar in this generic diagram where SC coexists with SDW according to reference. *Tp* marks the onset of 1D charge localization which ends around 15 kbar at the dashed double-dotted line. The (TL)1D to 2D deconfinement occurs at *T*⋆. The dashed line between 2D and 3D regimes defines the upper limit for low temperature 3D coherent domain.

In particular, $(TMTTF)_2PF_6$, although the most insulating compound of the phase diagram can be made superconducting at low temperatures under a pressure of 45 kbar. The study of this generic phase diagram has in turn greatly contributed to the experimental exploration of 1D physics and to the comparison with the theory.

$(TM)_2X$ conductors exhibit a band filling commensurate with the underlying 1D lattice due to the 2:1 stoichiometry imposing half a hole per TM molecule. However, non-uniformity of the intermolecular spacing had been noticed from the early structural studies of $(TMTTF)_2PF_6$ crystal due to the periodicity of the anion packing being twice the periodicity of the molecular packing. This non-uniformity provides a dimerization of the intrastack overlap quite significant in the sulfur series (prominent 1/2 filling Umklapp scattering) and still present in the members of the $(TMTSF)_2PF_6$ series. An important consequence of the commensurate situation for $(TM)_2X$ materials is the existence of two localization channels either due to electron-electron Umklapp scattering with momentum transfer $4kF$ (two particles scattering) or to $8kF$ (four particles scattering) corresponding to half or quarter-filled bands respectively. Consequently, both Umklapp mechanisms compete for the establishment of the 1D charge gap $\Delta\rho$, 1/4 and 1/2 filling Umpklapp scatterings leading to charge ordered and Mott insulators respectively.

On the left side of the generic diagram a phase transition towards a long range charge ordered (CO) insulating phase has been observed. This phase at low temperature has been ascribed, according to NMR data, to the onset of a charge disproportionation between molecules on the molecular chains. Increasing pressure, spins localized on dimers of molecules couple to the lattice and give rise to a spin-Peierls ground state evolving through a quantum critical point around 10 kbar into a Néel antiferromagnetic phase and subsequently a spin density wave state under pressure which have been extensively studied by transport and AFMR.

Moving toward the right side of the diagram, the onset of charge localization below $T\rho$ (i.e, 1D confinement) decreases under pressure as a result of an interplay between the Mott localization and the interchain coupling (increasing under pressure) which tends to bypass the correlation-induced localization. Once the localization gap vanishes, 1D confinement ceases and the existence of a quasi-1D Fermi surface becomes meaningful. In some respects the deconfinement observed under pressure around 15 kbar is the signature for a crossover from strong to weak coupling in the generic phase diagram. The Néel phase turns into an ordered weak antiferromagnetic spin density wave phase (*SDW*) above the deconfinement pressure of about 15 kbar.

The temperature T^* where the c^*-axis transport switches from an insulating to a metallic temperature dependence corresponds to a cross-over between two regimes, a high temperature regime in which the finite QP weight at Fermi energy is smaller than the regular metallic value. T^* is highly pressure dependent and can be as small as 20 K or so, a value much smaller than the one expected for the bare $t\perp$ of order 200 K. T^* would recover the value of the bare coupling only under infinite pressure.

Regarding the upper part of the Generic phase diagram transport in $(TM)_2X$ has been interpreted in terms of the Tomonaga-Luttinger liquid model for a commensurate conductor. A metal-like behaviour of the longitudinal resistance is still observed as long as T is larger than $\Delta\rho$ leading to a resistance displaying a power law $\rho(T)\approx T^\theta$ (θ>0). Experimental studies have revealed such a metallic behaviour for the resistance either in $(TMTTF)_2X$ under pressure or in $(TMTSF)_2X$ even at ambient pressure. What has been found is a sublinear exponent, namely θ=0.93 for the constant volume temperature dependence of the $(TMTSF)_2PF_6$ resistance. In the vicinity of T^*, this power law evolves toward a constant value in line with a Fermi liquid picture.

When quarter-filled Umklapp scattering prevails at high temperature $\theta=16K_\rho-3$ and consequently $K_\rho=0.23$ according to the data of $(TMTSF)_2PF_6$.

The temperature T^* where the c^*-axis transport switches from an insulating to a metallic temperature dependence corresponds to a cross-over between two regimes; a high temperature regime with no QP weight at Fermi energy in which the QP weight increases with decreasing temperature.

A three dimensional anisotropic coherent picture prevails at low temperature in $(TMTSF)_2X$ compounds according to the Kohler's rule and angular magnetoresistance oscillations observed in $(TMTSF)_2ClO_4$ and $(TMTSF)_2PF_6$ under pressure. In addition, optical reflectance data of light polarised along c^* support the existence for $(TMTSF)_2ClO_4$ of a weak Drude behaviour in the liquid helium temperature domain when $kBT<t\perp c$. However, the upper limit for the temperature domain of 3D coherence has been established comparing the temperature dependence of the resistivity along a and c^*. This 2D to 3D crossover domain is defined by the temperature above which the temperature dependencies of both components of transport are no longer identical.

Organic Superconductivity

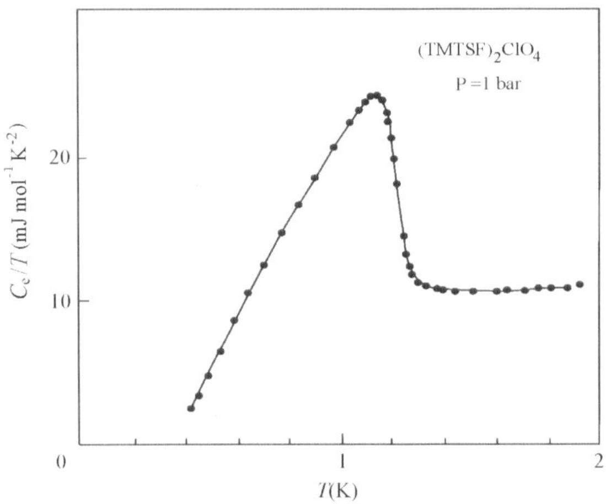

Figure: Electronic contribution to the specific heat of $(TMTSF)_2ClO_4$, plotted as Ce/T versus T, according to ref.

Although superconductivity in organic conductors has been first stabilized under pressure, more detailed investigations of this phenomenon have been conducted in $(TMTSF)_2ClO_4$, the only compound of the 1D-Bechgaard salts series exhibiting superconductivity at ambient pressure below 1.2 K. Evidences for superconductivity in $(TMTSF)_2ClO_4$ came out from transport, specific heat measurements and Meissner flux expulsion.

The electronic contribution to the specific heat of $(TMTSF)_2ClO_4$ in a C_e/T *vs* T plot, displays a very large anomaly around 1.2 K. The total specific heat obeys the classical relation in metals $C/T=\gamma+\beta T^2$, where the Sommerfeld constant for electrons γ = 10.5 mJ mol^{-1}K^{-2}, corresponding to a density of states at the Fermi level $N(EF)$ = 2.1 states eV^{-1} mol^{-1} for the two spin directions The specific heat jump at the transition amounts then to $\Delta C_e/\gamma T_c$=1.67, *i.e.* only slightly larger than the BCS ratio for a *s*-wave superconductor. Therefore, the specific heat data and the comparison between the value of the density of states derived specific heat Pauli susceptibility end support to a weak coupling Fermi liquid picture (at least in the low temperature range). The first critical field H_{c1} obtained from the Meissner magnetization curves at low temperature read, 0.2,1 and 10 Oe along axes a, $b'$and $c\star$ respectively. Following the values for the second critical fields H_{c2} derived either from the Meissner experiments and the knowledge of the thermodynamical field or from a

- Headaches, loss of coordination and nausea;
- Damage to liver, kidney and central nervous system;
- Some organics can cause cancer in animals, some are suspected or known to cause cancer in humans.

Key signs or symptoms associated with exposure to VOCs include:

- Conjunctival irritation;
- Nose and throat discomfort;
- Headache;
- Allergic skin reaction;
- Dyspnea;
- Declines in serum cholinesterase levels;
- Nausea;
- Emesis;
- Epistaxis;
- Fatigue;
- Dizziness.

The ability of organic chemicals to cause health effects varies greatly from those that are highly toxic, to those with no known health effect.

As with other pollutants, the extent and nature of the health effect will depend on many factors including level of exposure and length of time exposed. Among the immediate symptoms that some people have experienced soon after exposure to some organics include:

- Eye and respiratory tract irritation;
- Headaches;
- Dizziness;
- Visual disorders and memory impairment.

References

- Thomas Brock, Michael Groteklaes, Peter Mischke Lehrbuch der Lacktechnologie, Vincentz Network GmbH & Co KG, 2000, ISBN 3-87870-569-7 p. 67
- General-introduction-organic-compounds, organic-chemistry: toppr.com, Retrieved 18 July 2018
- Michael T. Musser "Cyclohexanol and Cyclohexanone" in Ullmann's Encyclopedia of Industrial Chemistry, Wiley-VCH, Weinheim, 2005.doi:10.1002/14356007.a08_217
- Alicyclic-compound, science: britannica.com, Retrieved 28 May 2018

- Martin, Richard (September 24, 2015). "New Technology to Capture, Convert Carbon Dioxide | MIT Technology Review". Retrieved 2015-09-27
- Diazonium-salt, science: britannica.com, Retrieved 11 April 2018
- Higuchi, T.; Zuck, Donald Anton (1951). "Behaviors of Several Compounds as Indicators in Lithium Aluminum Hydride Titration of Functional Groups". Journal of the American Chemical Society. 73 (6): 2676. doi:10.1021/ja01150a073
- Diazonium-salt-the-methods-of-preparation, chemistry: byjus.com, Retrieved 31 March 2018
- Marye Anne Fox; James K. Whitesell (2004). Organic Chemistry (3, illustrated ed.). Jones & Bartlett Learning. pp. 535–538. ISBN 978-0-7637-2197-8
- Volatile-organic-compounds-impact-indoor-air-quality, indoor-air-quality-iaq: epa.gov, Retrieved 22 April 2018
- Medwedew, S. S.; Alexejewa, E. N. (1932). "Organic peroxides II. Of the reaction between benzoyl hydroperoxide or benzoyl peroxide and triphenylmethyl". Berichte der deutschen chemischen Gesellschaft (A and B Series). 65 (2): 137. doi:10.1002/cber.19320650204

3

Organometallic Compounds

Compounds that contain one or more chemical bonds between a metal, such as alkaline, transition and alkaline earth metals, and a carbon atom of an organic molecule are called organometallic compounds. The topics elaborated in this chapter on s- and p-block organometallic compounds, organoboron compounds, metal hydrides, organoaluminium compounds, etc. will help in providing a better perspective about organometallic compounds.

Organometallic compound is a member of a class of substances containing at least one with carbon with very little polarization of an organic group. Organometallic compounds constitute a very large group of substances that have played a major role in the development of the science of chemistry. They are used to a large extent as catalysts (substances that increase the rate of reactions without themselves being consumed) and as intermediates in the laboratory and in industry. The class includes such compounds as ferrocene, a remarkably stable compound in which an iron atom is sandwiched between two hydrocarbon rings.

Organometallic compounds are typically discussed in terms of the metal as either main-group compounds or transition metal compounds. The main-group metals of organometallic compounds are typically considered to be those of the S-block and the heavier elements of the p-block in the periodic table of elements. The transition metals include those elements in the d- and f-blocks.

The physical and chemical properties of organometallic compounds vary greatly. Most are solids, particularly those whose hydrocarbon groups are ring-shaped or aromatic, but some are liquids and some are gases. Their heat and oxidation stability vary widely. Some are very stable, but a number of compounds of electropositive elements such as lithium, sodium, and aluminum are spontaneously flammable. Many organometallic compounds are highly toxic, especially those that are volatile.

The properties of the organometallic compounds depend in large measure on the type of carbon-metal bonds involved. Some are ordinary covalent bonds, in which pairs of electrons are shared between atoms. Others are multicentre covalent bonds, in which the bonding involves more than two atoms. A third type are ionic bonds, in which the bonding electron pair is donated by only one atom. In donor-acceptor bonds, the metal atom is connected to hydrocarbons with multiple bonds between carbon atoms.

Where metal atoms form covalent bonds with carbon atoms, the electrons are usually shared unequally. As a result, the bond is polarized—one end is more negative than the other. The extent of polarization depends on the strength with which the metal atom binds electrons. Organometallic compounds range in polar power from methylpotassium, in which the bond is almost like certain ionic bonds, to lead, which bonds with carbon with very little polarization.

Classification of Organometallic Compounds

A useful subdivision is by the type of M-C bond:

- Ionic - with most Group 1 elements;
- Covalent - with many Group 12, 13, 14 and 15 elements;
- Electron deficient - with Li, Be, Mg, B, Al.

Ionic

Ionic organometallic compounds are generally formed from elements such as sodium, potassium etc. where the metals are considered electropositive. If the organic groups are able to delocalise the negative charge over several carbon atoms then less electropositive elements like magnesium can also form ionic compounds, eg Cp_2Mg. In this case the charge is considered to be delocalised over each of the five carbon atoms in each ring.

Covalent

The simplest model of the M-C bond is where it consists of essentially a single covalent 2-electron bond. These compounds are often volatile and are comparable to typical organic compounds being soluble in organic solvents.

Electron deficient

Electron deficient organometallic compounds are generally associated with elements that have less than half-filled valence shells and are designated as such because of an insufficient number of valence electrons to allow all the atoms to be linked by traditional two-electron two-centre bonds. The compounds often have bridged or polymeric structures. The methyl derivatives of Li, Be and Al are found to be 3-D polymers, linear chains and dimeric respectively and despite the increase in RMM of the monomeric unit there is actually an increase in volatility.

Compound	RMM of monomeric unit	Structure	Volatility
LiMe	21.96	3D-polymer	infusible
BeMe2	39.08	Linear chain	sublimes at 473 K
AlMe3	72.08	Dimer	Melts at 288 K

Stability of Organometallic Compounds

The M-C bond energies for some methyl derivatives are shown in the table, plotting these values against Atomic Number of the metal shows that there is a decrease down a Group. This behaviour is expected since there should be better orbital overlap between similar valence orbitals and this would decrease for the larger more diffuse elements lower down a Group.

Bond dissociation energies and Heats of Formation											
Me_2M	**D / kJ mol^{-1}**	**ΔHf/kJ mol^{-1}**	**BP /K**	**Me_3M**	**D / kJ mol^{-1}**	**ΔHf/kJ mol^{-1}**	**BP /K**	**Me_4M**	**D / kJ mol^{-1}**	**ΔHf/kJ mol^{-1}**	**BP /K**
Me_2Be			490.2	Me_3B	364.0	-122.6	251.2	Me_4C	347.3	-167.4	283.2

Me_2Mg				Me_3Al	276.1	-129.7	399.2	Me_4Si	292.9	-238.5	300.2
Me_2Zn	175.7	54.8	317.2	Me_3Ga	246.9	-45	329.2	Me_4Ge	246.9	-71	316.2
Me_2Cd	138.1	109.6	379.2	Me_3In	171.5		409.2	Me_4Sn	217.6	-19.2	350.2
Me_2Hg	121.3	93.3	366.2	Me_3Tl			420.2	Me4Pb	154.8	136.4	383.2
				Me3As	230.1	15.5	325.2				
				Me3Sb	217.6	31.0	352.2				
				Me3Bi	142.3	192.9	383.2				

Concepts and Techniques

As in other areas of chemistry, electron counting is useful for organizing organometallic chemistry. The 18-electron rule is helpful in predicting the stabilities of metal carbonyls and related compounds. Most organometallic compounds do not however follow the 18e rule. Chemical bonding and reactivity in organometallic compounds is often discussed from the perspective of the isolobal principle.

As well as X-ray diffraction, NMR and infrared spectroscopy are common techniques used to determine structure. The dynamic properties of organometallic compounds is often probed with variable-temperature NMR and chemical kinetics.

Organometallic compounds undergo several important reactions:

- Oxidative addition and reductive elimination;
- Transmetalation;
- Carbometalation;
- Hydrometalation;
- Electron transfer;
- β-hydride elimination;
- Organometallic substitution reaction;
- Carbon-hydrogen bond activation;
- Cyclometalation;
- Migratory insertion;
- Nucleophilic abstraction.

Industrial Applications

Organometallic compounds find wide use in commercial reactions, both as homogeneous catalysis and as stoichiometric reagents For instance, organolithium, organomagnesium, and organoaluminium compounds, examples of which are highly basic and highly reducing, are useful stoichiometrically, but also catalyze many polymerization reactions.

Almost all processes involving carbon monoxide rely on catalysts, notable examples being described as carbonylations. The production of acetic acid from methanol and carbon monoxide is catalyzed via metal carbonyl complexes in the Monsanto process and Cativa process. Most synthetic aldehydes are produced via hydroformylation. The bulk of the synthetic alcohols, at least those larger than ethanol, are produced by hydrogenation of hydroformylation-derived aldehydes. Similarly, the Wacker process is used in the oxidation of ethylene to acetaldehyde.

Almost all industrial processes involving alkene-derived polymers rely on organometallic catalysts. The world's polyethylene and polypropylene are produced via both heterogeneously via Ziegler-Natta catalysis and homogeneously, e.g., via constrained geometry catalysts.

Most processes involving hydrogen rely on metal-based catalysts. Whereas bulk hydrogenations, e.g. margarine production, rely on heterogeneous catalysts, For the production of fine chemicals, such hydrogenations rely on soluble organometallic complexes or involve organometallic intermediates. Organometallic complexes allow these hydrogenations to be effected asymmetrically.

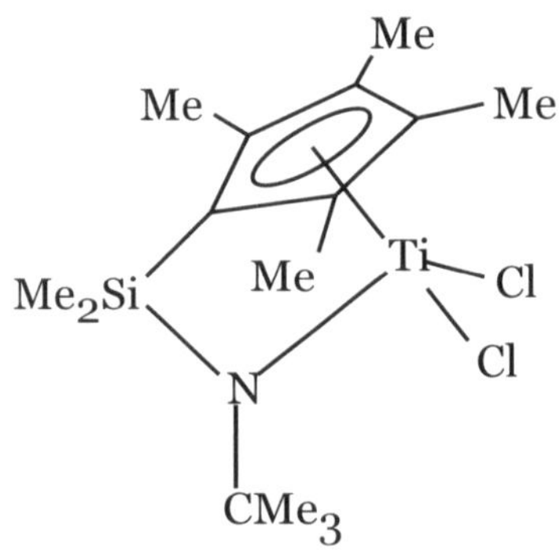

A constrained geometry organotitanium complex is a precatalyst for olefin polymerization.

Many semiconductors are produced from trimethylgallium, trimethylindium, trimethylaluminium, and trimethylantimony. These volatile compounds are decomposed along with ammonia, arsine, phosphine and related hydrides on a heated substrate via metalorganic vapor phase epitaxy (MOVPE) process in the production of light-emitting diodes (LEDs).

Organometallic Reactions

The synthesis of many organic molecules are facilitated by organometallic complexes. Sigma-bond metathesis is a synthetic method for forming new carbon-carbon sigma bonds. Sigma-bond metathesis is typically used with early transition-metal complexes that are in their highest oxidation state. Using transition-metals that are in their highest oxidation state prevents other reactions from occurring, such as oxidative addition. In addition to sigma-bond metathesis, olefin metathesis is used to synthesize various carbon-carbon pi bonds. Neither sigma-bond metathesis or olefin metathesis change the oxidation state of the metal. Many other methods are used to form new carbon-carbon bonds, including beta-hydride elimination and insertion reactions.

Catalysis

Organometallic complexes are commonly used in catalysis. Major industrial processes include hydrogenation, hydrosilylation, hydrocyanation, olefin metathesis, alkene polymerization, alkene oligomerization, hydrocarboxylation, methanol carbonylation, and hydroformylation. Organometallic intermediates are also invoked in many heterogeneous catalysis processes, analogues to those listed above. Additionally, organometallic intermediates are assumed for Fischer-Tropsch process.

Organometallic complexes are commonly used in small-scale fine chemical synthesis as well, especially in cross-coupling reactions that form carbon-carbon bonds, e.g. Suzuki-Miyaura coupling, Buchwald-Hartwig amination for producing aryl amines from aryl halides, and Sonogashira coupling, etc.

Environmental Concerns

Natural and contaminant organometallic compounds are found in the environment. Some that are remnants of human use, such as organolead and organomercury compounds, are toxicity hazards. Tetraethyllead was prepared for use as a gasoline additive but has fallen into disuse because of lead's toxicity. Its replacements are other organometallic compounds, such as ferrocene and methylcyclopentadienyl manganese tricarbonyl (MMT). The organoarsenic compound roxarsone is a controversial animal feed additive. In 2006, approximately one million kilograms of it were produced in the U.S alone.

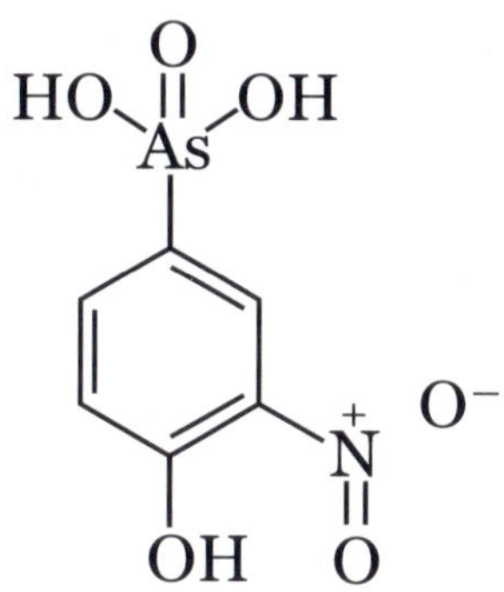

Roxarsone is an organoarsenic compound used as an animal feed.

s- and p-block Organometallic Compounds

The metal in main-group organometallic compounds can be any of the elements in the *s* block or any of the heavier elements in groups. The elements at the borderline between the *d* block and *p* block—namely, zinc, cadmium, and mercury will be discussed along with the *p*-block organometallics because of the similarity of their organometallic chemistry. In an internationally sanctioned system of nomenclature, the organic group is named first, followed by the metal, as in dimethylmercury. In writing the formula, this order is reversed, $Hg(CH_3)_2$. The organic groups, which are also called ligands, are named in the same way as for any organic compound. The number of carbon atoms on a group that are attached to the metal is indicated by the superscript in η^n. This convention is known as hapto nomenclature. A single point of attachment, η^1, is usually not explicitly indicated, as in the above formula for dimethylmercury, a monohapto species. The compound with the common name ferrocene has the systematic name bis(η^5-cyclopentadienyl)iron, where the number of cyclopentadienyl ligands (two) is indicated by the prefix *bis* and the number of sites of attachment (five) for each of these is indicated by η^5. Ferrocene is thus called a pentahapto compound. The number of sites of attachment are also indicated in the formula $Fe(\eta^5\text{-}C_5H_5)_2$.

The Stability and Reactivity of Organometallic Compounds

The stability and reactivity of organometallic compounds are associated with the nature of the organic ligands and the metal to which they are attached. In each of the main groups of the periodic table, the thermal stability of a given type of organometallic compound generally decreases from the lightest to the heaviest element in a group. For example, in compounds containing group-1 metals, methyllithium ($LiCH_3$) is much more stable than methylpotassium (KCH_3), and, in those with group-14 metals, tetramethylsilicon, $Si(CH_3)_4$, is stable in the absence of air at 500° C (932° F), whereas tetramethyllead, $Pb(CH_3)_4$, rapidly decomposes at that temperature. This trend in stability is a consequence in part of the decrease in M–C bond strength on going down within a group. The trend does not hold for the *d*-block elements, where M–C bond strengths and stability often increase going down a group.

The reactivities of organometallic compounds with water and air vary widely. The highly active main-group metals such as lithium (Li), sodium (Na), magnesium (Mg), and aluminum (Al) form highly air- and water-sensitive organometallic compounds. For example, $Al_2(CH_3)_6$ undergoes immediate and violent reaction with water to liberate methane (CH_4) gas, and it bursts immediately into flame on contact with air. For the elements toward the right of the periodic table, the organometallic compounds are not water-sensitive; tetramethylsilicon, for example, does not react with water or air at room temperature.

The Synthesis of *s*- and *p*-block Organometallic Compounds

Synthesis of *s*- and *p*-block organometallic compounds can often be accomplished by one of several general reaction types. The most important of these are outlined below.

Formation of Alkyllithium and Grignard Reagents

The highly active metals combine with a halogen-substituted hydrocarbon to produce simple organometallic compounds. For example, methyllithium, an important reagent in organic synthesis, is produced commercially by following the reaction:

$$2Li + CH_3Cl \rightarrow LiCH_3 + LiCl$$

With other active metals, such as magnesium, aluminum, and zinc, the reaction generally yields the organometallic halide. A common reaction of this type is the synthesis of a Grignard reagent, an alkylmagnesium halide that finds wide use in organic synthesis (the s indicates that the metal is in solid form).

$$Mg(s) + CH_3Br \xrightarrow{(C_2H_5)_2O} CH_3MgBr$$

Double Displacement

The synthesis of organometallic compounds by double displacement involves organometallic

(MR) and binary halide (EX, where E is a metal or metalloid and X is a halogen) starting materials. It provides a convenient synthetic procedure that is widely used in the laboratory and to a lesser extent on a commercial scale. As the following examples illustrate, the organic group on the more active metal is transferred to the less active metal or metalloid. In this context the most common highly active metals are lithium, aluminum, and magnesium.

$$4Li(CH_3) + SiCl_4 \rightarrow 4LiCl + Si(CH_3)_4$$

$$Al_2(CH_3)_6 + 2BF_3 \rightarrow 2AlF_3 + 2B(CH_3)_3$$

Redistribution

Double displacements involving the same central element are often referred to as redistribution reactions. A commercially important example is the redistribution of silicon tetrachloride and tetramethylsilicon (also known as tetramethylsilane) at elevated temperatures. $SiCl_4 + (CH_3)_4Si \rightarrow CH_3SiCl + (CH_3)_2SiCl_2 + (CH_3)_3SiH + ...$, the products from this reaction can be separated by distillation. This reaction is performed industrially where $(CH_3)_2SiCl_2$ is removed from the equilibrating mixture and then hydrolyzed to produce the intermediates for silicone polymers, which have the form $-(Si(CH_3)_2-O)-n$.

Hydrometallation

The addition of a metal hydride to a multiple bond is called hydrometallation, and it leads to the formation of a metal-carbon bond. $M-H + H_2C{=}CH_2 \rightarrow MH_2C-CH_3$. This reaction is driven mainly by the high C–H bond strength relative to most E–H bond strengths. Two important hydrometallation reactions are hydroboration and hydrosilation, illustrated, respectively, by the following examples:

$$H_3BOR_2 + \left(H_3C\right)_2 C{=}CH_2 \xrightarrow{R_2O} \left(H_3C\right)_2 CH\text{-}CH_2BH_2$$

$$SiH_4 + \left(CH_3\right)_2 C{=}CH_2 \xrightarrow{H_2PtCl_6} \left(CH_3\right)_2 CH\text{-}CH_2SiH_3$$

In the hydroboration and hydrosilation of an unsymmetrical alkene, the boron or silicon binds to the carbon atom that has less-bulky substituents, and the smaller hydrogen atom binds to the carbon atom that has bulky substituents—$(CH_3)_2C$ in the above equations. Hydroboration was discovered and developed in the United States by Herbert C. Brown, who shared the Nobel Prize for Chemistry in 1979 for this research. Both hydroboration and hydrosilation are widely used in the synthesis of complex organic molecules. In these applications, the B–C or Si–C bond is generally cleaved in a subsequent step to produce a product that is free of boron or silicon.

Reduction

All organometallic compounds are potential reducing agents, and those of the electropositive elements are very strong reducing agents because the metal gives up electrons to the carbon, resulting in a polar M–C bond with a partial positive charge on the metal and a negative charge on the carbon. Organometallic compounds of highly electropositive elements such as lithium, sodium, and aluminum ignite spontaneously and sometimes explode on contact with air or other oxidizing agents. The useful organometallic reagents $Li(CH_3)$, $Zn(CH_3)_2$, $B(CH_3)_3$, and $Al_2(CH_3)_6$ are spontaneously flammable in air (pyrophoric). Accordingly, techniques have been developed to handle these and other pyrophoric compounds. Glass reaction vessels sealed from the atmosphere and purged with nitrogen gas are commonly used for handling air-sensitive organometallic compounds in the laboratory. Large quantities of pyrophoric compounds such as $Al_2(C_2H_5)_6$ are routinely handled with ease in the chemical industry by using closed metal reactors for the production of these and other much less reactive compounds. Organometallic compounds have reduced reactivity when the metallic component is not highly electropositive and when the metal is completely surrounded by attached groups. For example, elevated temperatures are required to initiate combustion with $Si(CH_3)_4$ and $Sn(CH_3)_4$, and at room temperature they can be handled in air.

Carbanion Character

The partial negative charge of an organic group bonded to a highly active metal results in a distinctive pattern of reactivity that is frequently referred to as nucleophilic or carbanion character. Thus, organometallic compounds containing highly active (electropositive) metals, such as lithium, magnesium, aluminum, and zinc, react rapidly and completely with water, liberating a hydrocarbon in the process. For example, dimethylzinc liberates methane gas along with solid zinc hydroxide.

$$Zn(CH_3)_2 + 2H_2O \rightarrow Zn(OH)_2 + 2CH_4$$

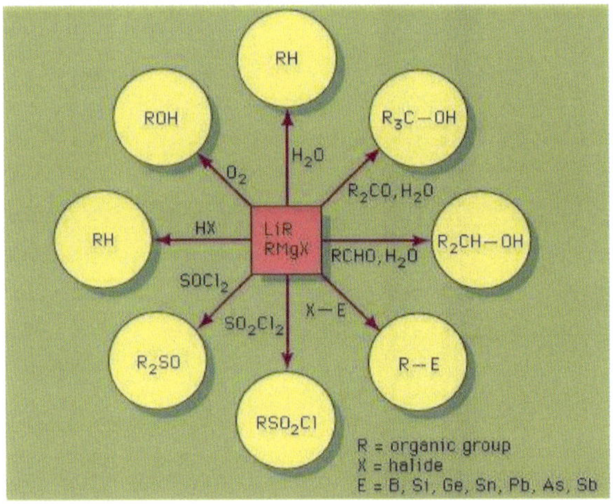

Highly reactive organometallic reagents, such as alkyllithium and Grignard reagents (LiR and RMgX, respectively), have multiple synthetic applications.

The above hydrolysis of dimethylzinc can be viewed as a transfer of a slightly acidic H^+ from water to the strongly basic carbanion CH_3^- in dimethylzinc.

Alkyllithium, alkylaluminum, and alkylmagnesium compounds are the most common carbanion reagents in laboratory-scale synthetic chemistry; carbanion character is greatly diminished for the less metallic elements boron and silicon. The nucleophilic character of organometallic compounds of active metals has many synthetic applications. For example, the organic group in organometallic compounds of active metals attacks the carbonyl carbon of a ketone, and upon hydrolysis a tertiary alcohol results. Similarly, aldehydes can be converted to secondary alcohols by reaction with an organometallic reagent followed by hydrolysis. Double displacement reactions can be used to prepare sulfones (R_2SO_2) and sulfoxides (R_2SO) by treating thionyl chloride ($SOCl_2$) or sulfuryl dichloride (SO_2Cl_2) with an alkyllithium or a Grignard reagent.

One consequence of the carbanion character of organometallic compounds containing active metals is the protolysis (proton-transfer) reaction that takes place with very weak protonic acids, including water. Alcohols react in a manner similar to the reaction of water, and this provides a convenient way of introducing an alkoxide substituent into an organometallic compound.

$$(C_2H_5)_3Ga + HOCH_3 \rightarrow [(C_2H_5)_3GaOHCH_3] \rightarrow (C_2H_5)_2Ga(OCH_3) + C_2H_6$$

The rate of reaction decreases with bulky organic groups on the alcohol. For example, *tert*-butyl alcohol,

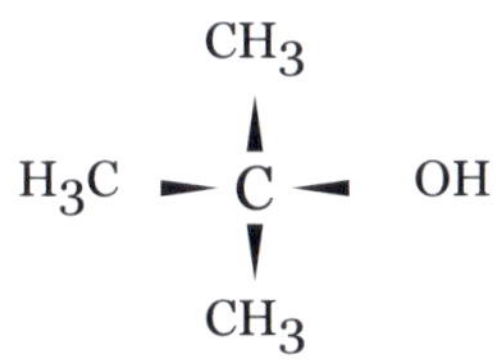

reacts slowly with most active organometallics, and it is therefore employed in the laboratory to safely destroy reactive organometallic wastes.

β-hydrogen Elimination

This is a common organometallic reaction in which a hydrogen atom on a carbon atom that is one position removed from the metal (the β position) transfers to the metal with the liberation of an alkene. The following example shows the formation of ethylene, C_2H_4.

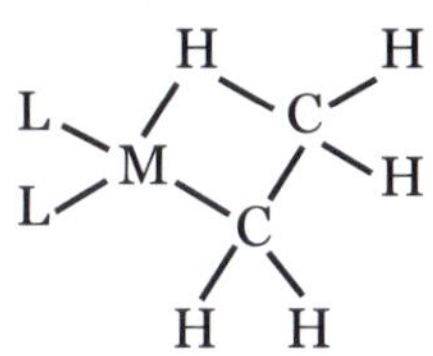

The transition state thought to exist in the β-hydrogen elimination reaction and the addition of M–H across a C=C double bond. The hydrogen atom forms a bridge between the carbon and metal (M) atoms. L represents a ligand.

Compounds with an accessible central metal atom tend to readily undergo β-hydrogen elimination. $M\text{-}\underset{\alpha}{CH_2}-\underset{\beta}{CH_3} \rightarrow M\text{-}H + H_2C{=}CH_2$

This reaction is the reverse of the addition of an M–H bond to an alkene, and under some conditions significant equilibrium concentrations of both reactants and products are observed. The β-hydrogen elimination reaction is thought to proceed through the transition state, and as might be expected from this reaction mechanism, compounds with an accessible central metal atom tend to undergo β-hydrogen elimination readily. For example, it occurs with trialkylaluminum compounds but not with tetraalkylsilicon compounds, in which reaction of the silicon atom is hindered by the bulky substituents.

Organoboron Compounds

Organoboron compounds have been playing an increasingly important role for organic synthesis, functional molecules, functional polymers, B carriers for neutron capture therapy, and biologically active agents.

There are several other types of organoboron derivative that display physical and chemical properties unshared by the previously described organoboron compounds. Such organoboron derivatives include allylboron , propargylboron and allenylboron derivatives. The carbon–boron bond in these derivatives is generally more labile than that in the more usual organoboron derivatives. For example, compounds tend to undergo allylic rearrangements causing their isomerization which can take place well below room temperature. These allylic organoboron compounds also tend to react readily with a variety of reagents, e.g. water, alcohols, ketones and alkynes, which are relatively inert to typical alkylboron compounds. Benzylboron derivatives may also be viewed as allylic organoboron compounds. Indeed, methanolysis of tribenzylborane has been shown to involve, at least in part, a cyclic transition state on the basis of deuteration studies:

1. Lewis Acidic Oxophilic Metal, many boron reagents provide for simultaneous activation of acceptor and donor portions of substrate, e.g., in conjugate addition reactions.

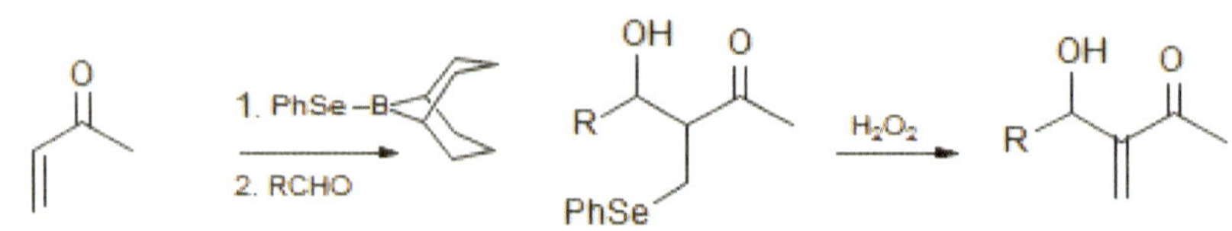

2. Boron hydrides can serve as both electrophilic and nucleophilic H^- donor. Borohydrides have powerful nucleophilic properties, boranes are weak electrophiles.

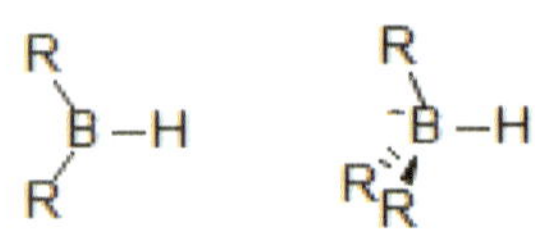

3. Carbanion donor, trialkyl boranes are not very reactive towards typical electrophiles (acids, iodine, bromine, etc). Reactivity is greatly increased if boronates are formed (e.g. by addition of hydroxide), or if complexation of the borane to the electrophile activates the boron in the same way. Thus enol, allyl and propargyl boranes will transfer the group on boron to Lewis basic electrophiles. Other types show little tendency to behave as carbanion sources.

4. Transmetalation of organoboron compounds to organocopper and organopalladium (Suzuki coupling) provides a powerful method for C-C bond formation.

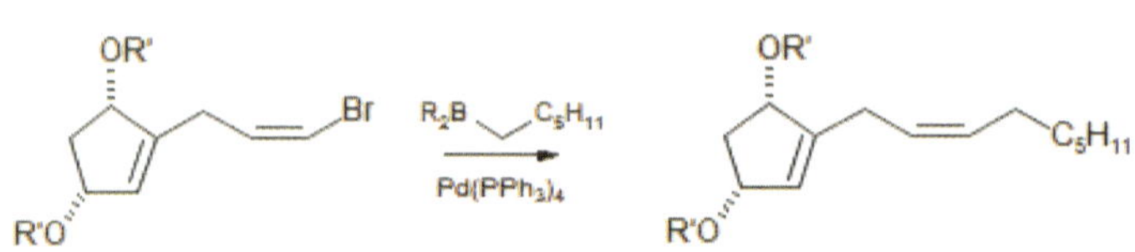

5. Organic groups on anionic boron readily migrate to electrophilic sites on adjacent atoms.

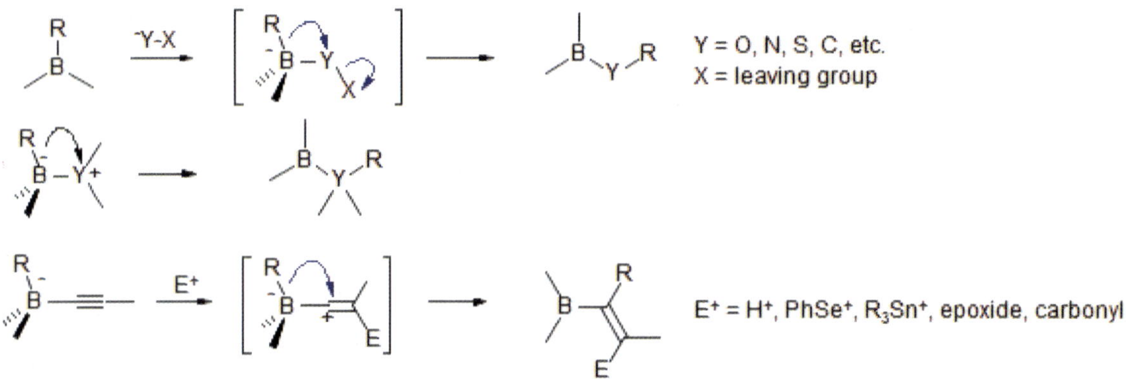

Properties of the B-C Bond

The B-C bond has low polarity (the difference in electronegativity 2.55 for carbon and 2.04 for boron), and therefore alkyl boron compounds are in general stable though easily oxidized.

In part because its lower electronegativity, boron often forms electron-deficient compounds, such as the triorganoboranes. Vinyl groups and aryl groups donate electrons and make boron less electrophilic and the B-C bond gains some double bond character. Like the parent borane, diborane, organoboranes are classified in organic chemistry as strong electrophiles because boron is unable to gain a full octet of electrons. Unlike diborane however, most organoboranes do not form dimers.

Reactions

Hydroboration-oxidation

In organic synthesis the hydroboration reaction is taken further to generate other functional groups in the place of the boron group. The hydroboration-oxidation reaction offers a route to alcohols by oxidation of the borane with hydrogen peroxide or to the carbonyl group with the stronger oxidizing agent chromium oxide.

Rearrangements

A second group of reactions that organoboron compounds are involved in create new carbon carbon bonds. Carbon monoxide is found to react very easily with a trialkylborane. What follows is a 1,2-rearrangement when an alkyl substituent on the anionic boron migrates to the adjacent electrophilic carbon of the carbonyl group. The carbonyl group can then be reduced to a hydroxyl group.

Allylboration

Asymmetric allylboration demonstrates another useful application of organoboranes in carbon–carbon bond formation. In this example from Nicolaou's synthesis of the epothilones, asymmetric allylboration (using an allylborane derived from chiral alpha-pinene) is used in conjunction with TBS protection and ozonolysis. Overall, this provides a two-carbon homologation sequence that delivers the required acetogenin sequence.

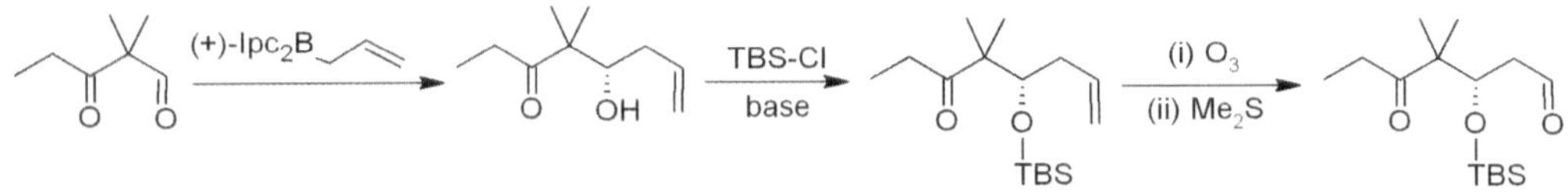

As Reducing Agent

Borane hydrides such as 9-BBN and L-selectride (lithium tri-sec-butylborohydride) are reducing agents. An example of an asymmetric catalyst for carbonyl reductions is the CBS catalyst. This catalyst is also based on boron, the purpose of which is coordination to the carbonyl oxygen atom.

Borates

Trialkylboranes, BR_3, can be oxidized to the corresponding borates, $B(OR)_3$. One method for the determination of the amount of C-B bonds in a compound is by oxidation of R_3B with trimethylamine oxide (Me_3NO) to $B(OR)_3$. The trimethylamine (Me_3N) formed can then be titrated.

Boronic acids $RB(OH)_2$ react with potassium bifluoride $K[HF_2]$ to form trifluoroborate salts $K[RBF_3]$ which are precursors to nucleophilic alkyl and aryl boron difluorides, $ArBF_2$.The salts are more stable than the boronic acids themselves and used for instance in alkylation of certain aldehydes:

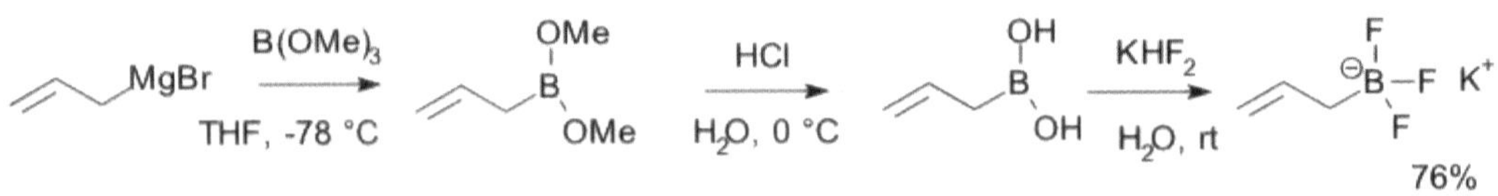

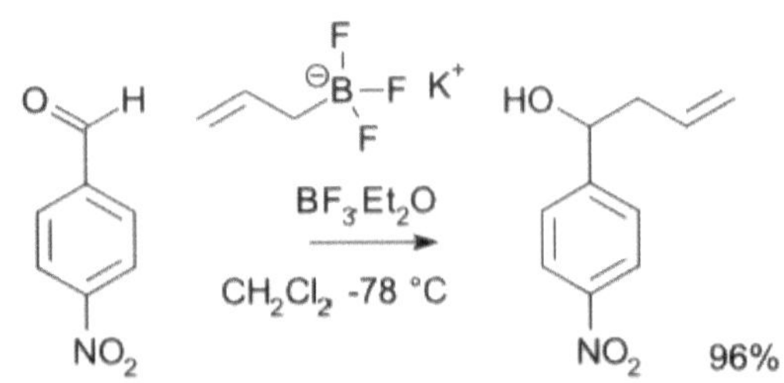

Suzuki Reaction and Related Reactions

Organoboron compounds also lend themselves to transmetalation reactions, especially with organopalladium compounds. This reaction type is exemplified in the Suzuki reaction, which involves coupling of aryl- or vinyl-boronic acid with an aryl- or vinyl-halide catalyzed by a palladium(0) complex,

$$R_1\text{-}BY_2 + R_2\text{-}X \xrightarrow[\textit{Base}]{\textit{Pd catelyst}} R_1\text{-}R_2$$

This reaction is an important method for making carbon-carbon bonds.

Classes of Organoboron Compounds

Triorganoboranes and Hydrides

Among the most studied classes of organoboron compounds have the formula BR_nH_{3-n}. As discussed above, these compounds are used as catalysts, reagents, and synthetic intermediates. The trialkyl and triaryl derivatives feature trigonal planar boron center that is typically only weakly Lewis acidic. Except for very bulky derivatives, the hydrides exist as dimers, reminiscent of the structure of diborane itself.

Borinic and Boronic Acids and Esters (BRn(OR)3-n)

Compounds of the type $BR_n(OR)_{3-n}$ are called borinic esters (n=2), boronic esters (n=1), and borates (n=0). Boronic acids are used in Suzuki reaction. Trimethyl borate, which is debatably not an organoboron compound, is an intermediate in the production of sodium borohydride.

Carboranes

Carboranes are cluster compounds with carbon and boron vertices. The best known is orthocarborane, with the formula $C_2B_{10}H_{12}$. Although they have few commercial applications, carboranes have attracted much attention as precursors to reagents and new materials. Anionic derivatives, dicarbollides, e.g., $[C_2B_9H_{11}]^{2-}$ are ligands that behave like cyclopentadienide.

Bora-substituted Aromatic Compounds

In borabenzene, one CH center in benzene is replaced by boron. These compounds are invariably isolated as adducts, e.g., C_5H_5B-pyridine. The cyclic compound borole, a structural analog of pyrrole, has not been isolated, but substituted derivatives known as boroles are known. The cyclic compound borepin is aromatic.

Boryl Compounds

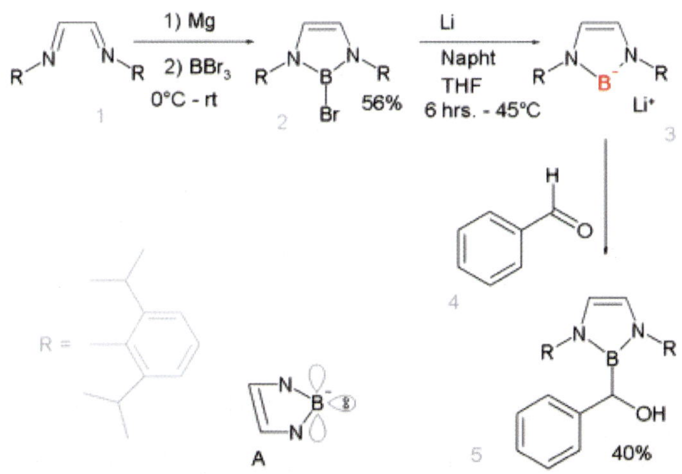

Boryl anions have the formula R_2B^-. Nucleophilic anionic boryl compounds have long been elusive but a 2006 study described a boryllithium compound, which reacts as a nucleophile: Organometallic compounds with metal to boron bonds, (i.e., M–BR_2), are known as boryl complexes. Related ligands are borylenes (M–B(R)–M).

The absence of lithium boryl compounds is notable because in other period 2 elements lithium salts are common e.g. lithium fluoride, lithium hydroxide lithium amide and methyllithium. The gap highlights the very low electronegativity of boron. Reaction of base with a borohydride R_2BH does not result in deprotonation to the boryl anion R_2B^- but to formation of the boryl anion $R_2B^-H(base)^+$. This reaction product has a complete octet. Instead the boryl compound is prepared by reductive heterolysis of a boron-bromide bond by lithium metal. The new boryl lithium compound is very similar to and isoelectronic with N-heterocyclic carbenes. It is designed to benefit from aromatic stabilization and from kinetic stabilization from the bulky 2,6-diisopropylphenyl groups. X-ray diffractionconfirms sp2 hybridization at boron and its nucleophilic addition reaction with benzaldehyde gives further proof of the proposed structure.

Alkylideneboranes

Alkylideneboranes of the type RB=CRR with a boron – carbon double bond are rarely encountered. An example is borabenzene. The parent compound is $HB=CH_2$ which can be detected at low temperatures. A fairly stable derivative is $CH_3B=C(SiMe_3)_2$ but is prone to cyclodimerisation.

NHC Adducts of Boron

NHCs and boranes form stable NHC borane adducts. Triethylborane adducts can be synthesised directly from the imidazolium salt and lithium triethylborohydride. Members of this compound class are investigated for use as reagent or catalyst.

Diborenes

Chemical compounds with boron to boron double bonds are rare. In 2007 the first neutral diborene (RHB=BHR) was presented by Gregory Robinson of the University of Georgia. Each boron atom has a proton attached to it and each boron atom is coordinated to a NHC carbene. The parent structure with the additional carbene ligands is diborane.

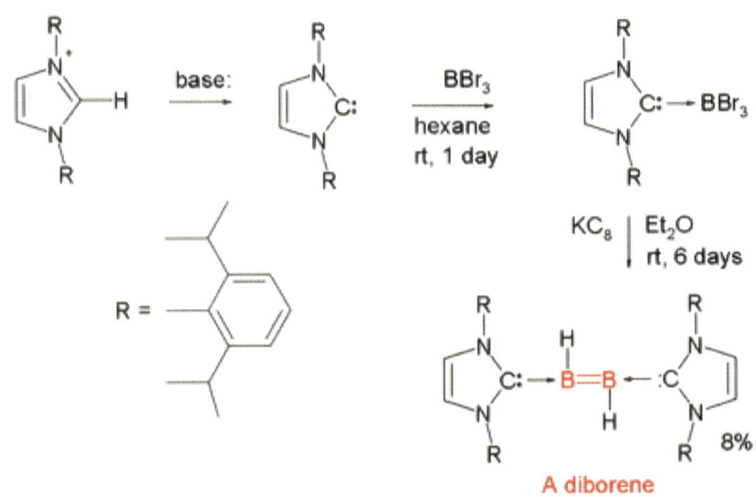

A reported diboryne is based on similar chemistry.

Organoaluminium Compounds

Organoaluminium compounds is organometallic compounds in which the aluminum is directly bonded with the carbon atom. Two types of organoaluminumcompounds are known: complete R_3A1 and incomplete R_2A1X or $RA1X_2$(where R stands for CH_3, C_2H_5, C_6H_5, and othersand X stands for halogen, OR, or H). Aluminum trialkyls are colorless fluids which are extremely sensitive to oxygen andmoisture; trimethylaluminum and triethylaluminum ignite spontaneously in the air, and they decompose explosively in water.Organoaluminum compounds are handled in an inert atmosphere (nitrogen, argon). All organoaluminum compounds producestable complexes with an ether or with amines—for example, $(CH_3)_3AlxO(CH_3)_2$ and $(CH_3)_3AlxN(CH_3)_3$. With NaR and LiR,complexes of the $Me[AIR_4]$ type are formed. All these complexes are less reactive, but they also ignite in the air.

Organoaluminum compounds are also obtained by the action of alkyl halides on an alloy of magnesium with aluminum—for example, $6C_2H_5Br + 2A1 + 3Mg = 2(C_2H_5)_3A1 + 3MgBr_2$.

Organoaluminum compounds are used in industry as catalysts of low-pressure olefin polymerization (for example, in producing low-pressure polyethylene) and for obtaining ultrapure aluminum.

With less bulky alkyl groups, dimerization occurs and one of the distinguishing features of alkyl bridge is the small Al-C-Al angle, which is ~ 75°.

The 3c,2e bonds are very weak and tend to dissociate in the pure liquid which increases with increase in the bulkiness of the alkyl group.

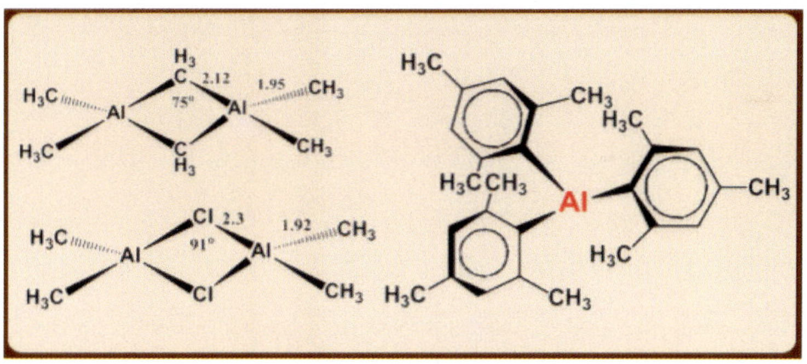

$$Al_2(CH_3)_6 \rightleftharpoons 2Al(CH_3)_3 \quad K = 1.52x10^{-8}$$

$$Al_2(C_4H_9)_6 \rightleftharpoons 2Al(C_4H_9)_3 \quad K = 2.3x10^{-4}$$

Perpendicular orientation of pheynl groups in Al_2Ph_6 Triphenylaluminium exists as a dimer with bridging η^1-phenyl groups lying in a plane perpendicular to the line joining the two Al atoms.

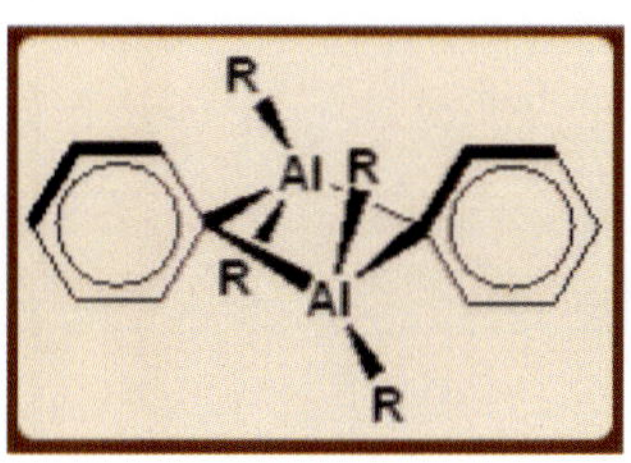

This structure is favored partly on steric grounds and partly by supplementation of the Al-C-Al bond by electron donation from the phenyl π-orbitals to the Al atoms.

Tendency for bridging: X > Ph > alkyl.

3c, 2e bonds formed by a symmetric combination of Al and C orbitals.

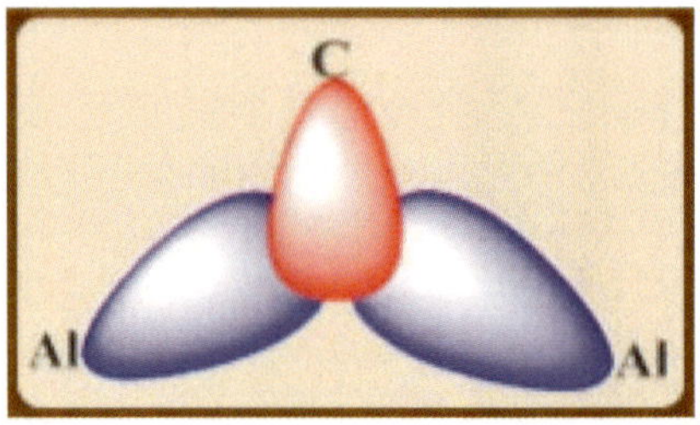

An additional interaction between the pπ orbital on C and an antisymmetric combination of Al orbitals.

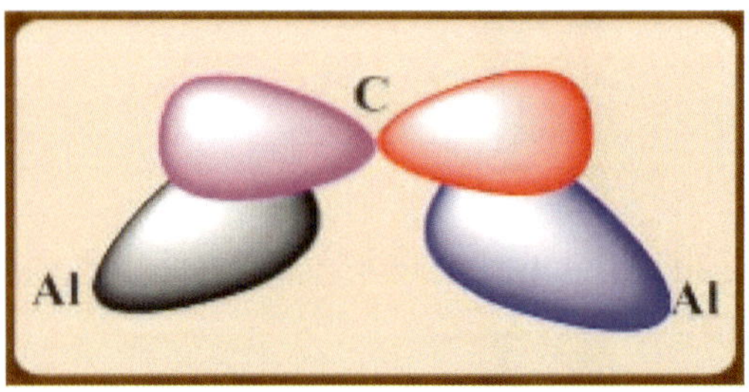

Synthesis

Very useful as alkene polymerization catalysts and chemical intermediates. Expensive carbanion reagents for the replacement of halogens organic groups by metathesis.

Laboratory scale preparations involves:

$$2Al + 3Hg(CH_3)_2 \rightarrow Al_2(CH_3)_6 + 3Hg$$

Commercial method:

$$2Al + CH_3Cl \rightarrow Al_2Cl_2(CH_3)_4$$

$$Al_2Cl_2(CH_3)_4 + 6Na \rightarrow Al_2(CH_3)_6 + 2Al + 6NaCl$$

Commercial method for ethylaluminium and higher homologs:

$$2Al + 3H_2 + 6RHC{=}CH_2 \xrightarrow[110-200\,atm]{60-110^\circ C} 2Al_2(CH_2CH_2R)_6$$

The reaction probably proceeds by the formation of a surface Al—H species that adds across the double bond of the alkene in a hydrometallation reaction.

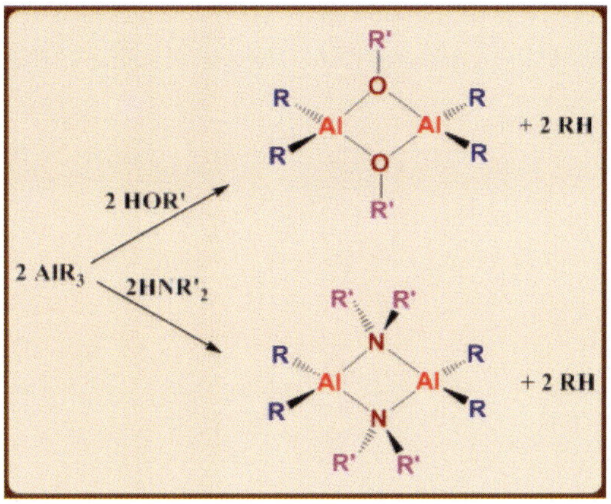

Reactions:

Alkylaluminum compounds are mild Lewis acids and form complexes with ethers, amines and anions. When heated, often β-hydrogen elimination is responsible for the decomposition of ethyl and higher alkylaluminium compounds. E.g. $Al(^iC_4H_9)_3$.

Tendency towards bridging structure is: $PR_2^- > X^- > H^- > Ph^- > R^-$.

d-block and f-block Metallocene

The nomenclature for organometallic compounds of the *d*-block elements, also known as transition metals, follows rules similar to those described for the main-group organometallics. Groups attached to the metal (ligands) are named first, followed by the metal. When a ligand is attached to the metal by more than one carbon atom, the number of sites of attachment is indicated by n in η^n.

Metal Carbonyls

Many of the early discoveries in *d*-block organometallic chemistry involved the metal carbonyls—i.e., compounds consisting of a metal atom bonded to one or more carbon monoxide (CO) ligands. Mond's discovery of the first simple metal carbonyl, tetracarbonylnickel, $Ni(CO)_4$, at the end of the 19th century was quickly followed by a series of discoveries in his laboratory and elsewhere showing that most of the *d*-block metals form neutral homoleptic carbonyls. (The term *homoleptic* refers to identical groups attached to a central atom.) The remarkable volatility of tetracarbonylnickel, whose boiling point is 43° C, prompted one of Mond's contemporaries to state that "Mond put wings on metals." The ease of formation of tetracarbonylnickel (from metallic nickel and carbon monoxide) and its high volatility led Mond to develop an efficient industrial process for separating nickel and cobalt, which often occur together in their ores. This process requires diligent safety measures because tetracarbonylnickel is extremely toxic.

The Structure of Metal Carbonyls

The carbonyl ligands in the tetracarbonylnickel molecule project toward the vertices of a tetrahedron, and thus the structure is referred to as tetrahedral. Similarly, the six carbonyl ligands in hexacarbonylchromium project toward the vertices of an octahedron:

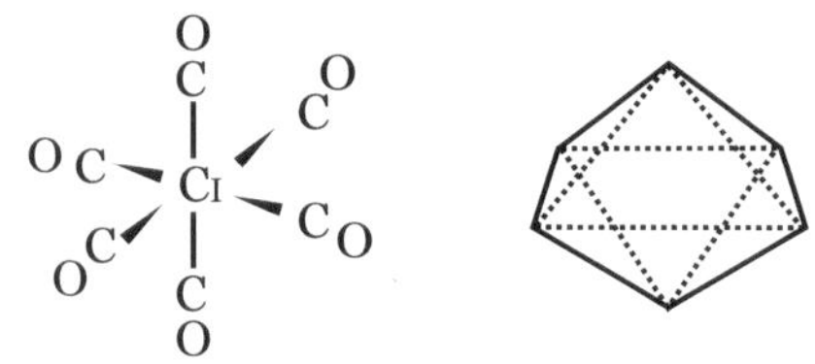

Many other metal carbonyls contain two or more metal atoms, such as decacarbonyldimanganese and octacarbonyldicobalt, shown here:

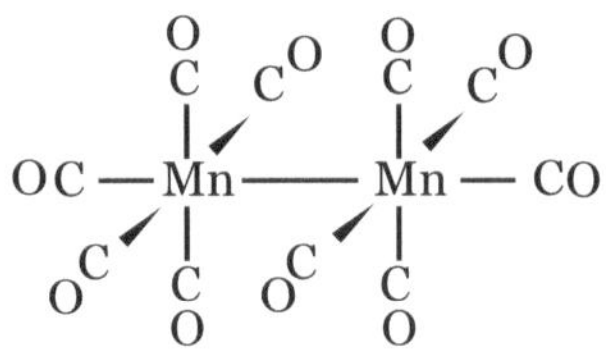

$Mn_2\,(CO)_{10}$

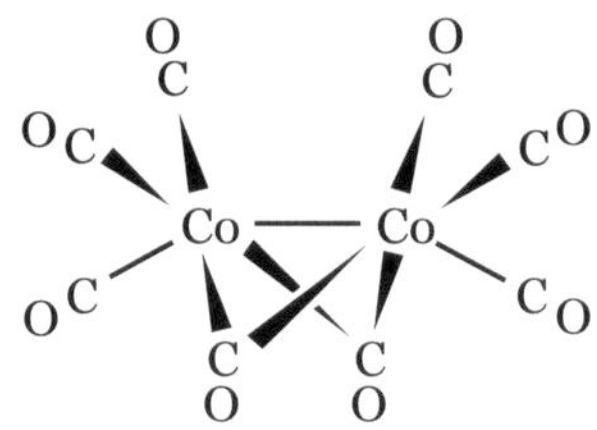

$Co_2\,(CO)_6$

In all these structures carbon monoxide is connected to the metal through its carbon atom. When more than one metal atom is present, as in octacarbonyldicobalt, the carbon of the carbonyl ligand may bridge between metal atoms.

The formulas of most homoleptic metal carbonyl compounds conform to the rule that each metal atom in the metal carbonyl molecule must have 18 valence electrons. The valence electrons represent the outer electrons on the metal plus those from the ligand. For example, the electron count for $Fe(CO)_5$ includes eight from the iron atom (it is in group 8 and hence has eight valence electrons) and two from each carbonyl ligand, giving a total of 18 electrons. This 18-electron rule applies to many organometallic compounds of the *d*-block metals other than carbonyls, but there are exceptions in the organometallic chemistry of the *d*-block metals. The most notable exceptions are metals on the far left of the *d* block and on the right of the *d* block where the platinum metals—rhodium (Rh), iridium (Ir), palladium (Pd), and platinum (Pt)—often exhibit a 16-electron count in their organometallic compounds.

Zero-oxidation-state Metal Carbonyls

The central metal in a neutral metal carbonyl, such as those described above, is assigned an oxidation state of zero, quite unlike the case in simple inorganic compounds in which positive oxidation states are the norm, as, for example, Fe^{3+} in $FeCl_3$ or Ni^{2+} in $NiBr_2$. Unlike the free metals, which also have a zero oxidation state, many carbonyls are soluble in a variety of simple organic solvents and are highly reactive. Because of these chemical and physical properties, the metal carbonyls are convenient starting materials for the synthesis of compounds with the metal atom in a zero or low oxidation state. One simple reaction is the substitution of other ligands such as triethylphosphine, $P(Et)_3$, for CO (Et is a common abbreviation for the ethyl group, $-C_2H_5$).

$$Cr(CO)_6 + P(Et)_3 \rightarrow Cr(CO)_5P(Et)_3 + CO$$

Because most other ligands are far bulkier than carbon monoxide and also because of differences in bonding properties, it is often not possible to replace all the carbonyl ligands by groups such as triethylphosphine, and a mixture of products with varying degrees of substitution can result.

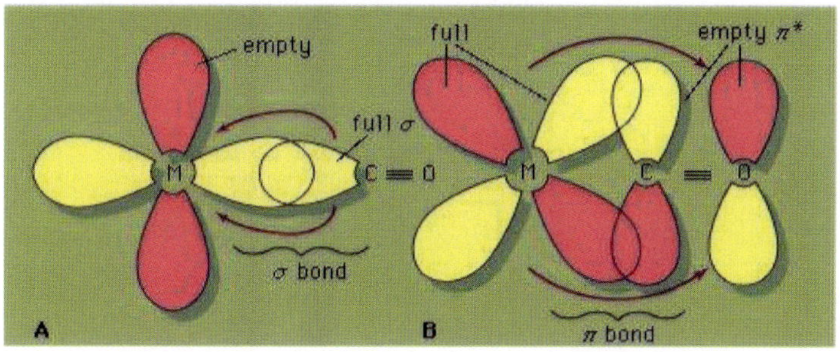

Bonding of the CO ligand to a metal atom(A) The σ bond formed by the partial donation of a pair of electrons from an orbital largely localized on the carbon atom to a *d* orbital on the metal atom. (B) Back π bonding from filled *d* orbitals on the metal atom into an empty π* orbital on the CO ligand. In most cases, the net back π bonding predominates, and electron density is transferred from the metal to the CO ligand.

The remarkable ability of the carbonyl ligand to stabilize compounds with metals in the zero oxidation state has led to detailed studies of M–CO bonds. A minor part of the M–CO bonding can be attributed to the tendency of the carbonyl ligand to donate a pair of electrons to the metal, which is the principal mode of interaction for most ligands with metal cations in classical coordination compounds such as $[Co(NH_3)_6]^{3+}$. The second mode of interaction with the metal is the simultaneous back-donation of electron density from the metal to the carbonyl ligand, which is called back π bonding.

Metal Carbonyl Anions

More remarkable than the formation of zero-oxidation-state metal carbonyls is the reduction of many of these carbonyl compounds to metal carbonyl anions, in which the metal has a negative oxidation state. The following example demonstrates that the two-electron reduction by sodium metal is accompanied by the loss of the two-electron donor carbonyl ligand, and so the 18-electron count on iron is preserved; the solvent is tetrahydrofuran (THF).

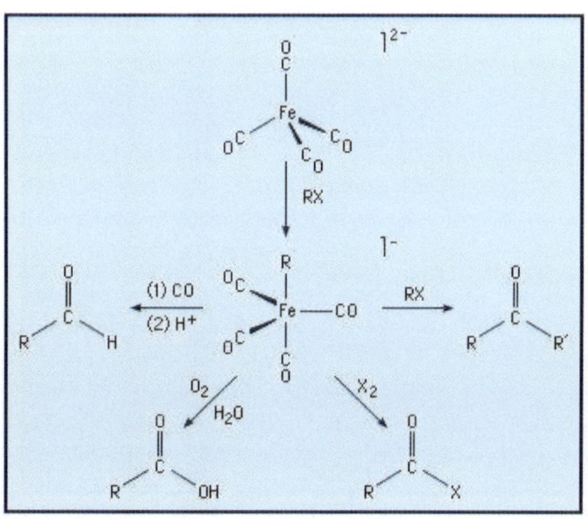

The organometallic reagent $Fe(CO)_4^{2-}$ can undergo multiple reactions in the synthesis of organic molecules.

$$2Na + Fe(CO)_5 \xrightarrow{THF} Na_2\left[Fe(CO)_4\right] + CO$$

The stabilization of very electron-rich complexes, such as $[Fe(CO)_4]^{2-}$, is attributed to the back π bonding that shifts electron density from the metal to the carbonyl ligands, and this view is substantiated by C–O bond distances and other physical data. The metal atom in these carbonyl anions is assigned a negative oxidation state. This formalism does not acknowledge the delocalization of electron density from the metal to the ligand, but the chemical properties of the carbonyl anions do suggest that some of the negative charge resides on the metal. For example, a metal carbonyl anion can be protonated with the H^+ ion, which generally attaches to the central metal and not a carbonyl ligand, as in the following example.

$$[Fe(CO)_4]^{2-} + HCl \rightarrow [HFe(CO)_4]^{-} + Cl^{-}$$

Owing to their high reactivity, carbonyl anions are useful starting materials for the synthesis of other organometallic compounds, and this accounts for their applications in organic synthesis. For example, $[Fe(CO)_4]^{2-}$ is used to extend a carbon chain by transfer of the carbonyl substituent, producing aldehydes, ketones, or carboxylic acids.

Aluminoxane

Aluminoxane is a class of organometallic compounds of aluminium having general formula R_2Al-O-AlR_2; they are used as catalysts in the polymerization of epoxides and oxetanes to form polyether elastomers.

Structure of an Aluminoxane

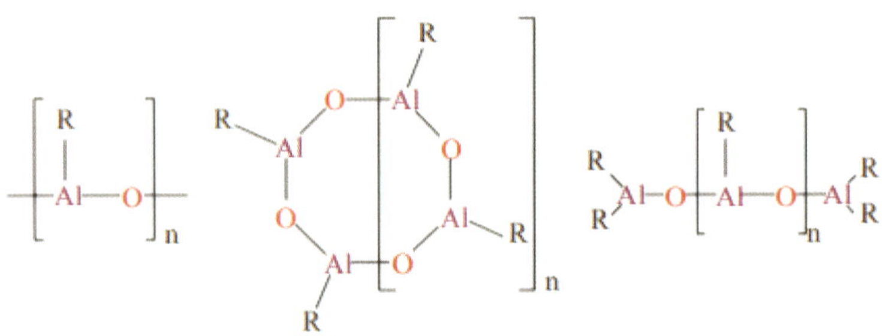

Traditional approach: "bottom-up"

$$AlX_3 \xrightarrow{H_2O} \left[Al(O)_x(OH)_y(X)_z\right]_n$$

hydrolysis of a small molecule but Aluminoxane have a core structure analogous to boehmite so can they be prepared from the mineral?

The Barron approach: "top-down"

$$\left[Al(O)_x(OH)_y(X)_z\right]_n \xleftarrow{HX} \left[Al(O)(OH)\right]_n$$

Synthesis of Carboxylate Aluminoxanes

$$[Al(O)(OH)]_n + RCO_2H\ (xs)$$

$$\downarrow$$

$$[Al(O)_x(OH)_y(O_2CR)_z]_n$$

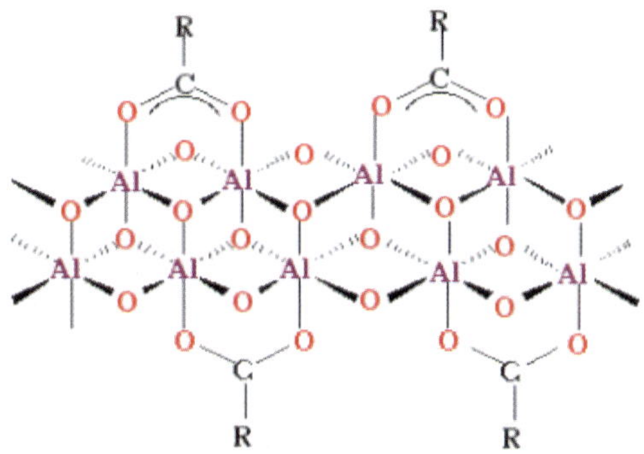

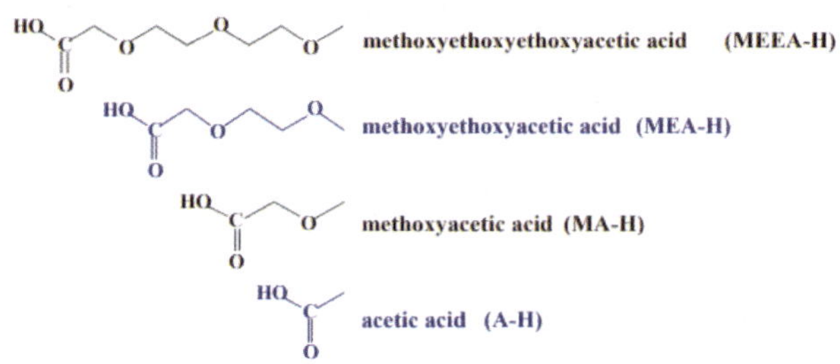

The Particle Size Change with pH

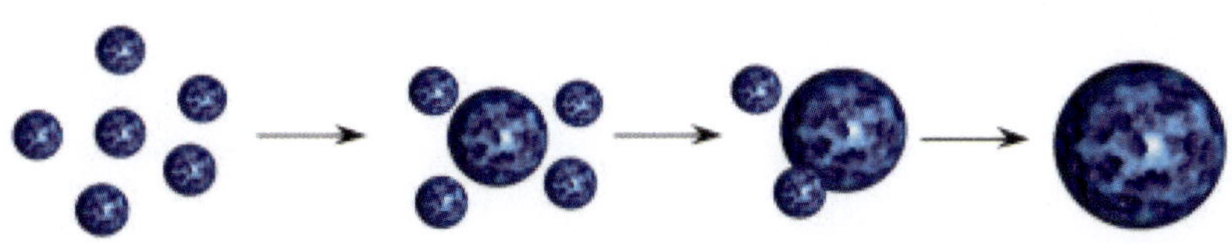

Seeded particle growth - several small alumoxane particles agglomerate to form a larger particle in a step-wise fashion.

The Structures Found for Catalyst Aluminoxanes

- Al-O cage structures;
- Aluminum 4-coordinate, oxygen 3-coordinate;

- Electron precise - not apparent as Lewis acids:
- No alkyl exchange o n = 6, 7, 8, 9, 12.

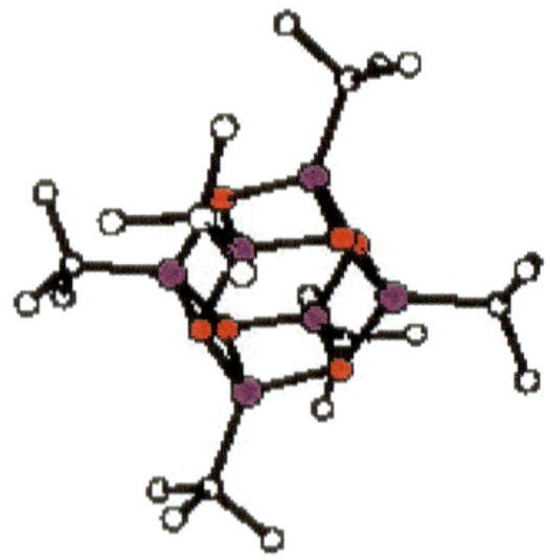

Molecular structure of $[(^tBu)AlO]_6$

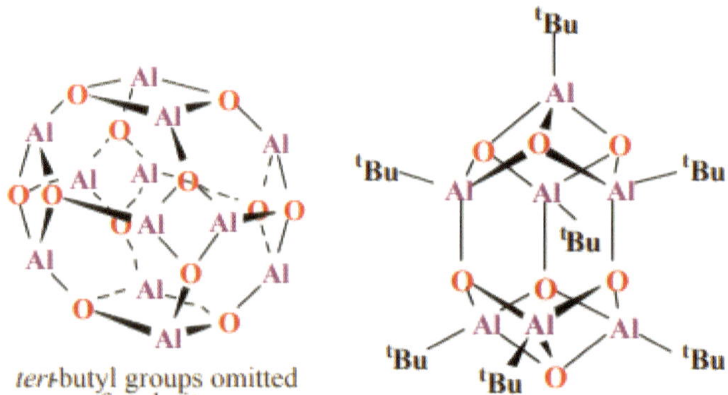

tert-butyl groups omitted for clarity

Metal Hydride

Metal hydrides occupy an important place in transition metal organometallic chemistry as the M–H bonds can undergo insertion reactions with a variety of unsaturated organic substrates yielding numerous organometallic compounds with M–C bonds. Not only the metal hydrides are needed as synthetic reagents for preparing the transition metal organometallic compounds but they also are required for important hydride insertion steps in many catalytic processes. The first transition metal hydride compound was reported by W. Heiber in 1931 when he synthesized $Fe(CO)_4H_2$. Though he claimed that the $Fe(CO)_4H_2$ contained Fe–H bond, it was not accepted until 1950s, when the concept of normal covalent M–H bond was widely recognized.

The metal hydride moieties are easily detectable in 1H NMR as they appear high field of TMS in the region between 0 to 60 ppm, where no other resonances appear. The hydride moieties usually couple with metal centers possessing nuclear spins. Similarly, the hydride moieties also couple with the adjacent metal bound phosphine ligands, if at all present in the complex, exhibiting characteristic cis (J = 15–30 Hz) and trans (J = 90–150 Hz) coupling constants. In the IR spectroscopy, the M–H frequencies appear between (1500–2200) cm^{-1} but their intensities are mostly weak. Crystallographic detection of metal hydride moiety is difficult as hydrogen atoms in general are poor scatterer of X–rays. Located adjacent to a metal atom in a M–H bond, the detection of hydrogen atom thus becomes challenging and as a consequence the X–ray crystallographic method systematically underestimates the M–H internuclear distance by ~ 0.1 Å. However, better data could be obtained by performing the X–ray diffraction studies at a low temperature in which the thermal motion of the atoms are significantly reduced. In light of these facts, the neutron diffraction becomes a powerful method for detection of the metal hydride moieties as hydrogen scatters neutrons more effectively and hence the M–H bond distances can be measured more accurately. A limitation of neutron diffraction method is that large sized crystals are required for the study.

Synthesis

Following reactions are employed for synthesizing metal hydrides:

- Protonation reactions

  For this reaction to occur the metal center has to be basic and electron rich.

  $$\left[Fe(CO)_4\right]^{2-} \xrightarrow{H^+} \left[HFe(CO)_4\right]^{-} \xrightarrow{H^+} H_2Fe(CO)_4$$

- From hydride donors

  Generally for this method, a main group hydride is reacted with metal halide.

  $$WCl_6 + LiBEt_3H + PR_3 \rightarrow WH_6(PR_3)_3$$

- Using dihydrogen (H_2) addition This method involves oxidative addition of H_2 and thus requires metal centers that are capable of undergoing the oxidative addition step.

  $$WMe_6 + PMe_2Ph \xrightarrow{H_2} WH_6(PMe_2Ph)_3$$

- From a ligand

  This method takes into account the β–elimination that occur in a variety of metal bound ligand moieties, thereby yielding a M–H bond.

  $$RuCl_2(PPh_3)_3 + KOCHMe_2 + PPh_3 \rightarrow RuH_2(PPh_3)_4 + Me_2CO + KCl$$

Metal hydrides are reactive species kinetically and thus participate in a variety of transformations like the ones discussed below.

- Deprotonation reactions

  The deprotonation reaction can be achieved by a hydride moiety resulting in the formation of H_2 gas as shown below:

  $$WH_6(PMe_3)_3 + NaH \rightarrow Na\left[WH_5(PMe_3)_3\right] + H_2$$

- Hydride transfer and insertion

  In this reaction a hydride transfer from a metal center to formaldehyde resulting in the formation of a metal bound methoxy moiety is observed as shown below:

  $$Cp_2{}^*ZrH_2 + CH_2O \rightarrow Cp_2{}^*Zr(OMe)_2$$

- Hydrogen atom transfer reaction

  An example of hydrogen atom transfer reaction is given below:

  $$[Co(CN)_5H]^{3-} + PhCH{=}CHCOOH \rightarrow [CO(CN)_5]^{3-} + PhC{\cdot}H\text{-}CH_2COOH$$

It is interesting to note that the nature of hydrogen atom in a M–H bond can vary from being protic in nature, when bound to electron deficient metal centers as in metal carbonyl compounds, to that of being hydridic in nature, when bound to more electropositive early transition metals. In the latter case, the hydride moieties tend to be basic and exhibit hydride transfer reactions with electrophiles like aldehydes or ketones. Furthermore, the protonation of these basic metal hydrides leads to the elimination of dihydrogen (H_2) gas along with the generation of a vacant coordination site at the metal center.

Bridging Hydrides

The metal hydrides usually show two modes of binding, namely terminal and bridging. In case of the bridging hydrides, the hydrogen atom can bridge between two or even more metal centers and thus, the bridging hydrides often display bent geometries.

σ–complexes

σ–complexes are rare compounds, in which the σ bonding electrons of a X–H bond further participate in bonding with a metal center (X = H, Si, Sn, B, and P). The σ complexes thus exhibit an askewed binding to a metal center with the hydrogen atom, containing no lone pair, being more close to the metal center and thereby resulting in a side–on structure. Many times if the metal center is electron rich, then further back donation to the σ* orbital of the metal bound X–H moiety may occur resulting in a complete cleavage of the X–H bond.

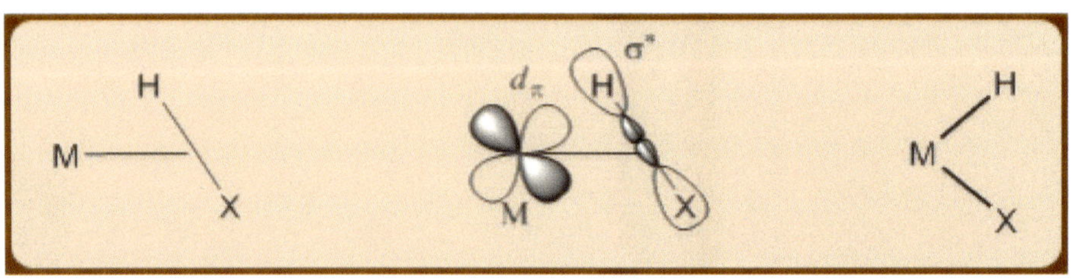

Metal Dihydrogen Complexes

The simplest variant of a σ–complex contains a dihydrogen ligand. The first dihydrogen complex was isolated by Kubas, after which many new ones were reported.

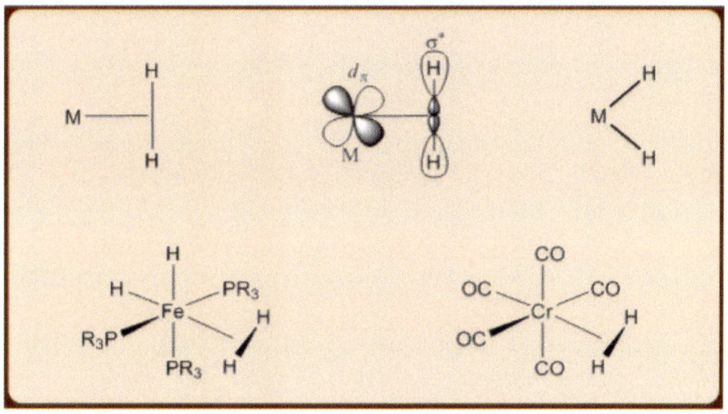

Quite expectedly, the dihydrogen moiety bound to a metal in a σ–complex is found to be more acidic (pKa = 0 – 20) when compared to the free dihydrogen molecule (pKa = 35). It is interesting to note that the pKa change associated with the binding of dihydrogen to a metal in a σ–complex relative to that of the free H_2 molecule is significantly larger than the change associated with binding of H_2O to metal. Owing to this inherent acidity, the deprotonation of the metal bound dihydrogen moiety by a base can thus be appropriately employed for heterolytic activation of the dihydrogen moiety as illustrated below:

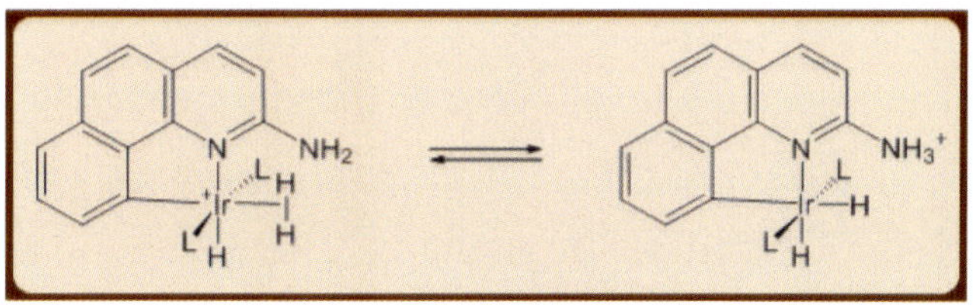

The dihydrogen complexes of metals are often referred to as nonclassical hydrides. The electron rich π basic metals are anticipated to split the metal bound dihydrogen moieties resulting in classical dihydride complexes. Along the same line of thinking, the electron deficient and less π basic metal would tend to stabilize a dihydrogen complex. The dihydrogen complexes can also be characterized by the X ray diffraction as well as neutron diffraction methods. In IR spectrum, the metal bound H–H stretch appear in the range (2300 – 2900) cm^{-1} while in the 1H NMR spectrum the same appear between 0 to –10 ppm as a broad peak. The dihydrogen complexes are often characterized by isotopic labeling studies of metal bound H–D moiety that shows a coupling constant of 20 – 34 Hz as supposed to 43 Hz observed in case of the free H–D molecule.

Organocadmium Compound

Organocadmium compounds are organometallic compounds that contain a carbon-cadmium chemical bond. The description of the synthesis, physical properties, reactions and use of organocadmium compounds is under the domain of organocadmium chemistry. Such compounds are characterized by their sensitivity to light, air and moisture. There is much similarity in this field and the chemistries corresponding to that of zinc and mercury, which are group 12 elements.

The electrolysis of bromoethane, 1-bromopropane and chlorobenzene has been conducted in a cell consisting of an inert cathode and a sacrificial cadmium anode by passing the nitrogen gas. The electrolysis cell can be represented as:

$$Cd_{(+)} \left| \begin{array}{c} RX/Ar\text{-}X+CH_3CN+ \\ +BU_4NCI+N_2 \end{array} \right| Pt_{(-)}$$

Where, RX is alkyl halides
Ar-X is aryl halide

The efforts to prepare their coordination complexes by refluxing them with ligand in different solvents were also failed. Therefore, their adducts have also been prepared electrochemically by adding 1.0 g of ligand 1,10-phenanthroline or 2,2'-bipyridyl to above reaction mixture and the solid products are isolated in the anode compartment as the electrolysis proceeded.

All the above synthesized electrochemical products are more sensitive to air or moisture. All these compounds are insoluble in most of the commonly used organic solvents like methanol, ethanol, acetone, benzene, chloroform, carbon tetrachloride, N,N-dimethyl formamide, dimethyl sulfoxide and carbon disulfide. It has also been observed that all these isolated complexes do not melt upto 300° C, however, decompose in a temperature range of 200° C to 240° C. The decomposition of the products is indicated from the change in colour of these compounds.

Microanalysis and IR data of organocadmium compounds. The composition of the electrochemical synthesized complexes was determined by analytical and elemental analysis data (Cd, Br, Cl, C and H). From the analytical and elemental analysis data, the complexes are proposed to have the general molecular formula RCdX. These compounds are highly hygroscopic and decompose in air. Therefore, these compounds have been handled in a dry box.

However, major vibrational bands of these compounds appeared in the region of 674-660 cm^{-1}, 467-455 cm^{-1} and 430-425 cm^{-1}, 349-330 cm^{-1}, 227-222 cm^{-1} and 152-150 cm^{-1}.

It is reported in literature that ν (M-C) absorption bands appear in the range of 700-450 cm^{-1}. It is also reported in literature that ν (M-C) peaks observed in the region of 435-420 cm^{-1}. Thus, the band observed in the present products in the range of 674-660 cm^{-1}, 467-455 cm^{-1} and 430-425 cm^{-1} were assigned to ν (Cd-C) stretching vibrations. Appearance of the absorption band due to ν (Cd-C) shows that these compounds may be organocadmium compounds. Moreover, these compounds are highly hygroscopic and decompose in air like the organometallic compounds.

In literature it is reported that metal-halogen stretching vibrations for inorganic complexes are occur in the far infrared in the 500-200 cm^{-1} region. Survey of literature reveals that for polymeric metal complexes ν (M-Cl) stretches are observed in the region of 356-330 cm^{-1}. So, in the present study, medium intensity bands observed at 349 cm^{-1} and 330 cm^{-1} are due to ν (Cd-Cl) stretches of chlorobenzene cadmium complex.

The metal bromides ν (M-Br) stretching vibrations in the far infrared region are reported229 in the region of 254-210 cm^{-1}. In literature the ν (M-Br) peaks are also reported230 in the region of 165- 150 cm^{-1}. In the present products, the bands appeared at 227 cm^{-1}, 150 cm-1 and 222 cm^{-1}, 152

cm^{-1} are due to ν (Cd-Br) stretching vibrations of bromoethane and 1- bromopropane cadmium complex respectively. These peaks are comparatively broad indicating polymeric nature.

So, physical characteristics and infrared spectra of the present electrochemical products reveal that these compounds are organocadmium compounds. All these organocadmium are having high melting point and insoluble nature in all the commonly used organic solvents thereby indicating their polymeric nature.

The probable structure of the organocadmium may be through halide bridges are shown below:

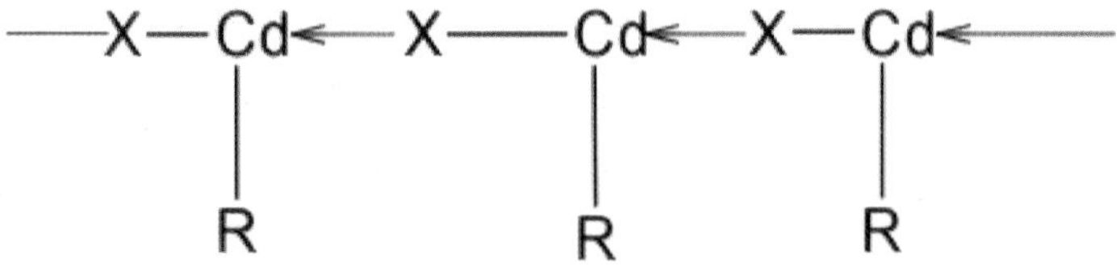

An interesting feature of these reactions is that current efficiencies of all these systems are very high (more than 100% in all these cases).

At cathode:

$$RX + e^{-} \rightarrow R^{\cdot} + x^{-}$$

The halide ion formed at cathode migrates to the anode compartment under the influence of the applied potential and undergo the following sequence of reactions:

$$X^{-} \rightarrow X^{\cdot} + e^{-}$$

$$\dot{X} + Cd_{(+)} \rightarrow Cd\dot{X}$$

$$Cd\dot{X} + RX \rightarrow RCdX + \dot{X}$$

The reaction ($\dot{X} + Cd_{(+)} \rightarrow Cd\dot{X}$) and ($Cd\dot{X} + RX \rightarrow RCdX + \dot{X}$) constitute a chain process and proceed without the consumption of any current from the source and thus explain the high current efficiencies of these systems.

Microanalysis and IR data of coordination complexes of organocadmium compounds. The isolated solid complexes are subjected to elemental analysis of Cd, Br, Cl, C, H and N to identify their tentative formulae. This data is best suited to the molecular formula RCdX.L. which shows that one ligand is coordinated to metal. All these compounds are insoluble in commonly used organic solvents indicate that these may be polymeric in nature.

IR spectra of these metal complexes show peaks in the range of 1545-1468 cm^{-1}, 649-638 cm^{-1}, 470-450 cm^{-1}, 426-419 cm^{-1}, 359-332 cm^{-1}, 231-225 cm^{-1} and 158-154 cm^{-1}.

The bands appearing in the regions of 649-638 cm^{-1}, 470-450 cm^{-1} and 426-419 cm^{-1} in the present products can be assigned to ν (Cd–C) stretching vibrations. In literature the ν (M-Br) peaks are also reported230 at 165- 150 cm^{-1}. So, in present electrochemical synthesized products, the peaks in the region of 359-332 cm^{-1} due to ν (Cd-Cl) and in the region of 231- 225 cm^{-1} and 158-154 cm^{-1} are due ν (Cd-Br) stretching vibrations.

Comparisons of the present infrared data to those of the parent products show a shift of these bands towards the higher regions. The additional bands appearing in the region of 1545 – 1468 cm^{-1} in these products may be due to ν (C···C) and ν (C···C) stretching vibrations of the ligand molecules. However, these bands appear at slightly lower region as compared to those reported in the pure ligand molecules.

The shift of stretching vibration of ν (Cd–C), ν (Cd-Cl) and ν (Cd-Br) towards higher wave number side than those of the parent organocadmium compounds and appearance of absorption bands due to ligand molecules confirmed the attachment of the ligand which was also supported by the analytical data of these adducts.

Current efficiencies of all these systems were determined which are found to be very high. High current efficiencies indicate that the reactions leading to the formation of these coordination compounds are the predominant reactions of the systems.

System	Potential Applied (Volts)	Product	Colour	Elemental analysis Found(Calculated)%				Current efficiency (Gram Equivalents per Faraday)
				Cd	C	H	X	
Bromoethane	25	C_2H_5CdBr	White	49.00 (50.79)	9.89 (10.84)	2.20 (2.26)	35.00 (36.10)	**1.2**
1-Bromopropane	20	C_3H_7CdBr	Cream	47.75 (47.77)	15.35 (15.30)	3.00 (2.98)	33.90 (33.96)	**2.0**
Chlorobenzene	25	C_6H_5CdCl	Grey	49.95 (49.98)	33.00 (32.01)	2.28 (2.22)	15.05 (15.78)	**0.92**

Table: Electrolysis characteristics, elemental analysis and other related data of electrolysis of alkyl/aryl halides at Cadmium anode.

System	Absorption Bands (cm^{-1}) Possible Assignment	
Bromoethane	425(m), 467 (s), 674 (s)	ν(Cd-C)
	227(m), 150(m)	ν(Cd-Br)
1-Bromopropane	430(m),466 (w), 670 (s)	ν(Cd-C)
	222(m), 150(m)	ν(Cd-Br)
Chlorobenzene	428(w), 455 (m), 660(s)	ν(Cd-C)
	330(m), 349(m)	ν(Cd-Cl)

Table: Selected infrared absorption bands of products of electrolysi of alkyl/aryl halides at Cadmium anode.

The term (s), (m) and (w) represents the strong, medium and week intensity bands.

System	Potential Applied (Volts)	Product	Colour	Elemental analysis Found(Calculated)%					Current efficiency (Gram Equivalents per Faraday)
				Cd	C	H	X	N	
Bromoethane + 1,10-phenanthroline	30	$C_2H_5CdBrC_{12}H_8N_2$	Cream	28.00 (27.99)	41.85 (41.84)	4.00 (3.24)	19.91 (19.90)	6.00 (6.97)	**1.32**
1-Bromopropane+ 1,10-phenanthroline	35	$C_3H_7CdBrC_{12}H_8N_2$	Cream	27.00 (27.05)	42.32 (43.32)	3.70 (3.61)	19.00 (19.23)	7.00 (6.74)	**2.20**
Chlorobenzene+ 1,10-phenanthroline	40	$C_6H_5CdClC_{12}H_8N_2$	Light Brown	27.02 (27.44)	52.62 (53.32)	3.20 (3.21)	8.70 (8.76)	7.00 (6.91)	**1.30**
Bromoethane+ 2,2'-bipyridyl	25	$C_2H_5CdBrC_{10}H_8N_2$	Cream	29.16 (29.78)	39.00 (38.15)	3.55 (3.44)	21.20 (21.16)	7.45 (7.42)	**1.36**
1-Bromopropane+ 2,2'-bipyridyl	30	$C_3H_7CdBrC_{10}H_8N_2$	Cream	28.70 (28.71)	40.00 (39.85)	3.80 (3.83)	20.45 (20.41)	7.10 (7.15)	**1.48**
Chlorobenzene + 2,2'-bipyridyl	40	$C_6H_5CdClC_{10}H_8N_2$	Grey	30.00 (29.49)	51.35 (50.38)	3.40 (3.41)	8.50 (9.31)	7.05 (7.35)	**0.98**

Table: Electrolysis characteristics, elemental analysis and other related data of electrolysis of alkyl/aryl halide + Ligand systems at Cadmium anode.

System	Absorption Bands (cm^{-1}) Possible Assignment	
Bromoethane + 1,10-phenanthroline	419(w), 469(s), 638(s) 225 (m), 156(m) 1515, 1539	ν(Cd-C) ν(Cd-Br) ν(C---C)/ν(C---N)
1-Bromopropane+ 1,10-phenanthroline	422(m), 470(s), 637(s) 231(m), 154(m) 1513, 1545	ν(Cd-C) ν(Cd-Br) ν(C---C)/ν(C---N)
Chlorobenzene+ 1,10-phenanthroline	420(s), 453(s), 639(w) 336(m), 359(m) 1515,1532	ν(Cd-C) ν(Cd-Cl) ν(C---C)/ν(C---N)
Bromoethane+ 2,2'-bipyridyl	422(s), 465(s), 648(m) 231(m), 158(m) 1472, 1520	ν(Cd-C) ν(Cd-Br) ν(C---C)/ν(C---N)
1-Bromopropane+ 2,2'-bipyridyl	424(w), 468(s), 642(w) 228 (m), 158(m) 1468, 1535	ν(Cd-C) ν(Cd-Br) ν(C---C))/ν(C---N)
Chlorobenzene + 2,2'-bipyridyl	426(w), 450(s),649(s) 332 (m), 350(m) 1473, 1535	ν(Cd-C) ν(Cd-Cl) ν(C---C)/ν(C---N)

Table: Selected infrared absorption bands of products of electrolysis of alkyl/aryl halide+Ligand systems at Cadmium anode.

The term (s), (m) and (w) represents the strong, medium and week intensity bands.

Electrochemical synthesis of organoaluminium compounds and their coordination complexes. Electrochemical oxidation of bromoethane, 1-bromopropane, chlorobenzene and cyclopentadiene at sacrificial aluminium anode and inert platinum cathode has been done in an H-type cell under the similar reaction conditions as in the preparation of organocadmium compounds. The cell representation is:

$$Al_{(+)} \left| \begin{matrix} RX/Ar\text{-}X \text{ or } Cp+ \\ CH_3CN+Bu_4NCl+N_2 \end{matrix} \right| Pt_{(-)}$$

Where, RX is alkyl halides

Ar-X is aryl halide

Cp is cyclopentadiene

Solid compound separated out in the anode compartment as the electrolysis proceeded.

In the similar way, the adducts of these compounds have also been prepared electrochemically by adding 1.0 g of ligand 1,10-phenanthroline or 2,2'-bipyridyl to the above reaction mixture and all these compounds have been characterized.

The parent compounds and their adducts are insoluble in various organic solvents. All these resulting products are more sensitive to air or moisture and do not melt up to a temperature of 240° C.

Microanalysis and IR data of organoaluminium compounds. All these complexes were analyzed by analytical data and elemental analysis to establish the composition. On the basis of this data, we can say that the products of general molecular formula $(R)_2AlX$ have been isolated in these complexes and molecular formula AlCp3 has been isolated in case of cyclopentadiene aluminium complex. Infrared spectra of all these organometallic compounds have been taken. The band positions in the spectral region 4000-400 cm^{-1} and 400-30 cm^{-1}.

A glance of table above reveals that the products obtained from the electrochemical reactions of alkyl halides show characteristic absorption bands in the region of 638-563 cm^{-1} corresponding to ν (Al-C) stretching vibrations. 222-223 the presence of ν (Al-C) absorption bands in these complexes show that these products may be organoaluminium halides. The moisture sensitivity of these products also supports that these are organoaluminium halides. The bands centered at 238 cm^{-1}, 156 cm^{-1} and 230 cm^{-1}, 158 cm^{-1} in the present products are corresponding to ν(Al-Br) vibrations and present at the 352 cm^{-1} are assignable to ν(Al-Cl) stretching vibrations of organoaluminium compounds. Infrared spectrum of the product of the electrochemical reaction of cyclopentadiene show absorption bands at 2976 cm^{-1}, 1412 cm^{-1}, 1020 cm^{-1}, 832 cm^{-1} which can be assigned to ν(C-H) deformation 231 (out of plane and in plane) of the cyclopentadienyl ring. Another medium intensity band assignable to ν (Al-ring) vibrations has been observed at 410 cm^{-1} in this compound.

The current efficiencies of all these systems have been found to be more than 100% except cyclopentadiene and chlorobenzene systems. The reason for this is same as discussed earlier. High current efficiency of all these systems indicating the reaction leading to formation of $AlCp_3$ is the predominant reaction of this system.

The mode of this reaction is as follow:

At cathode,

At anode,

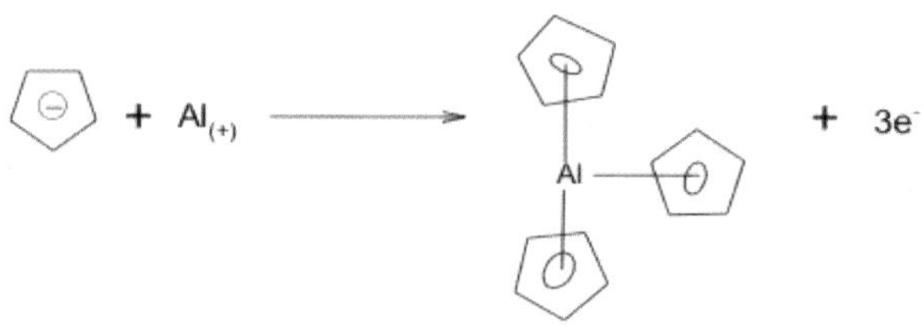

Microanalysis and IR data of coordination complexes of organoaluminium compounds. The compositions of the resulting products were established by analytical/elemental analysis and this data is best suited to the molecular formula $(R)_2$ AlX.L in the case of alkyl/aryl halides and AlCp3.L in the case of cyclopentadiene.

Infrared spectral studies of these products have been carried out. The absorption bands in the regions of 632-598 cm^{-1} and 448-423 cm^{-1} in the resulting products can be assigned to ν (Al–C) stretching vibrations. In the present study, the medium intensity absorption bands in the region of 359-335 cm^{-1} are assignable to ν (Al–Cl) stretching vibrations and in the region of 242-224 cm^{-1} and 162-152 cm^{-1} are corresponding to ν (Al–Br) stretching vibrations of organoaluminium compounds respectively. In the case of cyclopentadiene -aluminium complex the persistence of bands at 3030 cm^{-1}, 1412 cm^{-1}, 1028 cm^{-1}, 840 cm^{-1} are due to cyclopentadienyl rings.

The attachment of the ligand molecules is confirmed by the appearance of these absorption bands at a little higher region than the parent compounds and the additional bands appearing in these products due to ν $(C \equiv C)$ and ν $(C \equiv N)$ of the ligand molecules give the another evidence for the attachment of the ligand with aluminium.

Current efficiencies of the alkyl/aryl halide systems are found to be more than 100% (except cyclopentadiene and chlorobenzene). High current efficiencies indicate the mechanism of the formation of these coordination compounds at sacrificial aluminium anode and inert cathode as is given below:

At cathode,

$$RX + e^{-} \rightarrow R^{\cdot} + X^{-}$$

The halide ion formed at cathode migrates to the anode compartment under the influence of the applied potential and undergoes the following sequence of reactions:

At anode,

$$X^{-} \rightarrow X^{\cdot} + e^{-}$$

$$\dot{X} + Al_{(+)} \rightarrow Al\dot{X}$$

$$Al\dot{X} + 2RX \rightarrow (R)_2\ AlX + 2\dot{X}$$

The reaction ($\dot{X} + Al_{(+)} \rightarrow Al\dot{X}$) and ($Al\dot{X} + 2RX \rightarrow (R)_2\ AlX + 2\dot{X}$) constitute a chain process and proceed without the consumption of any current from the source and thus explain the high current efficiencies of these systems. The present technique provides a new synthetic method for the laboratory preparation of organocadmium and organoaluminium compounds in a single step and the method is associated with high current efficiencies.

System	Potential Applied(Volts)	Product	Colour	Elemental analysis Found(Calculated)%				Current efficiency (Gram Equivalents per Faraday)
				Al	C	H	X	
Bromoethane	42	$(C_2H_5)_2AlBr$	Grey	15.90 (16.37)	30.00 (29.11)	6.02 (6.06)	49.00 (48.45)	**1.3**
1-Bromopropane	40	$(C_3H_7)_2AlBr$	Grey	13.05 (13.99)	37.81 (37.33)	7.09 (7.26)	40.00 (41.42)	**2.0**
Chlorobenzene	35	$(C_6H_5)_2AlCl$	Grey	12.15 (12.47)	66.28 (66.51)	4.28 (4.62)	15.35 (16.40)	**0.84**
Cyclopentadiene	30	$(C_5H_5)_2Al$	Dark Brown	17.35 (17.20)	76.35 (76.43)	6.22 (6.37)	----	**0.81**

Table: Electrolysis characteristics, elemental analysis and other related data of electrolysis of alkyl/aryl halides at Aluminium anode.

System	Absorption Bands (cm^{-1})	Possible Assignment
Bromoethane	560(s), 638(m) 230(m), 156(m)	ν(Al-C) ν(Al-Br)
1-Bromopropane	572(s), 625(m) 238(m), 158(m)	ν(Al-C) ν(Al-Br)
Chlorobenzene	563(s), 614(m) 352 (m)	ν(Al-C) ν(Al-Cl)
Cyclopentadiene	2976(s) 1412(s) 1020(m) 832(m) 410(m)	ν(C-H) of Cp ν(C-C) of Cp ν(C-H) of Cp ν(C-H) of Cp ν(Al-Cp) ring

Table: Selected infrared absorption bands of products of electrolysis of alkyl/aryl halides at aluminium anode.

The term (s) and (m) represents the strong and medium intensity bands.

System	Potential Applied (Volts)	Product	Colour	Elemental analysis Found(Calculated)%					Current efficiency (Gram Equivalents per Faraday)
				Al	C	H	N	X	
Bromoethane+ 1,10-phenanthroline	45	$(C_2H_5)_2AlBrC_{12}H_8N_2$	Brown	7.42 (7.82)	55.00 (55.63)	5.18 (5.22)	8.10 (8.11)	21.90 (23.15)	**2.0**
1-Bromopropane + 1,10-phenanthroline	25	$(C_3H_7)_2AlBrC_{12}H_8N_2$	Light Brown	7.16 (7.24)	57.02 (57.89)	5.89 (5.90)	7.28 (7.50)	21.39 (21.41)	**1.4**
Chlorobenzene+ 1,10-phenanthroline	55	$(C_6H_5)_2AlClC_{12}H_8N_2$	Light Brown	7.18 (6.81)	71.53 (72.59)	4.49 (4.54)	7.42 (7.06)	8.94 (8.95)	**0.82**
Cyclopentadiene+ 1,10-phenanthroline	40	$(C_5H_5)_2AlC_{12}H_8N_2$	Dark brown	7.99 (8.00)	77.26 (78.28)	5.34 (5.34)	8.29 (8.30)	.....	**0.76**
Bromoethane+ 2,2'-bipyridyl	40	$(C_2H_5)_2AlBrC_{10}H_8N_2$	Light Grey	8.33 (8.41)	52.28 (52.32)	5.59 (5.61)	8.76 (8.72)	23.78 (24.88)	**1.6**
1-Bromopropane+ 2,2'-bipyridyl	30	$(C_3H_7)_2AlBrC_{10}H_8N_2$	Light Grey	7.36 (7.73)	54.80 (55.00)	6.15 (6.30)	8.00 (8.02)	22.80 (22.89)	**2.0**
Chlorobenzene+ 2,2'-bipyridyl	45	$(C_6H_5)_2AlClC_{10}H_8N_2$	Light Grey	7.42 (7.24)	7.00 (7.08)	4.82 (4.83)	7.60 (7.51)	9.25 (9.53)	**0.79**
Cyclopentadiene+ 2,2'-bipyridyl	45	$(C_5H_5)_2AlC_{10}H_8N_2$	Dark Brown	8.51 (8.62)	76.00 (76.63)	5.45 (5.75)	8.90 (8.94)	-----	**0.82**

Table: Electrolysis characteristics, elemental analysis and other related data of electrolysis of alkyl/aryl halide + Ligand systems at Aluminium anode.

System	Absorption Bands (cm^{-1})	Possible Assignment
Bromoethane + 1,10-phenanthroline	428(s), 598(m) 230(m), 162(m) 1464, 1600	ν(Al-C) ν(Al-Br) ν(C---C) /ν(C---N)
1-Bromopropane+ 1,10-phenanthroline	433(s), 602(m) 234 (m), 160(m) 1460, 1602	ν(Al-C) ν(Al-Br) ν(C---C)/ν(C---N)
Chlorobenzene + 1,10-phenanthroline	423(s), 625 (m) 355(m) 1459, 1605	ν(Al-C) ν(Al-Cl) ν(C---C) /ν(C---N)
Cyclopentadiene+ 1,10-phenanthroline	3030(s) 1412(s) 1028(m) 840(m) 430(m) 1456,1611	ν(C-H) of Cp ν(C-C) of Cp ν(C-H) of Cp ν(C-H) of Cp ν(Al-Cp) ring ν(C---C)/ν(C---N)

Table: Selected infrared absorption bands of products of electrolysis of alkyl/aryl halide +Ligand systems at Aluminium anode:

The term (s) and (m) represents the strong and medium intensity bands.

Bromoethane+ 2,2'-bipyridyl	432(s), 610(m) 224(m) 1440, 1608	ν(Al-C) ν(Al-Br) ν(C---C)/ν(C---N)
1-Bromopropane+ 2,2'-bipyridyl	448(s), 622(m) 242(m) 1448, 1600	ν(Al-C) ν(Al-Br) ν(C---C) /ν(C---N)
Chlorobenzene + 2,2'-bipyridyl	429(s), 632 (m) 359(m) 1452, 1605	ν(Al-C) ν(Al-Cl) ν(C---C)/ν(C---N)
Cyclopentadiene+ 2,2'-bipyridyl	3028(s) 1416(s) 1020(m) 830(m) 420(m) 1456,1600	ν(C-H) of Cp ν(C-C) of Cp ν(C-H) of Cp ν(C-H) of Cp ν(Al-Cp) ring ν(C---C) /ν(C---N)

References

- Oliveira, José; Elschenbroich, Christoph (2006). Organometallics (3., completely rev. and extended ed.). Weinheim: Wiley-VCH-Verl. ISBN 978-3-527-29390-2
- Organometallic-compound, science: britannica.com, Retrieved 25 April 2018
- Zeise, W. C. (1831). Von der Wirkung zwischen Platinchlorid und Alkohol, und von den dabei entstehenden neuen Substanzen. Annalen der Physik. 97(4): 497–541. Bibcode:1831AnP....97..497Z. doi:10.1002/andp.18310970402
- Organoboron-compound, chemistry: sciencedirect.com, Retrieved 15 March 2018
- Crabtree, Robert H. (2009). The Organometallic Chemistry of the Transition Metals (5th ed.). New York, NY: John Wiley and Sons. pp. 2, 560, and passim. ISBN 0470257628. Retrieved 23 May 2016
- Organoaluminum-Compounds: encyclopedia2.thefreedictionary.com, Retrieved 02 July 2018
- Berg, Jeremy M.; Lippard, Stephen J. (1994). Principles of bioinorganic chemistry ([Pbk. ed.]. ed.). Mill Valley: University Science Books. ISBN 0-935702-73-3

4

Organic Reactions

Chemical reactions involving organic compounds are called organic reactions. All the diverse types of organic reactions such as addition reactions, elimination reactions, redox reactions and substitution reactions have been carefully analyzed in this chapter.

Organic reactions encompass a wide assortment of reactions involving compounds that have carbon as the main element in their molecular structure. The reactions in which an organic compound may take part are largely defined by its functional groups.

There is no limit to the number of possible organic reactions and mechanisms. However, certain general patterns are observed that can be used to describe many common or useful reactions. Each reaction has a stepwise reaction mechanism that explains how it happens. Organic reactions can be organized into several basic types, with some reactions fitting into more than one category.

Some of the basic types of organic chemical reactions are noted below:

- Addition reactions, including such reactions as halogenation, hydrohalogenation, and hydration. Based on the mechanism, the main addition reactions are classified as electrophilic, nucleophilic, or radical addition.
- Elimination reactions, including processes such as dehydration.
- Substitution reactions are divided into several types: nucleophilic aliphatic substitution with SN1, SN2 and SNi reaction mechanisms; nucleophilic aromatic substitution; nucleophilic acyl substitution; electrophilic substitution; electrophilic aromatic substitution; and radical substitution.
- Organic redox reactions are redox reactions specific to organic compounds and are very common.
- Rearrangement reactions are divided into 1, 2-rearrangements, pericyclic reactions, and metathesis reactions.
- In condensation reactions, two reactants combine, and a small molecule (usually water) is split off. The opposite reaction, when water is consumed in a reaction, is called hydrolysis. Many polymerization reactions are derived from organic reactions. They are divided into addition polymerizations and step-growth polymerizations.

Addition Reaction

Addition reaction is a class of chemical reactions in which an atom or group of atoms is added to a molecule.

Addition reactions are typical of unsaturated organic compounds—i.e. alkenes which contain a carbon-to-carbon double bond, and alkynes, which have a carbon-to-carbon triple bond and aldehydes and ketones, which have a carbon-to-oxygen double bond. An addition reaction may be visualized as a process by which the double or triple bonds are fully or partially broken in order to accommodate additional atoms or groups of atoms in the molecule. Addition reactions to alkenes and alkynes are sometimes called saturation reactions because the reaction causes the carbon atoms to become saturated with the maximum number of attached groups.

A typical addition reaction may be illustrated by the hydrochlorination of propene (an alkene), for which the equation is:

$$CH_3CH = CH_2 + HCl \rightarrow CH_3C{+}HCH_3 + Cl^- \rightarrow CH_3CHClCH_3.$$

The reaction proceeds in two stages: first, the hydrogen ion, H+, of hydrogen chloride (the positively charged component) adds to one of the pair of carbon atoms joined by double bonds—in this case, the less alkylated carbon atom—followed by addition of the chloride ion, Cl– (the negatively charged component), to the other carbon atom.

In addition reactions to aldehydes and ketones, the sequence of events is reversed; i.e., the initial step is addition of the negatively charged component of the reagent to the carbon atom, followed by addition of the positively charged component to the oxygen atom. Thus, the reaction of a methyl ketone ($CH_3(C = O)R$, where, R is an alkyl group) with hydrogen cyanide proceeds as follows:

$$O \parallel H_3CCR + HCN \parallel O{-}OH \parallel H_3CCCN + H{+} \parallel H_3CCCN \parallel RR.$$

The hydrochlorination of propene or, in general, the addition to alkenes is said to be initiated by electron-seeking (electrophilic) reagents, while the additions to alkynes, aldehydes, and ketones are said to be initiated by electron-rich (nucleophilic) reagents. Other forms of addition reactions include: catalyzed addition reactions, such as the self-addition of alkenes (catalyzed by acids) or the hydrogenation of alkenes, aldehydes, and ketones (catalyzed by metals); addition reactions in which cyclic compounds are formed; and addition reactions that proceed by chain mechanisms.

Aldol Addition

An aldol addition is a reaction that occurs between the enolate of an aldehyde or ketone and the alpha-carbon of another aldehyde or ketone to form a beta-hydroxy aldehyde or ketone.

An enolate is an intermediate in organic chemistry that contains a formal negative one charge on either a carbon atom or an oxygen atom.

If the negative charge is on the carbon atom there's a double bond between carbon and oxygen, but if it's on the oxygen atom there's a double bond between two carbon atoms. This type of scenario is called resonance when the negative charge can be placed on more than one atom. Consider as an example the enolate of butanone.

Figure: An enolate is an intermediate in organic chemistry derived from an aldehyde or ketone.

Enolates form when a base reacts with an aldehyde or ketone and pull off one of the hydrogen atoms directly adjacent to the carbon-oxygen double bond (carbonyl group).

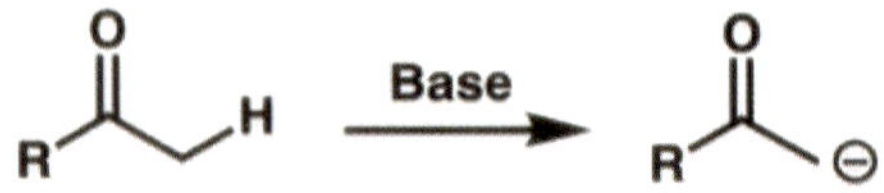

An enolate results when an aldehyde or ketone reacts with a base

The aldol addition in a nutshell is when the newly formed enolate reacts with another aldehyde or ketone to form a new carbon-carbon bond and thus a larger organic compound. Kind of like building railroad tracks. The general form of the aldol addition is enolate + aldehyde or ketone = aldol product.

The general form of the aldol addition reaction

Mechanism of the Aldol Addition

A three step mechanism:

Step 1: Enolate formation

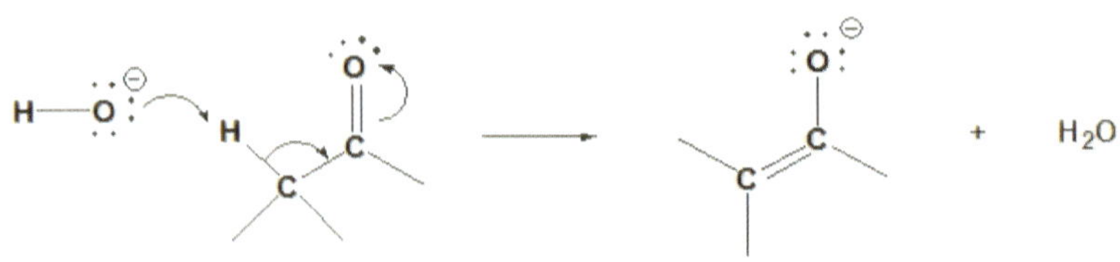

Step 2: Nucleophilic attack by the enolate

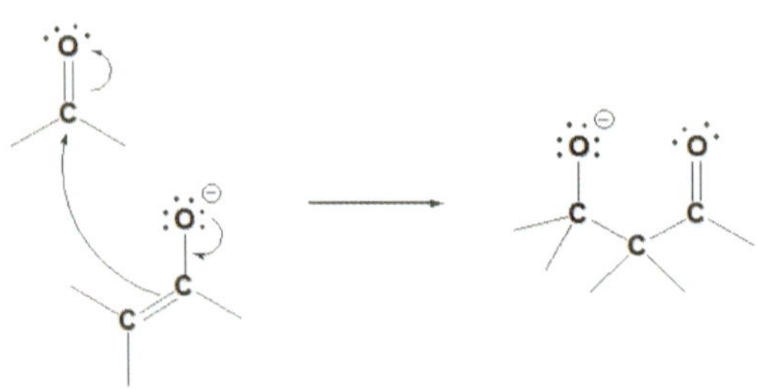

Step 3: Protonation

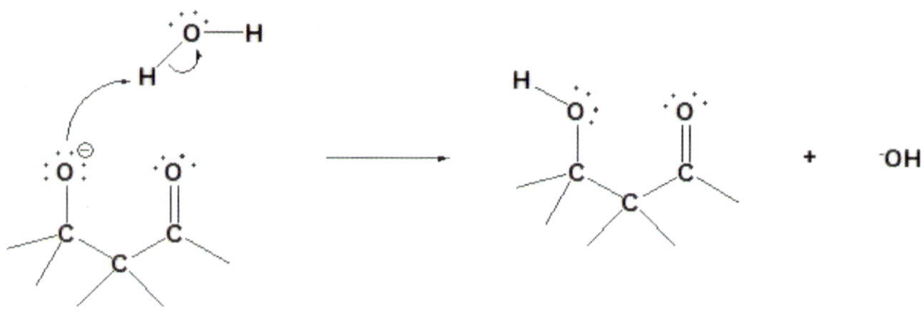

Aldol Condensation: The Dehydration of Aldol Products to Synthesize α, β Unsaturated Carbonyl (Enones)

The products of aldol reactions often undergo a subsequent elimination of water, made up of an alpha-hydrogen and the beta-hydroxyl group. The product of this β-elimination reaction is an α, β-unsaturated aldehyde or ketone. Base-catalyzed elimination occurs with heating. The additional stability provided by the conjugated carbonyl system of the product makes some aldol reactions thermodynamically and mixtures of stereoisomers (E & Z) are obtained from some reactions. Reactions in which a larger molecule is formed from smaller components, with the elimination of a very small by-product such as water, are termed Condensations. Hence, the following examples are properly referred to as aldol condensations. Overall the general reaction involves a dehydration of an aldol product to form an alkene.

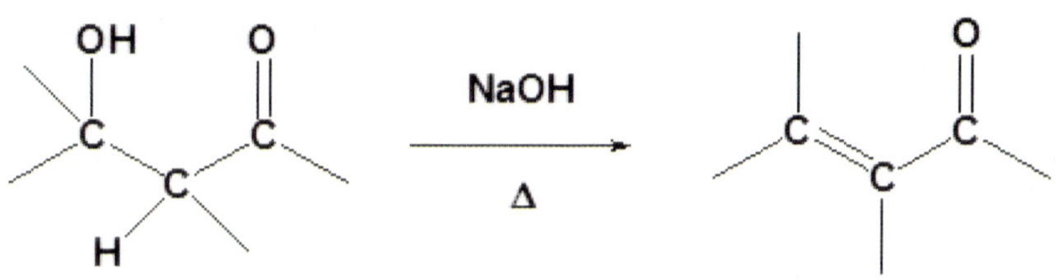

Figure: General reaction for an aldol condensation.

Going from reactants to products simply,

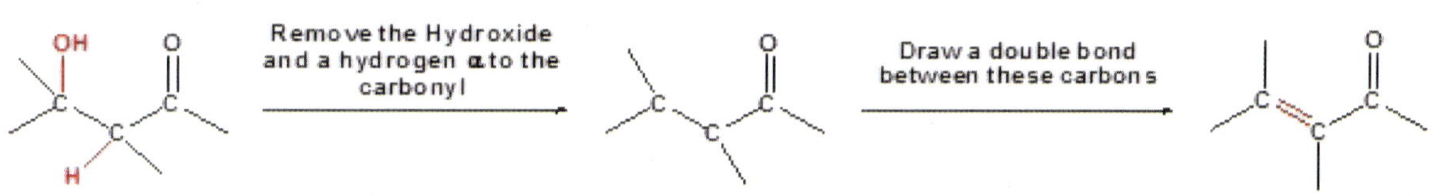

Figure: The aldol condensation example.

Example: Aldol Condensation

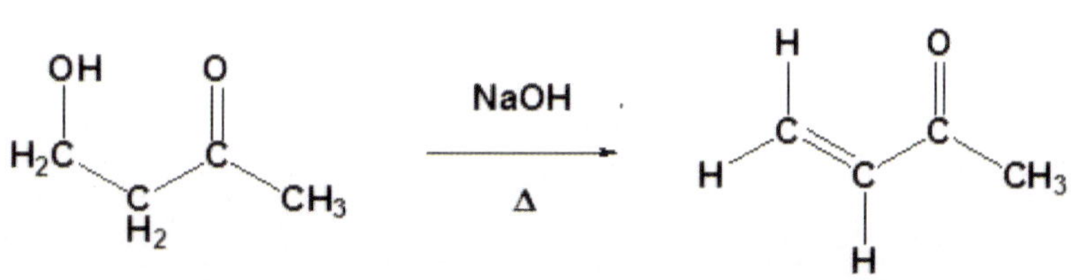

Aldol Condensation Mechanism

- From enolate,

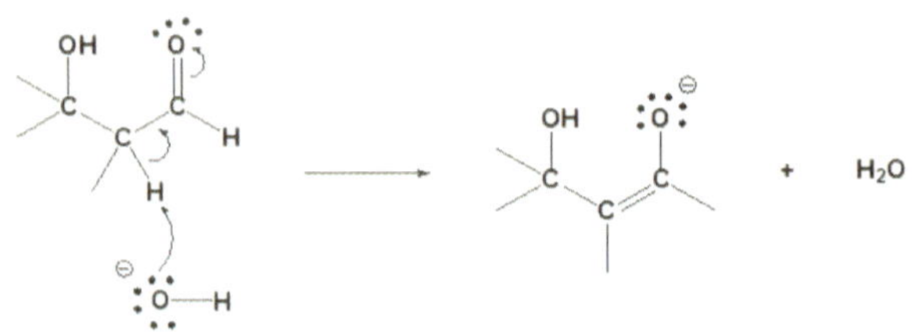

- Form enone,

When performing both reactions together always consider the aldol product first then convert to the enone. Note, the double bond always forms in conjugation with the carbonyl.

Example:

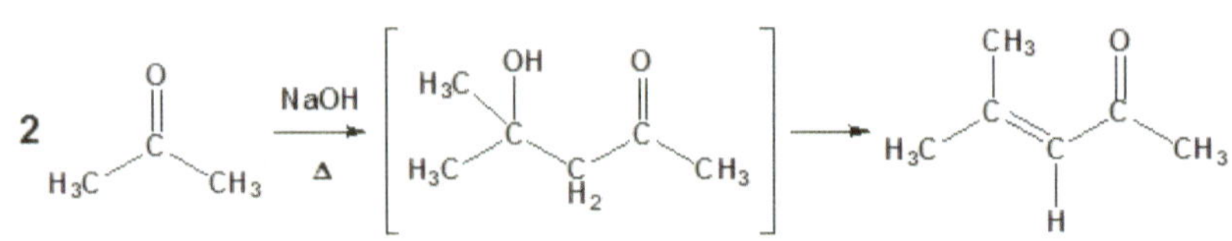

Intramolecular Aldol Reaction

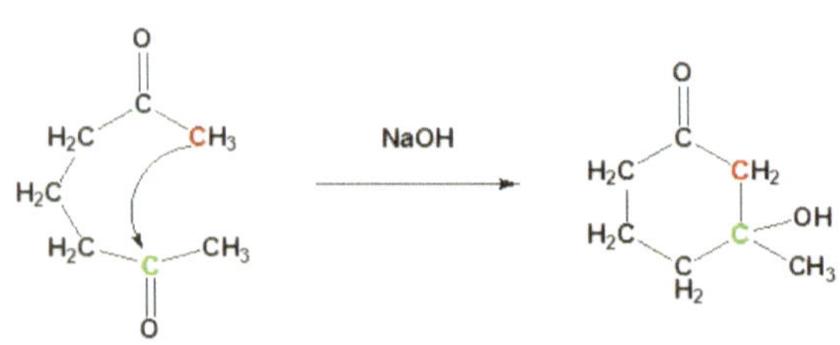

Molecules which contain two carbonyl functionalities have the possibility of forming a ring through an intramolecular aldol reaction. In most cases two sets of α hydrogen need to be considered. As with most ring forming reaction five and six membered rings are preferred.

As with other aldol reaction the addition of heat causes an aldol condensation to occur.

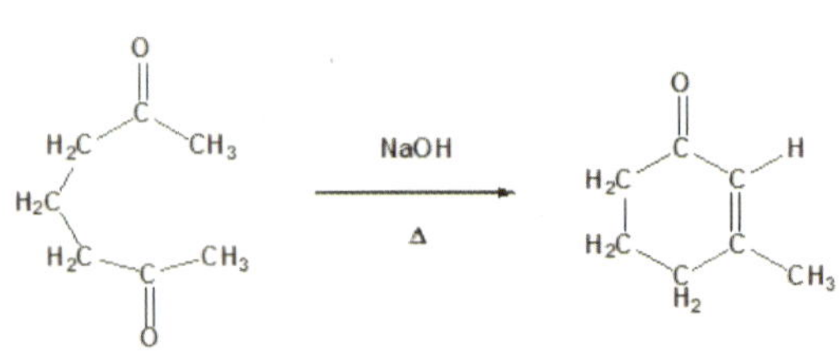

Mixed Aldol Reaction and Condensations

The previous examples of aldol reactions and condensations used a common reactant as both the enolic donor and the electrophilic acceptor. The product in such cases is always a dimer of the reactant carbonyl compound. Aldol condensations between different carbonyl reactants are called crossed or mixed reactions, and under certain conditions such crossed aldol condensations can be effective.

Example: Mixed Aldol Reactions

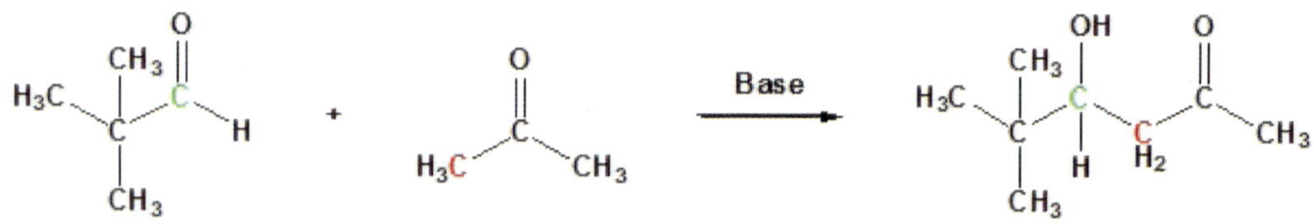

The success of these mixed aldol reactions are due to two factors. First, aldehydes are more reactive acceptor electrophiles than ketones, and formaldehyde is more reactive than other aldehydes. Second, aldehydes lacking alpha-hydrogens can only function as acceptor reactants, and this reduces the number of possible products by half. Mixed aldols in which both reactants can serve as donors and acceptors generally give complex mixtures of both dimeric (homo) aldols and crossed aldols. Because of this most mixed aldol reactions are usually not performed unless one reactant has no alpha hydrogens.

The following abbreviated formulas illustrate the possible products in such a case, red letters representing the acceptor component and blue the donor. If all the reactions occurred at the same rate, equal quantities of the four products would be obtained. Separation and purification of the components of such a mixture would be difficult:

$$AACH_2CHO + BCH_2CHO + NaOH \rightarrow A\text{–}A + B\text{–}B + A\text{–}B + B\text{–}A$$

The aldol condensation of ketones with aryl aldehydes to form α,β-unsaturated derivatives is called the Claisen-Schmidt reaction.

Example: Claisen-Schmidt Reaction

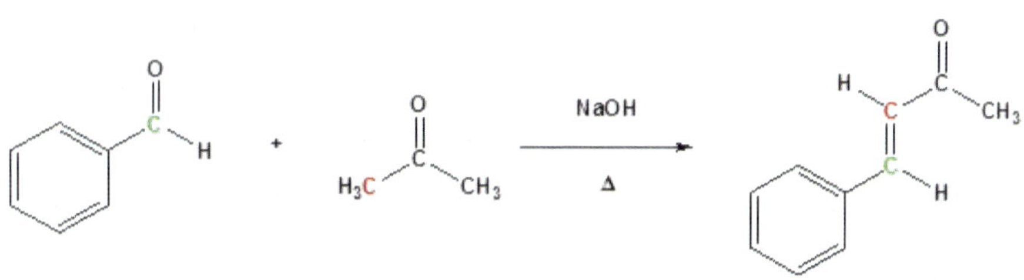

Stereoselectivity

The aldol reaction is particularly useful because two new stereogenic centers are generated in one reaction. Extensive research has been performed to understand the reaction mechanism and improve the selectivity observed under many different conditions. The syn/anti convention is commonly used to denote the relative stereochemistry at the α- and β-carbon.

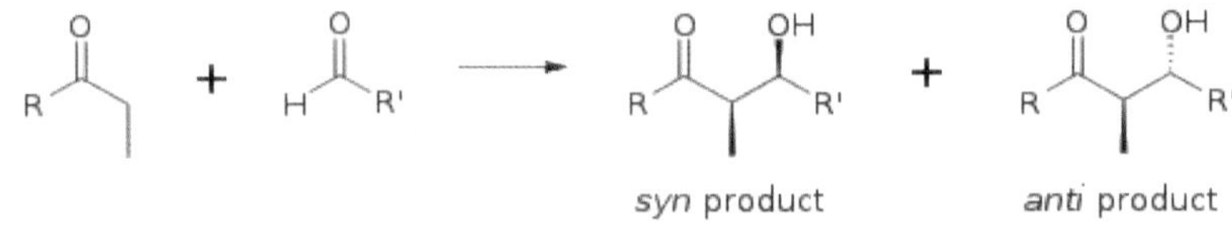

The convention applies when propionate (or higher order) nucleophiles are added to aldehydes. The R group of the ketone and the R' group of the aldehyde are aligned in a "zig zag" pattern in the plane of the paper (or screen), and the disposition of the formed stereocenters is deemed syn or anti, depending if they are on the same or opposite sides of the main chain.

Older papers use the erythro/threo nomenclature familiar from carbohydrate chemistry.

Enolate Geometry

There is no significant difference between the level of stereoinduction observed with E and Z enolates. Each alkene geometry leads primarily to one specific relative stereochemistry in the product, E giving anti and Z giving syn:

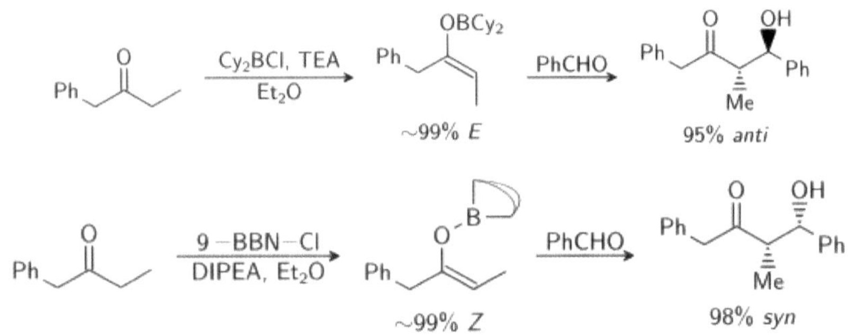

Metal Ion

The enolate metal cation may play a large role in determining the level of stereoselectivity in the aldol reaction. Boron is often used because its bond lengths are significantly shorter than that of other metals such as lithium, aluminium, or magnesium.

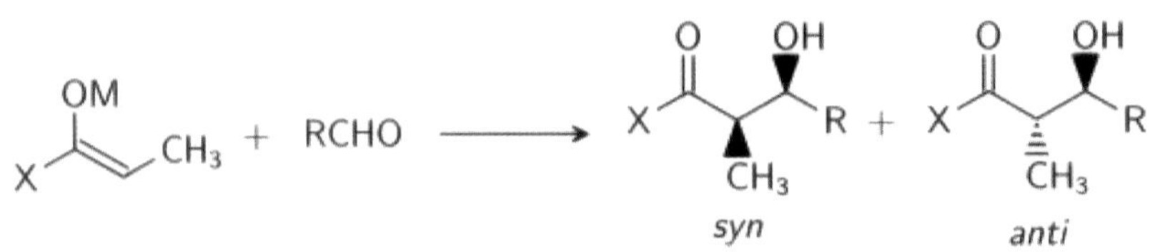

For example, boron–carbon and boron–oxygen bonds are 1.4–1.5 Å and 1.5–1.6 Å in length, respectively, whereas typical metal-carbon and metal-oxygen bonds are 1.9–2.2 Å and 2.0–2.2 Å in length, respectively. The use of boron rather than a metal "tightens" the transition state and gives greater stereoselectivity in the reaction. Thus the above reaction gives a syn:anti ratio of 80:20 using a lithium enolate compared to 97:3 using a bibutylboron enolate.

Alpha Stereocenter on the Enolate

The aldol reaction may exhibit "substrate-based stereocontrol", in which existing chirality on either reactant influences the stereochemical outcome of the reaction. This has been extensively studied, and in many cases, one can predict the sense of asymmetric induction, if not the absolute level of diastereoselectivity. If the enolate contains a stereocenter in the alpha position, excellent stereocontrol may be realized.

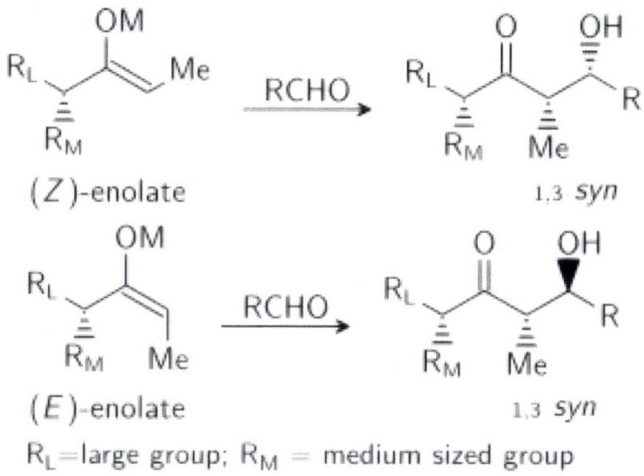

In the case of an E enolate, the dominant control element is allylic 1,3-strain whereas in the case of a Z enolate, the dominant control element is the avoidance of 1,3-diaxial interactions. The general model is presented below:

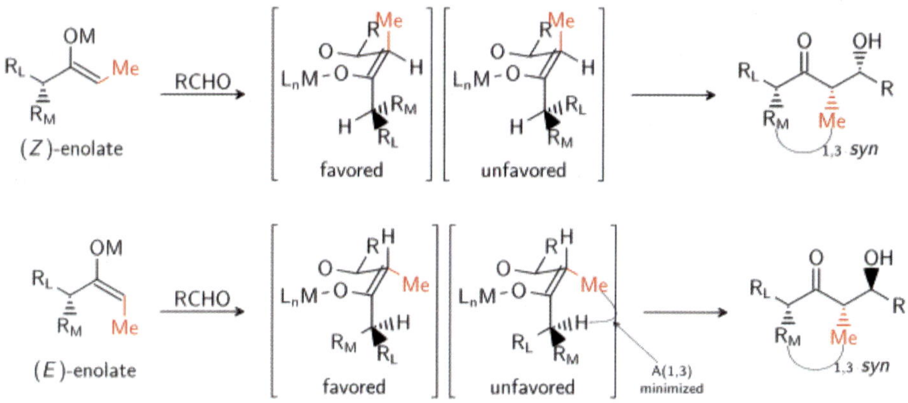

For clarity, the stereocenter on the enolate has been epimerized; in reality, the opposite diastereoface of the aldehyde would have been attacked. In both cases, the 1, 3-syn diastereomer is favored. There are many examples of this type of stereocontrol:

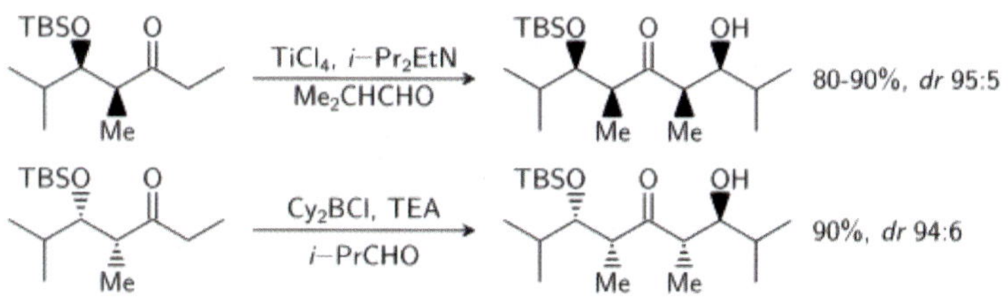

Alpha Stereocenter on the Electrophile

When enolates attacks aldehydes with an alpha stereocenter, excellent stereocontrol is also possible. The general observation is that E enolates exhibit Felkin diastereoface selection, while Z enolates exhibit anti-Felkin selectivity. The general model is presented below:

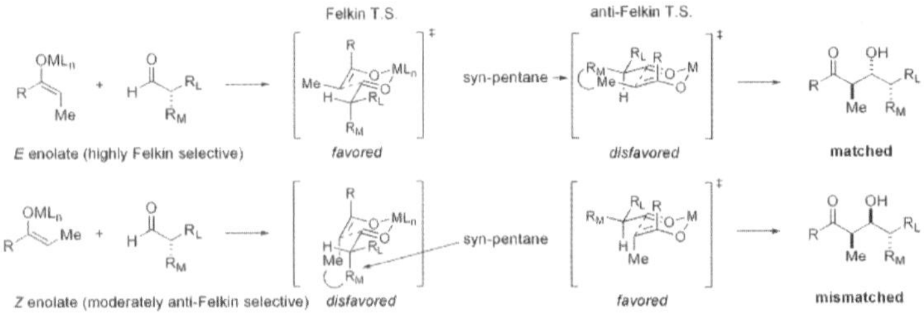

Since Z enolates must react through a transition state that contains either a destabilizing syn-pentane interaction or an anti-Felkin rotamer, Z-enolates exhibit lower levels of diastereoselectivity in this case.

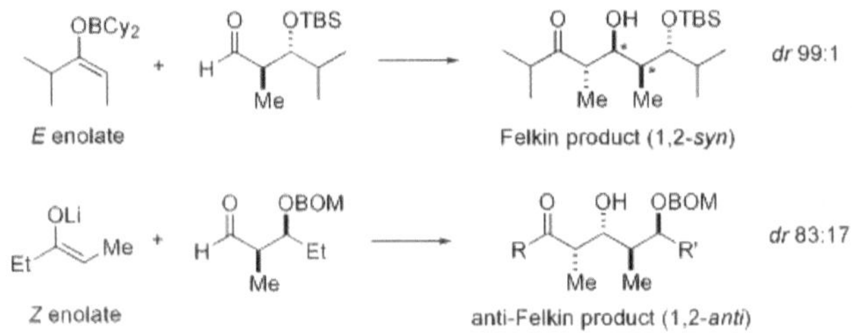

Unified Model of Stereoinduction

If both the enolate and the aldehyde contain pre-existing chirality, then the outcome of the "double stereodifferentiating" aldol reaction may be predicted using a merged stereochemical model that takes into account the enolate facial bias, enolate geometry, and aldehyde facial bias. Several examples of the application of this model are given below:

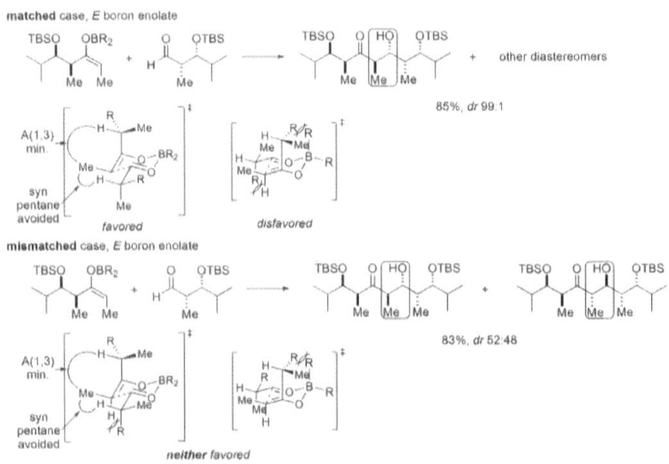

Evans' Oxazolidinone Chemistry

Modern organic syntheses now require the synthesis of compounds in enantiopure form. Since the aldol addition reaction creates two new stereocenters, up to four stereoisomers may result:

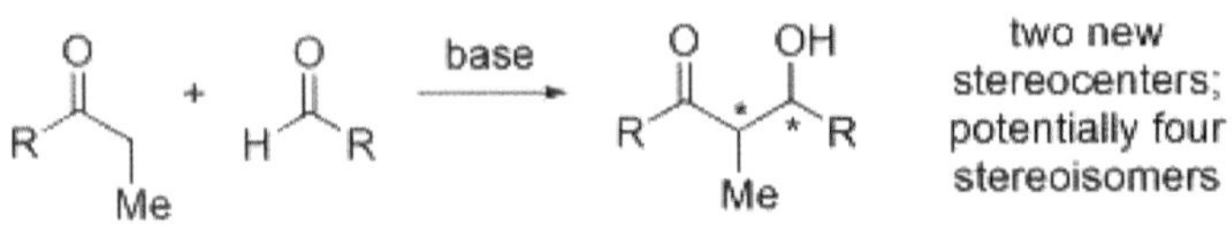

Many methods which control both relative stereochemistry and absolute stereochemistry have been developed.

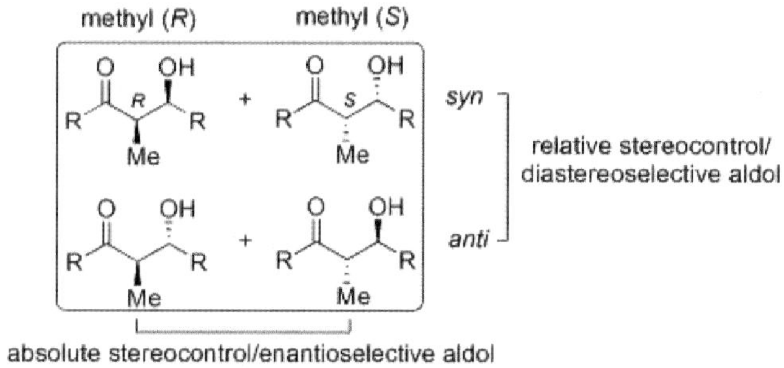

A widely used method is the Evans' acyl oxazolidinone method. Developed in the late 1970s and 1980s by David A. Evans and coworkers, the method works by temporarily creating a chiral enolate by appending a chiral auxiliary. The pre-existing chirality from the auxiliary is then transferred to the aldol adduct by performing a diastereoselective aldol reaction. Upon subsequent removal of the auxiliary, the desired aldol stereoisomer is revealed.

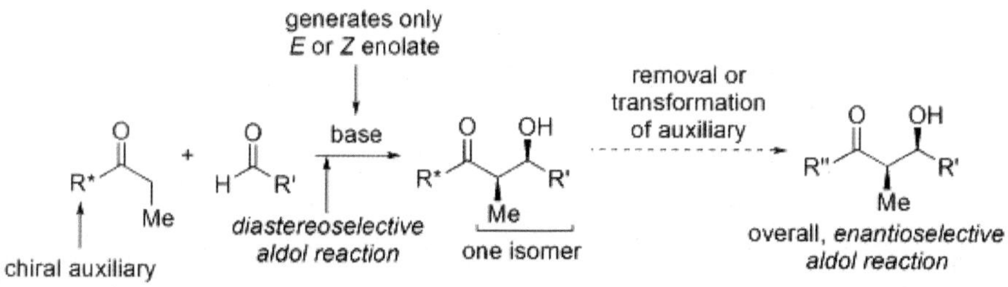

In the case of the Evans' method, the chiral auxiliary appended is an oxazolidinone, and the resulting carbonyl compound is an imide. A number of oxazolidinones are now readily available in both enantiomeric forms. These may cost roughly $10–$20 US dollars per gram, rendering them relatively expensive. However, enantiopure oxazolidinones are derived in 2 synthetic steps from comparatively inexpensive amino acids, which means that large-scale syntheses can be made more economical by in-house preparation. This usually involves borohydride mediated reduction of the acid moiety, followed by condensation/cyclisation of the resulting amino alcohol with a simple carbonate ester such as diethylcarbonate.

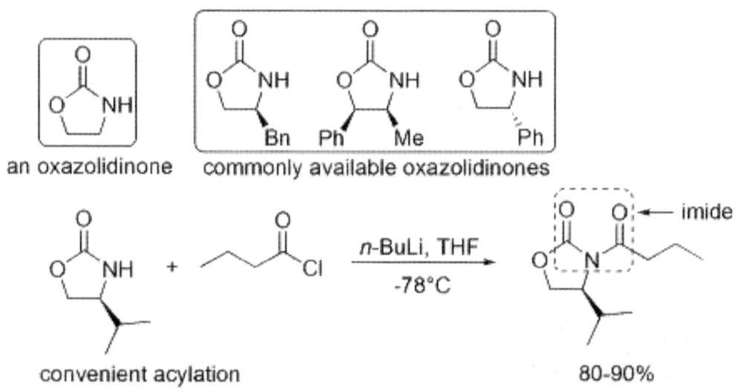

The acylation of an oxazolidinone is a convenient procedure, and is informally referred to as "loading done". Z-enolates, leading to syn-aldol adducts, can be reliably formed using boron-mediated soft enolization:

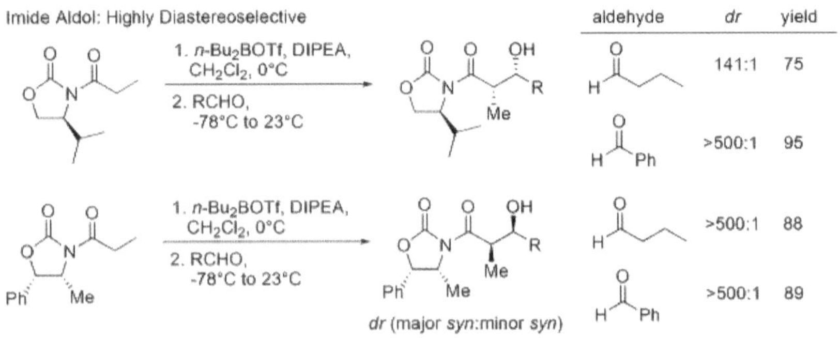

Often, a single diastereomer may be obtained by one crystallization of the aldol adduct. However, anti-aldol adducts cannot be obtained reliably with the Evans method. Despite the cost and the limitation to give only syn adducts, the method's superior reliability, ease of use, and versatility render it the method of choice in many situations. Many methods are available for the cleavage of the auxiliary:

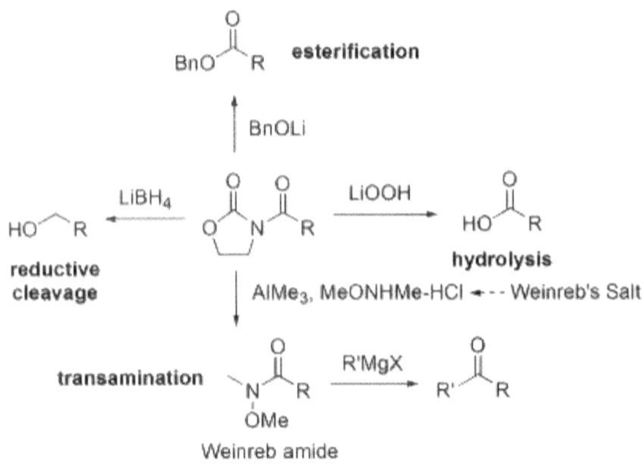

Upon construction of the imide, both syn- and anti-selective aldol addition reactions may be performed, allowing the assemblage of three of the four possible stereoarrays, syn-selective and anti-selective:

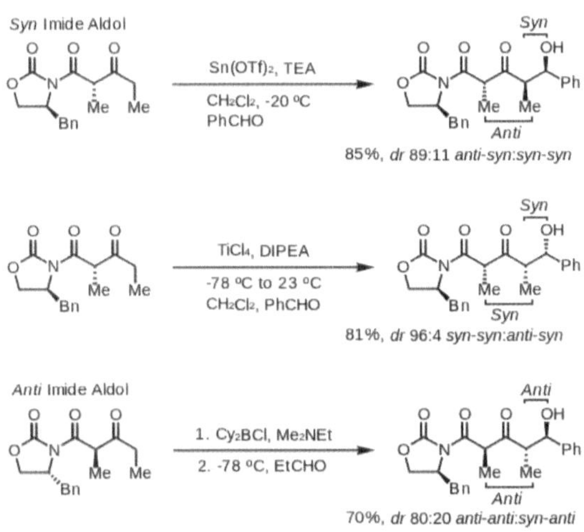

In the syn-selective reactions, both enolization methods give the Z enolate, as expected; however, the stereochemical outcome of the reaction is controlled by the methyl stereocenter, rather than the chirality of the oxazolidinone. The methods described allow the stereoselective assembly of polyketides, a class of natural products that often feature the aldol retron.

Modern Variations and Methods

Recent methodology now allows a much wider variety of aldol reactions to be conducted, often with a catalytic amount of chiral ligand. When reactions employ small amounts of enantiomerically pure ligands to induce the formation of enantiomerically pure products, the reactions are typically termed “catalytic, asymmetric”; for example, many different catalytic, asymmetric aldol reactions are now available.

Acetate Aldol Reactions

A key limitation to the chiral auxiliary approach described previously is the failure of N-acetyl imides to react selectively. An early approach was to use a temporary thioether group:

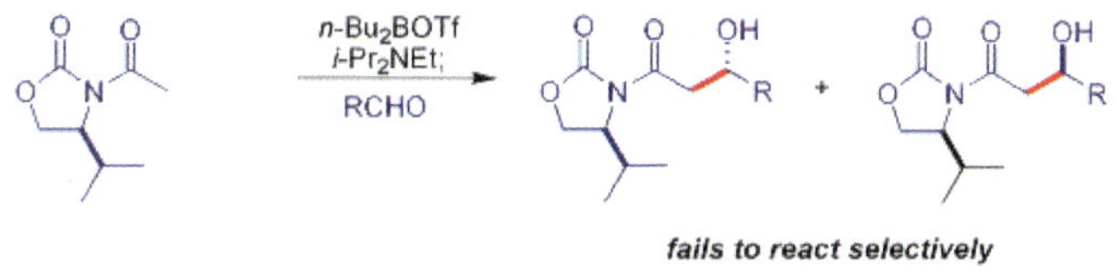

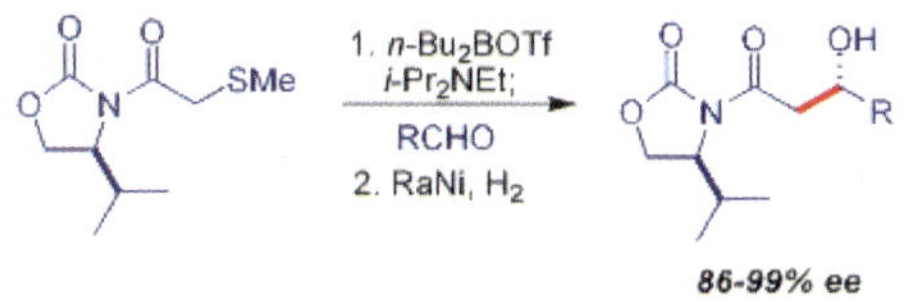

Mukaiyama Aldol Reaction

The Mukaiyama aldol reaction is the nucleophilic addition of silyl enol ethers to aldehydes catalyzed by a Lewis acid such as boron trifluoride (as boron trifluoride etherate) or titanium tetrachloride. The Mukaiyama aldol reaction does not follow the Zimmerman-Traxler model. Carreira has described particularly useful asymmetric methodology with silyl ketene acetals, noteworthy for its high levels of enantioselectivity and wide substrate scope.

The method works on unbranched aliphatic aldehydes, which are often poor electrophiles for catalytic, asymmetric processes. This may be due to poor electronic and steric differentiation between their enantiofaces.

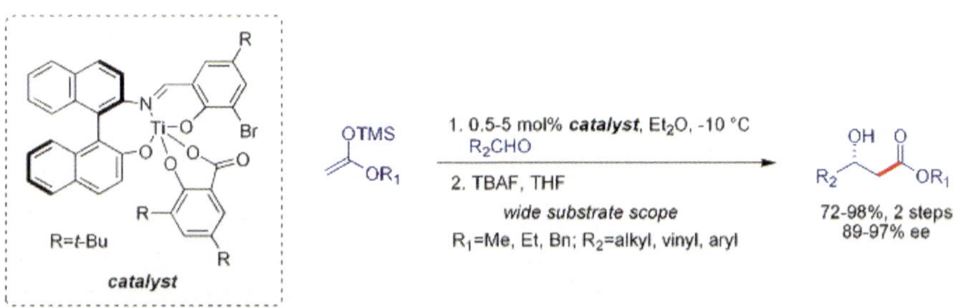

via sequential condensation:
amino alcohol + bromosalicaldehyde;
Ti(O-i-Pr)4 + di-tert-butyl-salicylic acid

The analogous vinylogous Mukaiyama aldol process can also be rendered catalytic and asymmetric. The example shown below works efficiently for aromatic (but not aliphatic) aldehydes and the mechanism is believed to involve a chiral, metal-bound dienolate.

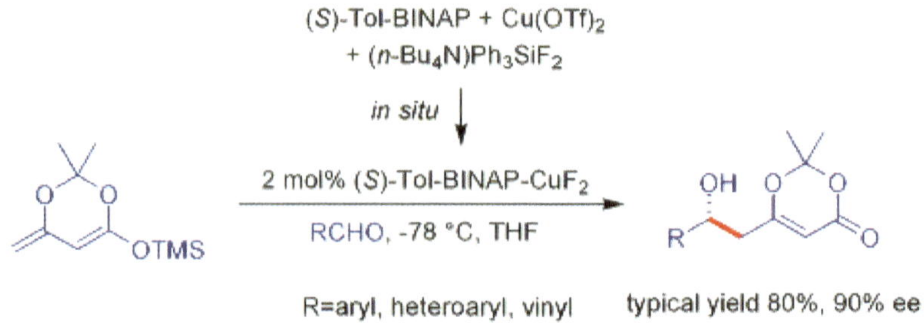

Crimmins Thiazolidinethione Aldol

A more recent version of the Evans' auxiliary is the Crimmins thiazolidinethione. The yields, diastereoselectivities, and enantioselectivities of the reaction are, in general, high, although not as high as in comparable Evans cases. Unlike the Evans auxiliary, however, the thiazoldinethione can perform acetate aldol reactions and can produce the "Evans syn" or "non-Evans syn" adducts by simply varying the amount of sparteine. The reaction is believed to proceed via six-membered, titanium-bound transition states, analogous to the proposed transition states for the Evans auxiliary.

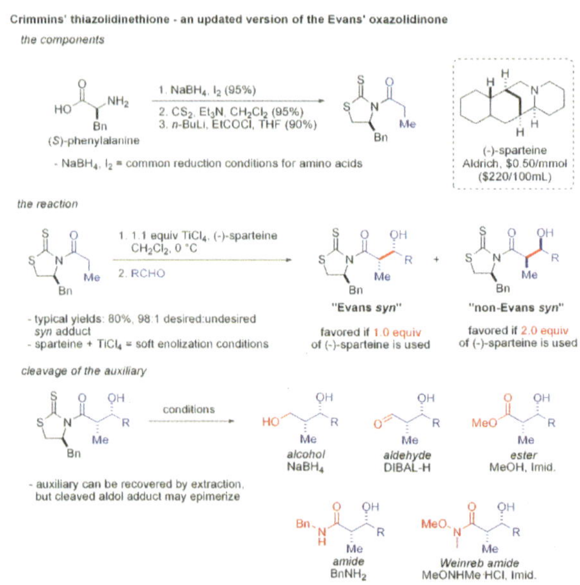

Organocatalysis

A more recent development is the use of chiral secondary amine catalysts. These secondary amines form transient enamines when exposed to ketones, which may react enantioselectively with suitable aldehyde electrophiles. The amine reacts with the carbonyl to form an enamine, the enamine acts as an enol like nucleophile, and then the amine is released from the product all the amine itself is a catalyst. This enamine catalysis method is a type of organocatalysis, since the catalyst is entirely based on a small organic molecule. In a seminal example, proline efficiently catalyzed the cyclization of a triketone:

direct measurements of transport, superconductivity can be considered to be in the extreme type II limit. The Ginzburg-Landau parameter κ can even overcome 1000 when the field is along a due to the weak interchain coupling in these Q1D conductors making the field penetration very easy for the parallel configuration of the external field. An interpretation for the critical fields assuming the pure limit of type II superconductors has been suggested in 1985. This proposal was based on the calculation of the microscopic expressions for the effective mass tensor in the Ginzburg-Landau equation near T_c.

Given the experimental determination of the critical field derivatives near T_c of $(TMTSF)_2ClO_4$, dH^a_{c2}/dT = 67.5kOe/K, $dH^{b'}_{c2}/dT$ = 36.3kOe/K, and dH^c_{c2}/dT = 1.39kOe/K, Eqs. lead to band parameters t_a:t_b:t_c = 1200, 300 and 6K respectively are derived when critical fields derivatives are determined from $T_{c,onset}$ for the three principal axes after.

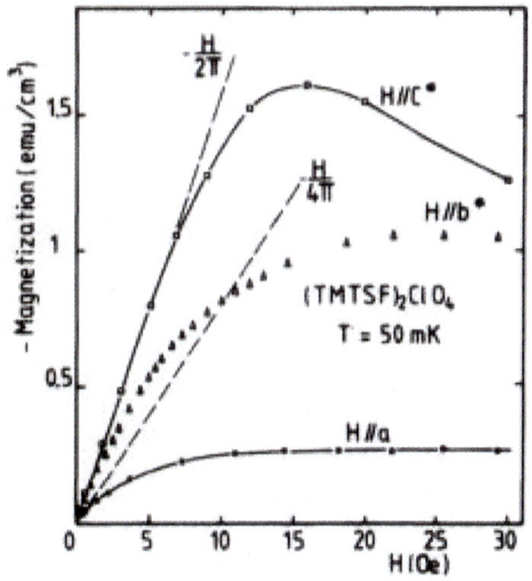

Figure: Diamagnetic shielding of $(TMTSF)_2ClO_4$ at T=0.05K for magnetic fields oriented along the three crystallographic axes.

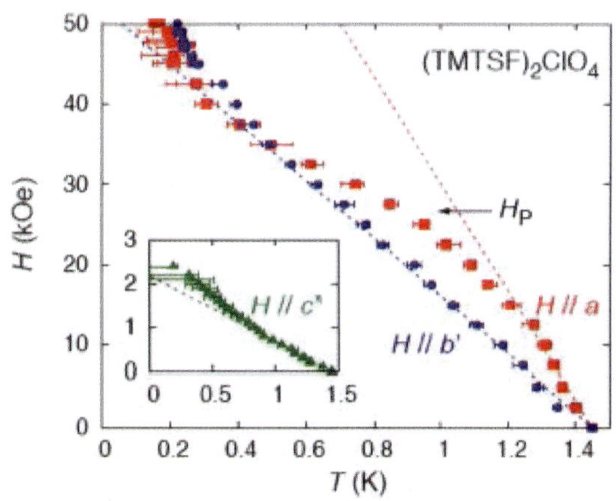

Figure: Critical fields of $(TMTSF)_2ClO_4$ determined from $T_{c,onset}$ for the three principal axes with an indication for the Pauli limit at low temperature after.

Pairing Mechanism in Organic Superconductivity

For several reasons one is entitled to believe that the pairing mechanism in organic superconductivity may differ from the regular electron-phonon driven pairing in traditional superconductors.

First, superconductivity of quasi one dimensional Bechgaard salt shares a common border with magnetism as displayed on the generic diagram. Second, strong antiferromagnetic fluctuations in the normal state above T_c in the vicinity of the *SDW* phase provide the dominant contribution to the nuclear hyperfine relaxation and are also controlling the linear temperature dependence of electronic transport. Some experimental results point to the existence of a non conventional pairing mechanism.

Non Magnetic Defects

A basic property of the BCS superconducting *s*-wave function is the time reversal symmetry of the Cooper pairs. Hence no pair breaking is expected from the scattering of electrons against spinless impurities. Experimentally, this property has been verified in non-magnetic dilute alloys of *s*-wave superconductors and brought a strong support for the BCS model of conventional *s*-wave superconductors. However, the condition for time reversal symmetry is no longer met for the case of *p*-wave pairing. Consequently, *Tc* for these superconductors should be strongly affected by any non-magnetic scattering. It is actually the extreme dependence of the critical temperature of Sr_2RuO_4 on non-magnetic disorder which has provided a strong support in favour of triplet superconductivity in this compound. It is also the remarkable sensitivity of organic superconductivity to irradiation detected in the early years which led Abrikosov to suggest the possibility of triplet pairing in these materials.

A more recent investigation of the influence of non magnetic defects on organic superconductivity has been conducted following a procedure which rules out the addition of possible magnetic impurities as it can be the case for X-ray irradiated samples. Non magnetic defects can be introduced in a controlled way either by fast cooling preventing the complete ordering of the tetrahedral ClO_4 anions or by slow cooling the solid solution $(TMTSF)_2(ClO_4)_{(1-x)}ReO_{4x}$.

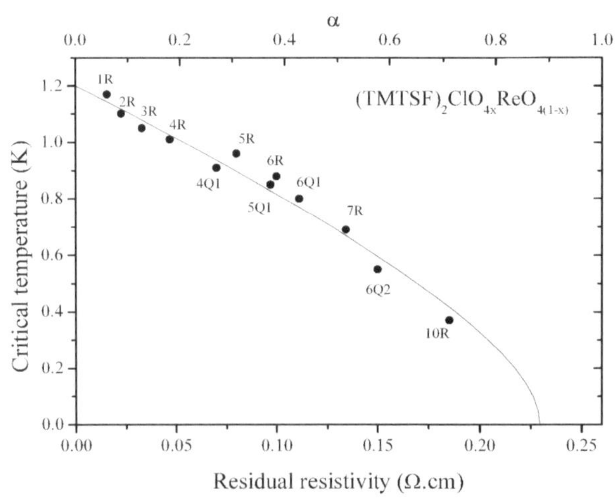

Figure: Phase diagram of $(TMTSF)_2(ClO4)_{(1-x)}ReO_{4x}$ governed by non magnetic disorder.

All circles refer to very slowly cooled samples in the R-state (the so-called relaxed state) with different ReO_4^- contents. The sample with $\rho o=0.27(\Omega cm)^{-1}$ is metallic down to the lowest temperature of the experiment. The continuous line is a fit of the data with the Digamma function and $T_c^o=1.23K$.

As displayed on figure above, T_c in the solid solution is suppressed and the suppression can be related to the enhancement of the elastic scattering in the normal state. The data on figure above show that the relation T_c versus ρ_0 follows the theory proposed for the pair breaking by magnetic impurities in conventional superconductors, namely

$$ln(\frac{T_c^o}{T_c}) = \psi(\frac{1}{2} + \frac{\alpha T_c^o}{2\pi T_c}) - \psi(\frac{1}{2}),$$

with $\psi(x)$ being the Digamma function, $\alpha = \hbar / 2\tau k_B T_c^o$ the depairing parameter τ. The experimental data follow the latter law with a good accuracy with T_c^o =1.23*K*.

The conventional pair breaking theory for magnetic impurities in usual superconductors has been generalized to the case of non-magnetic impurities in unconventional materials and the correction to T_c obeys the above relation. Since the additional scattering cannot be ascribed to magnetic scattering according to the EPR checks showing no additional traces of localized spins in the solid solution, the data in figure cannot be reconciled with the picture of a superconducting gap keeping a constant sign over the whole ($\pm k_F$) Fermi surface. They require a picture of pair breaking in a superconductor with some unconventional gap symmetry.

Symmetry	s	p	d	f	g
Sensitivity to impurities		√	√	√	√
NMR Knight shift and $1/T_1$	√		√		√
Doppler shift			√	(√)	√

Figure: Possible gap symmetries agreeing with the different experimental results. The *d*-wave (or *g*-wave) symmetry is the only symmetry agreeing with all experiments, (yellow column on line).

The influence of non-magnetic impurities on the superconducting phase implies the existence of positive as well as negative values for the *SC* order parameter. It precludes the usual case of *s*-symmetry but is still unable to discriminate between two possible options namely, singlet-*d (g)* or triplet.

Spin Susceptibility in the Superconducting Phase

Regarding the spin part of the *SC* wave function, a triplet pairing was first claimed in $(TMTSF)_2PF_6$ from a divergence of the critical field H_{c2} exceeding the Pauli limiting value reported at low temperature in $(TMTSF)_2PF_6$ under pressure when *H* is applied along the *b'* or *a* axes and from the absence of any change in the ^{77}Se Knight shift at T_c.

However more recent experiments performed at fields lower than those used in reference for the work on $(TMTSF)_2PF_6$ did reveal a clear drop of the ^{77}Se Knight shift below T_c in the compound $(TMTSF)_2ClO_4$. These new data provided a conclusive evidence in favour of singlet pairing. In addition, a steep increase of the spin lattice relaxation rate versus magnetic field for both field orientations ∥ *a* and *b'* provided evidence for a sharp cross-over or even a phase transition occurring at low temperature under magnetic field between the low field *d*-wave singlet phase and a high field regime exceeding the paramagnetic limit H_p being either a triplet-paired state or an inhomogenous Fulde-Ferrell-Larkin-Ovchinnikov state.

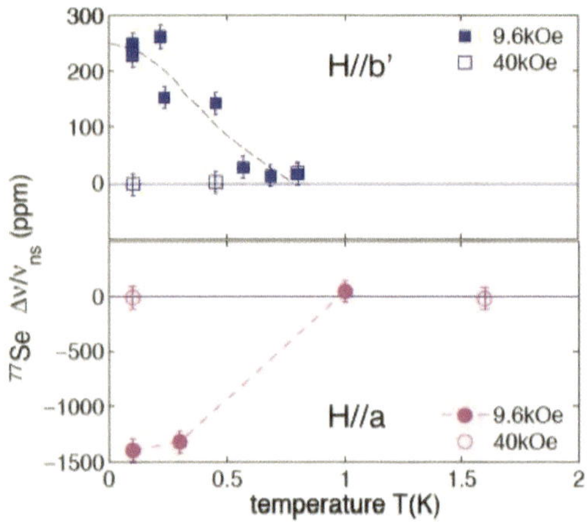

Figure: ^{77}Se Knight shift vs T for $(TMTSF)_2ClO_4$, for $H//b'$ and a, according to reference.

The sign of the variation of Knight shift at T_c depends on the sign of the hyperfine field. A variation is observed only under low magnetic field. The difference of T_c for the two field orientations is related to the anisotropy of H_{c2}. The field of $4T$ is still lower than the critical field derived from the onset of the resistive transition ($H_{c2} \approx 5T$).

Magneto Calorimetric Studies

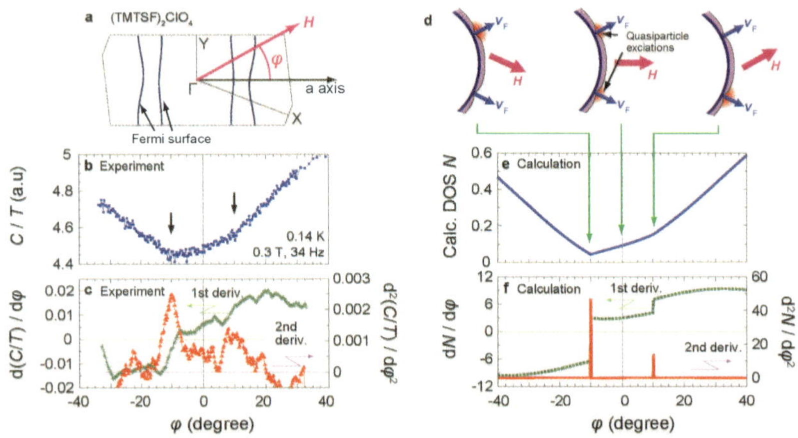

The above figure, a- Fermi surface of $(TMTSF)_2ClO_4$ at low temperature in the presence of anion ordering. b- Angular dependence of $C(\phi)$ at 0.14 K and 0.3 T according to ref. c- First and second derivatives of the data displayed on a. d- Sketch of the influence of a magnetic field on the quasi particle DOS. The contribution of Doppler-shifted quasi-particles to the DOS is minimum when the magnetic field is parallel to v_F at one of the nodes (left or right). e and f- Simulations of the specific heat in the vicinity of the angle $\phi=0$ using the band structure on figure. Arrows correspond to the different orientations of the magnetic field as shown on figure.

The response of superconductivity to non magnetic impurities and the loss of spin susceptibility at T_c should be sufficient to qualify the d-wave alternative as the likely one among the various possibilities displayed on figure. In such a case, nodes of the gap should exist on the Fermi surface and could be located. This has been made possible via a measurement of the quasi particle density of states in the superconducting phase performed in an oriented magnetic field. This experiment is based on a Volovik's important remark that for superconductors with line nodes, most of the density of states in their mixed state under magnetic field comes from the superfluid density far outside the vortex cores. The energy of QP's in the superfluid density rotating around the vortex core with the velocity vs($\perp H$) is Doppler-shifted by the magnetic field and reads,

$$\delta\omega \propto \mathrm{v}_s \cdot \mathrm{v}_F(k)$$

where, $\mathrm{v}_F(k)$ is the Fermi velocity at the k point on the Fermi surface. The Doppler shift can contribute to the DOS in the superconducting state as long as it remains smaller than the superconducting energy gap, *i.e* only for those k states located in the vicinity of gap nodes on the Fermi surface. Hence, the contribution of the Doppler shift is minimum when the Fermi velocity at a node and the magnetic field are parallel according to eqution ($\delta\omega \propto \mathrm{v}_s \cdot \mathrm{v}_F(k)$), and should contribute to a kink in the rotation pattern of the electronic specific heat $C(\phi)$.

The Doppler shift has already been used extensively in the case of 2D superconductors expected to reveal line nodes probing the thermal conduction or the specific heat while the magnetic field is rotated in the basal plane. For such 2D conductors $\mathrm{v}_F(\mathrm{k})$ is usually colinear with k. Therefore, the angular resolved specific heat (or thermal conductivity) enables to reveal the positions of the gap nodes according to the angles corresponding to minima of specific heat (thermal conductivity).

The situation is somewhat peculiar for Q1D Fermi surface. There, anomalies in the rotation pattern of C_v are expected at angles ϕ_n between H and $\mathrm{v}_F(\mathrm{k}_n)$ where k_n corresponds to a nodal position on the Fermi surface.

A simple model has been used to understand the data of angular resolved C_v of $(TMTSF)_2ClO_4$. The rotation pattern has been modeled by,

$$N(\phi) \propto \sqrt{H / H_{c2}(\phi) \sum_n A_n \,|\, sin(\phi - \phi n) \,|}$$

where the first factor reproduces the anisotropic character of the critical field in the basal plane while the weighted summation over angles accounts for the existence of nodes at angles ϕ_n between the magnetic field and the special k_n points where the Fermi velocity is parallel to H. The experimental observation of a rotation pattern at low field and low temperature which is non symmetrical with respect to the inversion of ϕ and the existence of kinks for $C(\phi)$ at $\phi=\pm 10^o$ on figure (b) have been taken as the signature of line nodes. According to the band structure calculation the location of the nodes on the Fermi surface becomes $k_y \sim \pm 0.25 b^*$. As far as the Pauli limitation is concerned, the superconducting phase diagram of $(TMTSF)_2ClO_4$ established by thermodynamic experiments indicates a value for the thermodynamic H_{c2} along a of 2.5 T for $H \| a$ much smaller than the expected value of 7.7 T for the orbital limitation derived from the measurement of the temperature derivative of H_{c2} close to T_c for a clean type II superconductor. On the other hand, along the two transverse directions orbital limitation is at work. The domain of field above the paramagnetic limit up to H_{c2} derived from resistivity data where the specific heat recovers its normal state value requires further experimental investigations but it might be the signature of some *FFLO* phase as suggested by A. Lebed recently.

Antiferromagnetic Fluctuations, Spin and Charge Sectors

The metallic phase of $(TMTSF)_2PF_6$ in the 3D coherent regime when pressure is in the neighbourhood of the critical pressure P_c behaves in way far from what is expected for a Fermi liquid.

The canonical Korringa law , $1/T_1T \propto \chi^2(q=0,T)$, is well obeyed at high temperature say, above 25 K, but the low temperature behaviour deviates strongly from the standard relaxation in paramagnetic metals. As shown on figure an additional contribution to the relaxation rate emerges in top of the usual Korringa relaxation. This additional contribution rising at low temperature has been attributed to the onset of antiferromagnetic fluctuations in the vicinity of P_c. On the other hand in the low temperature regime the relaxation rate follows a law such as $T_1T=C(T+\Theta)$ as shown on figure. This is the Curie-Weiss behaviour for the relaxation which is to be observed in a 2D fluctuating antiferromagnet. The positive Curie-Weiss temperature Θ which provides the energy scale of the fluctuations is zero when pressure is equal to P_c (the quantum critical conditions). When Θ becomes large comparable to T, the standard relaxation mechanism is recovered even at low temperature in agreement with the very high pressure results.

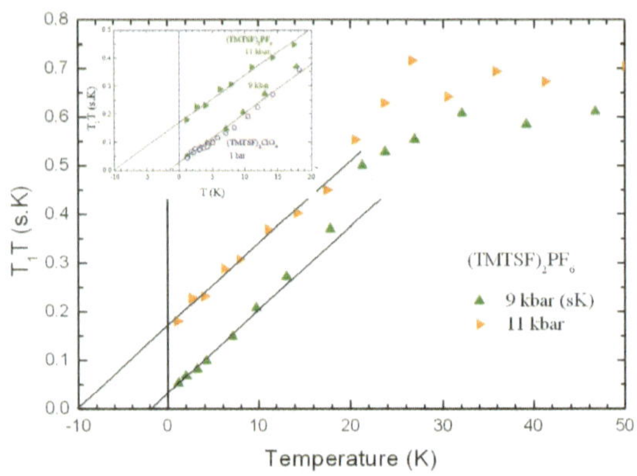

Figure: Plot of the nuclear relaxation versus temperature. A Korringa regime, T_1T = const is observed down to 25K.

The 2D AF regime is observed below ≈15K and the small Curie-Weiss temperature of the 9 kbar run is the signature of the contribution of quantum critical fluctuations to the nuclear relaxation. The Curie-Weiss temperature becomes zero at the QCP. These data show that the QCP should be slightly below 9 kbar with the present pressure scale. The inset shows that the organic superconductor $(TMTSF)_2ClO_4$ at ambient pressure is very close to fulfill quantum critical conditions.

The inelastic scattering in transport reveals at once a strong linear term at low temperature evolving to quadratic in the high temperature regime with a general tendency to become quadratic at all temperatures when pressure is way above the critical pressure P_c.

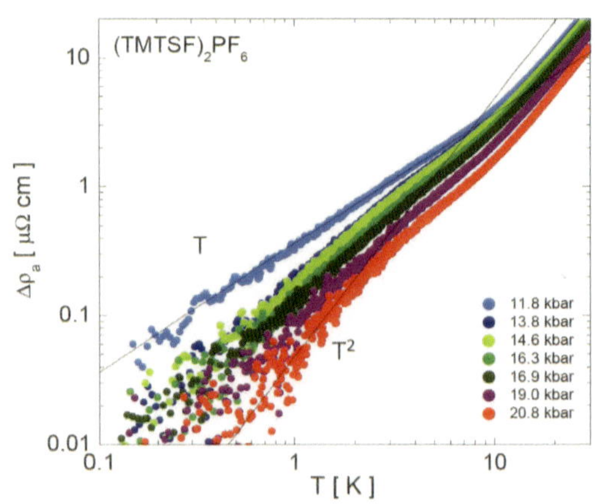

Figure: A log-log plot of the inelastic longitudinal resistivity of $(TMTSF)_2PF_6$ below 20 K.

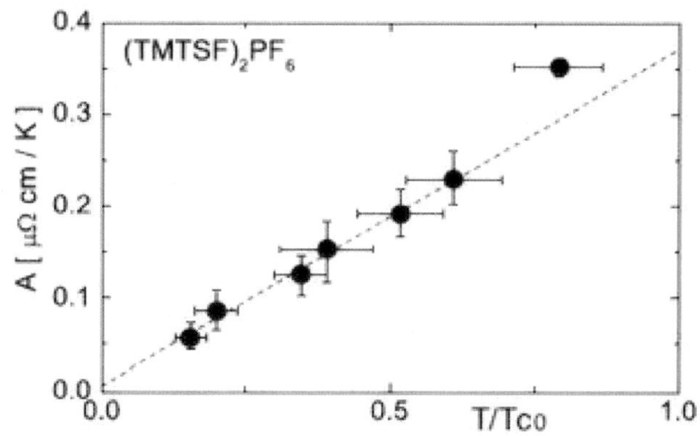

Figure: Coefficient A of linear resistivity as a function of T_c plotted versus T_c/T_{co} for $(TMTSF)_2PF_6$.

T_c is defined as the midpoint of the transition and the error bars come from the 10%and 90% points with T_{co}=1.23K under the pressure of 8 kbar which provides the maximum T_c in the *SDW/SC* coexistence regime.The dashed line is a linear fit to all data points except that at T_c=0.87K.

The existence of a linear temperature dependence of the resistivity is *at variance* with the sole T^2 dependence expected from the electron-electron scattering in a conventional Fermi liquid. This is clearly seen on a log-log plot of the resistivity versus T. Furthermore, the investigation of both transport and superconductivity under pressure in $(TMTSF)_2PF_6$ has established a correlation between the amplitude of the linear in temperature dependence of the resistivity and the value of T_c as displayed on figure suggesting a common origin for the inelastic scattering of the metallic phase and pairing in the *SC* phase $(TMTSF)_2PF_6$.

Discussions and Perspectives

The close proximity between antiferromagnetism and superconducting ground states of $(TM)_2X$ superconductors and the deviation of the metallic phase from the traditional Fermi liquid behaviour have been recognized as early as the beginning eighties. The possibility for a pairing mechanism involving carriers on neighbouring chains in these quasi 1D conductors avoiding the Coulomb repulsion has been proposed by V. Emery in the context of the exchange phonon mechanism. Shortly after, Emery and coworkers introduced the possibility that antiferromagnetic fluctuations play a role in the pairing mechanism but concluded that superconductivity could not emerge from pairing on the same organic chain. The exchange of spin fluctuations between carriers on neighbouring chains was thus proposed to provide the necessary glue for pairing in analogy with the exchange of charge density waves proposed by Kohn and Luttinger in the context of a new pairing mechanism in low dimensional conductors.

L. Caron and C. Bourbonnais extending their theory for the generic $(TM)_2X$ phase diagram to the metallic domain made the proposal for a gap equation with singlet superconductivity based on an interchain magnetic coupling with an attraction deriving from an interchain exchange interaction overcoming the on-stack Coulomb repulsion. Taking into account the interference between the diverging Cooper and Peierls channels the renormalizationtreatment of Q-1-D conductors received more recently a significant improvement.

Within the framework of a weak-coupling limit the problem of the interplay between antiferromagnetism and superconductivity in the Bechgaard salts has been worked out using the renormalization group (RG) approach as summarized below taking into account only the 2D problem. The RG integration of high energy electronic degrees of freedom was carried out down to the Fermi level, and leads to a renormalization of the couplings at the temperature *T*. The RG flow superimposes the 2*kF* electron-hole (density-wave) and Cooper pairing many-body processes which combine and interfere at every order of perturbation. As a function of the 'pressure' parameter $t'_{\perp}$ *i.e* the unnesting interstacks coupling, a singularity in the scattering amplitudes signals an instability of the metallic state toward the formation of an ordered state at some characteristic temperature scale. At low $t'_{\perp}$, nesting is sufficiently strong to induce a *SDW* instability in the temperature range of experimental T_{SDW}~10–20K. When the antinesting parameter approaches the threshold $t'^{*}_{\perp}$ from below ($t'^{*}_{\perp} \approx 25.4K$, using the above parameters), T_{SDW} sharply decreases and as a result of interference, *SDW* correlations ensure Cooper pairing attraction in the superconducting *d*-wave (*SCd*) channel. This gives rise to an instability of the normal state for the onset of *SCd* order at the temperature T_c with pairing coming from antiferromagnetic spin fluctuations between carriers of neighbouring chains. Such a pairing model is actually supporting the conjecture of interchain pairing in order for the electrons to avoid the Coulomb repulsion made by V. Emery.

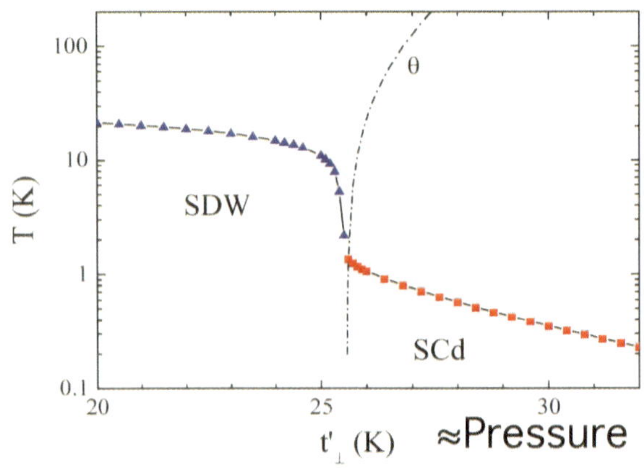

Figure: Calculated phase diagram of the quasi-one dimensional electron gas model from the renormalization group method at the one-loop level. Θ and the dash-dotted line defines the temperature region of the CW behavior for the inverse normalized *SDW* response function

The calculated phase diagram with reasonable parameters taking $g_1=g_2/2\approx0.32$ for the backward and forward scattering amplitudes respectively and $g3\approx0.02$ for the longitudinal Umklapp scattering term captures the essential features of the experimentally-determined phase diagram of $(TMTSF)_2PF_6$.

Sedeki *et-al* have proceeded to an evaluation of the imaginary part of the one-particle self-energy which provides access to the electron-electron scattering rate τ^{-1}. In addition to the regular Fermi-liquid component which goes as T^2 low frequency spin fluctuations yield $\tau^{-1}=aT\xi$, where *a* is constant and the antiferromagnetic correlation length $\xi(T)$ increases according to $\xi=c(T+\Theta)^{-1/2}$ as $T\rightarrow T_c$, where Θ is the temperature scale for spin fluctuations. It is then natural to expect the Umklapp resistivity to contain (in the limit $T\ll\Theta$) a linear term AT besides the regular BT^2, whose magnitude would presumably be correlated with T_c, as both scattering and pairing are caused by the same antiferromagnetic correlations. The observation of a *T*-linear law for the resistivity up to 8K in $(TMTSF)_2PF_6$ under a pressure of 11.8 kbar as displayed on figure is therefore consistent with the value of $\Theta=8K$ determined from NMR relaxation at 11 kbar on figure.

To conclude, both experimental and theoretical views point the contribution of electron correlations to the superconducting pairing problem. The extensive experimental evidence in favor of the emergence of superconductivity in the $(TM)_2X$ family next to the stability threshold for antiferromagnetism has shown the need for a unified description of all electronic excitation that lies at the core of both density-wave and superconducting correlations. In this matter, the recent progresses achieved by the renormalization group method for the 1D-2D electron gas model have resulted in predictions about the possible symmetries of the superconducting order parameter when a purely electronic mechanism is involved, predictions that often differ from phenomenologically based approaches to superconductivity but are in fair agreement with the recent experimental findings. In this respect reexaminations of the superconducting state, specific heat, thermal conductivity,etc, are highly demanded.

What is emerging from the work on these prototype 1D organic superconductors is their very simple electronic nature with only a single band at Fermi level, no prominent spin orbit coupling and extremely high chemical purity and stability.

They should be considered in several respects as model systems to inspire the physics of the more complex high T_c superconductors, especially for pnictides and electron-doped cuprates. Most concepts discovered in these simple low dimensional conductors should be very useful for the study of other low dimensional systems such as carbon nanotubes or artificial 1D structures, etc.

The pairing mechanism behind organic superconductivity is likely different from the proposal made by Little but it is nevertheless a phonon-less mechanism, at least in $(TM)_2X$ superconductors.

In conclusion, organic superconductivity has been discovered in the prototype Bechgaard salt $(TMTSF)_2PF_6$ but several other fascinating aspects of the physics of 1D conductors are worth mentioning namely, the magnetic field confinement discovered in the 1D's leading to the phenomenon of field induced spin density wave phases and the quantization of the Hall effect in these phases. Furthermore, the angular dependent magnetoresistance closely related to these anisotropic conductors, the so-called Lebed-Osada-Danner-Chaikin oscillations in $(TMTSF)_2X$ provide a nice illustration for the new features of the quasi-1D conductors.

Volatile Organic Compound

Volatile organic compounds are compounds that have a high vapor pressure and low water solubility. Many VOCs are human-made chemicals that are used and produced in the manufacture of paints, pharmaceuticals, and refrigerants. VOCs typically are industrial solvents, such as trichloroethylene; fuel oxygenates, such as methyl tert-butyl ether (MTBE); or by-products produced by chlorination in water treatment, such as chloroform. VOCs are often components of petroleum fuels, hydraulic fluids, paint thinners, and dry cleaning agents. VOCs are common ground-water contaminants.

Volatile organic compounds (VOCs) are emitted as gases from certain solids or liquids. VOCs include a variety of chemicals, some of which may have short- and long-term adverse health effects. Concentrations of many VOCs are consistently higher indoors (up to ten times higher) than outdoors. VOCs are emitted by a wide array of products numbering in the thousands. Examples include: paints and lacquers, paint strippers, cleaning supplies, pesticides, building materials and furnishings, office equipment such as copiers and printers, correction fluids and carbonless copy paper, graphics and craft materials including glues and adhesives, permanent markers, and photographic solutions.

Organic chemicals are widely used as ingredients in household products. Paints, varnishes, and wax all contain organic solvents, as do many cleaning, disinfecting, cosmetic, degreasing, and hobby products. Fuels are made up of organic chemicals. All of these products can release organic compounds while you are using them, and, to some degree, when they are stored.

Sources of VOCs

VOCs arise from a broad range of different sources. Their presence in atmospheric air can be accounted for by biological processes, for example, metabolic processes in plants, putrefaction or decomposition. Further sources of VOCs in atmospheric air include technical processes in which substances are generated through incomplete combustion (especially in vehicular emissions), or as volatile by-products of industrial or commercial processes. Possible sources of VOCs in indoor air include products and materials for use in construction and interior decoration (for example floor, wall, and ceiling coverings, paints, varnishes, adhesives, or materials used in furniture and decorating). Also of significance are body care products, cleaning materials, and hobby products, as well as tobacco smoke, and even food preparation and human metabolic processes.

In comparison to the sources of VOCs in atmospheric air, those to be found in indoor air in central Europe are generally of substantially greater significance in respect of their impact on health, as most people in this region spend much of their time indoors. In addition, there is mostly greater proximity to the sources. VOCs from outside can also penetrate into buildings. As a rule, however, ventilation leads to a reduction of the original interior concentrations.

VOCs into the Air

When solvents or liquid fuels evaporate, and liquid or paste-like products dry out, VOCs are set free in large amounts into the surrounding air. A less obvious phenomenon is the release of various associated substances, which are not strongly bound to the products. They can slowly disperse into the air from the product surface and are delivered in a constant stream from the inside of the product to its surface (material emission). This applies for example to residual solvents and building blocks in plastics (monomers), excipients such as plasticisers, solubilisers, antioxidants, stabilisers, and catalysts from production processes, as well as to associated substances such as aromatics, flame retardants, and biocides. Terpenes are typical VOCs: they are released into the air from materials and products of natural origin, for example, wood. VOCs also arise as reaction products, for example, between oxygen, ozone or water and natural substances of content, such as those to be found in wood and vegetable oils.

Health Effects

Health effects may include:

- Eye, nose and throat irritation;

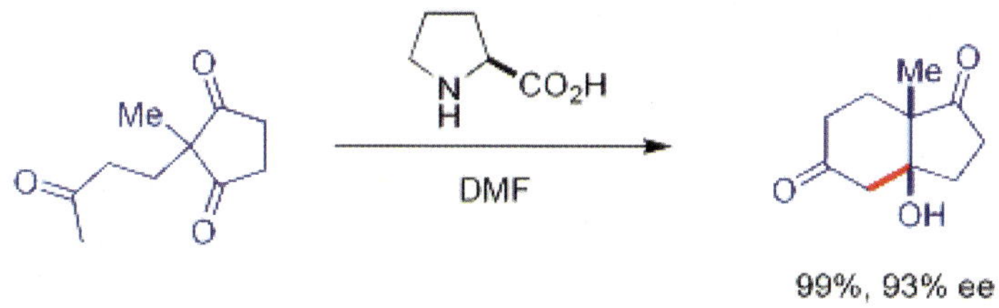

This reaction is known as the Hajos-Parrish reaction (also known as the Hajos-Parrish-Eder-Sauer-Wiechert reaction, referring to a contemporaneous report from Schering of the reaction under harsher conditions). Under the Hajos-Parrish conditions only a catalytic amount of proline is necessary (3 mol%). There is no danger of an achiral background reaction because the transient enamine intermediates are much more nucleophilic than their parent ketone enols. This strategy offers a simple way of generating enantioselectivity in reactions without using transition metals, which have the possible disadvantages of being toxic or expensive.

Proline-catalyzed aldol reactions do not show any non-linear effects (the enantioselectivity of the products is directly proportional to the enantiopurity of the catalyst). Combined with isotopic labelling evidence and computational studies, the proposed reaction mechanism for proline-catalyzed aldol reactions is as follows:

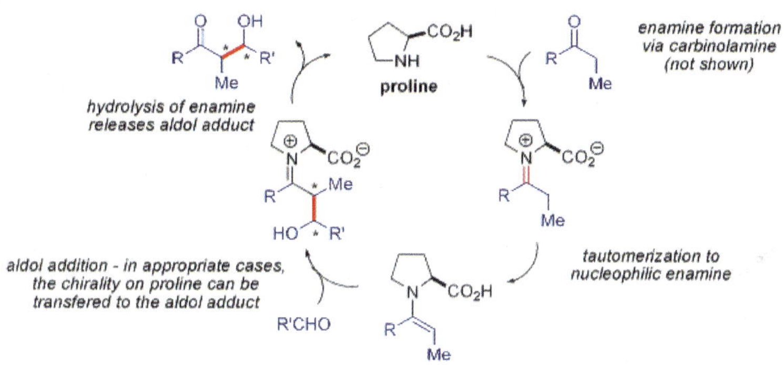

This strategy allows the otherwise challenging cross-aldol reaction between two aldehydes. In general, cross-aldol reactions between aldehydes are typically challenging because they can polymerize easily or react unselectively to give a statistical mixture of products. The first example is shown below:

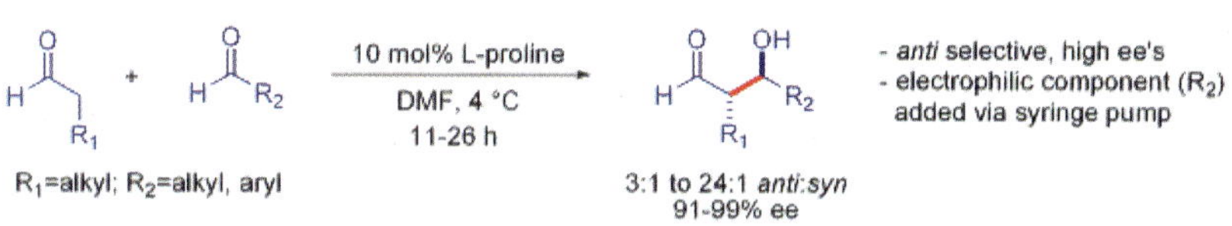

In contrast to the preference for syn adducts typically observed in enolate-based aldol additions, these organocatalyzed aldol additions are anti-selective. In many cases, the organocatalytic conditions are mild enough to avoid polymerization. However, selectivity requires the slow syringe-pump controlled addition of the desired electrophilic partner because both reacting partners typically have enolizable protons. If one aldehyde has no enolizable protons or alpha- or beta-branching, additional control can be achieved.

An elegant demonstration of the power of asymmetric organocatalytic aldol reactions was disclosed by MacMillan and coworkers in 2004 in their synthesis of differentially protected carbohydrates.

While traditional synthetic methods accomplish the synthesis of hexoses using variations of iterative protection-deprotection strategies, requiring 8–14 steps, organocatalysis can access many of the same substrates using an efficient two-step protocol involving the proline-catalyzed dimerization of alpha-oxyaldehydes followed by tandem Mukaiyama aldol cyclization.

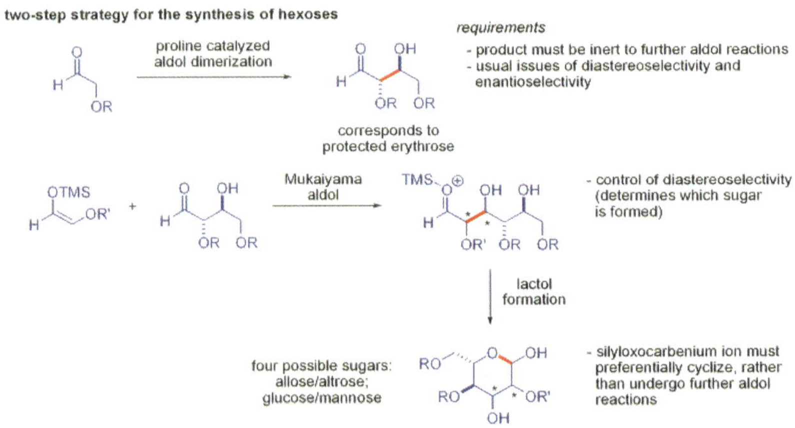

The aldol dimerization of alpha-oxyaldehydes requires that the aldol adduct, itself an aldehyde, be inert to further aldol reactions. Earlier studies revealed that aldehydes bearing alpha-alkyloxy or alpha-silyloxy substituents were suitable for this reaction, while aldehydes bearing Electron-withdrawing groups such as acetoxy were unreactive. The protected erythrose product could then be converted to four possible sugars via Mukaiyama aldol addition followed by lactol formation. This requires appropriate diastereocontrol in the Mukaiyama aldol addition and the product silyloxycarbenium ion to preferentially cyclize, rather than undergo further aldol reaction. In the end, glucose, mannose, and allose were synthesized:

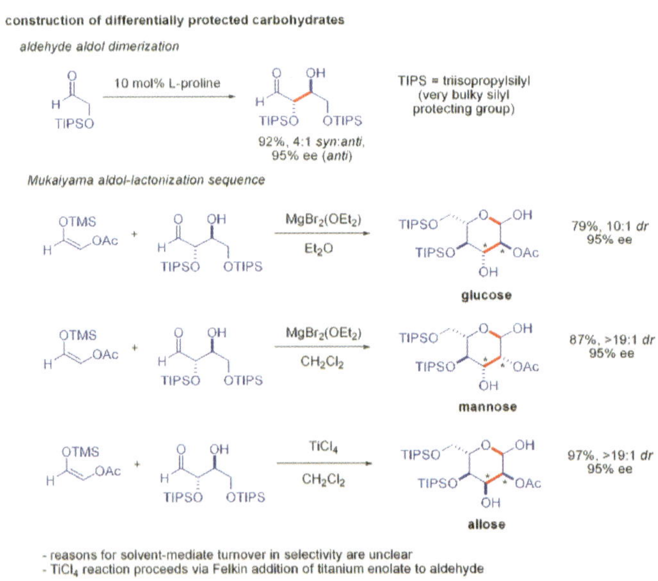

"Direct" Aldol Additions

In the usual aldol addition, a carbonyl compound is deprotonated to form the enolate. The enolate is added to an aldehyde or ketone, which forms an alkoxide, which is then protonated on workup. A superior method, in principle, would avoid the requirement for a multistep sequence in favor of a "direct" reaction that could be done in a single process step. One idea is to generate the enolate using a metal catalyst that is released after the aldol addition mechanism. The general problem is that the addition generates an alkoxide, which is much more basic than the starting materials. This product binds tightly to the metal, preventing it from reacting with additional carbonyl reactants.

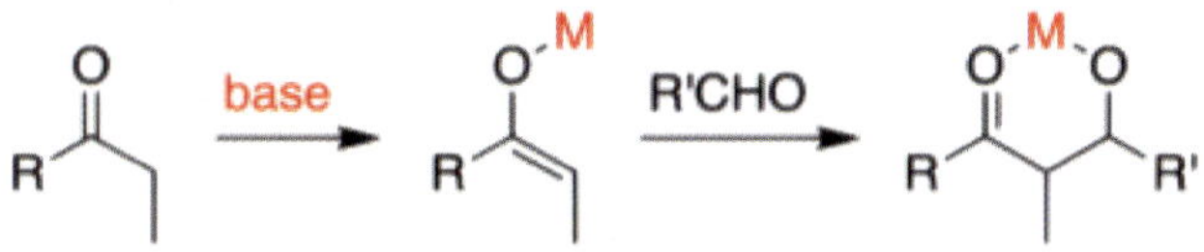

One approach, demonstrated by Evans, is to silylate the aldol adduct. A silicon reagent such as TMSCl is added in the reaction, which replaces the metal on the alkoxide, allowing turnover of the metal catalyst. Minimizing the number of reaction steps and amount of reactive chemicals used leads to a cost-effective and industrially useful reaction.

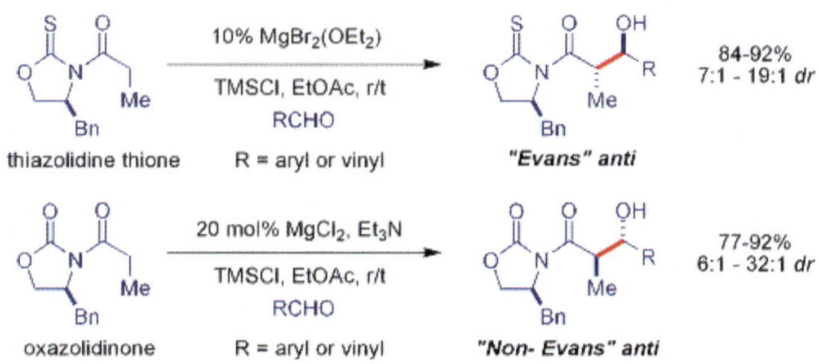

A more recent biomimetic approach by Shair uses beta-thioketoacids as the nucleophile. The ketoacid moiety is decarboxylated in situ. The process is similar to the way malonyl-CoA is used by Polyketide synthases. The chiral ligand is case is a bisoxazoline. Aromatic and branched aliphatic aldehydes are typically poor substrates.

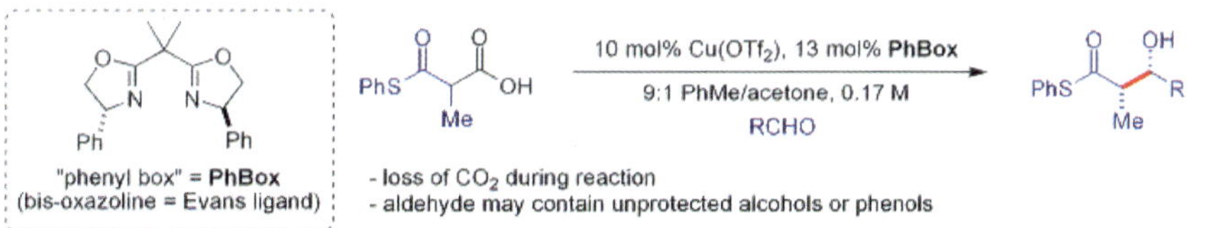

Biological Aldol Reactions

Examples of aldol reactions in biochemistry include the splitting of fructose-1, 6-bisphosphate into dihydroxyacetone and glyceraldehyde-3-phosphate in the fourth stage of glycolysis, which is an example of a reverse ("retro") aldol reaction catalyzed by the enzyme aldolase A (also known as fructose-1, 6-bisphosphate aldolase).

In the glyoxylate cycle of plants and some prokaryotes, isocitrate lyase produces glyoxylate and succinate from isocitrate. Following deprotonation of the OH group, isocitrate lyase cleaves isocitrate into the four-carbon succinate and the two-carbon glyoxylate via an aldol cleavage reaction. This cleavage is very similar mechanistically to the aldolase A reaction of glycolysis.

Mukaiyama Aldol Addition

educts
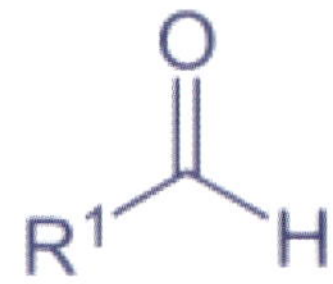 aldehyde (R1 = Alkyl, Aryl) or formate (R1 = OR)
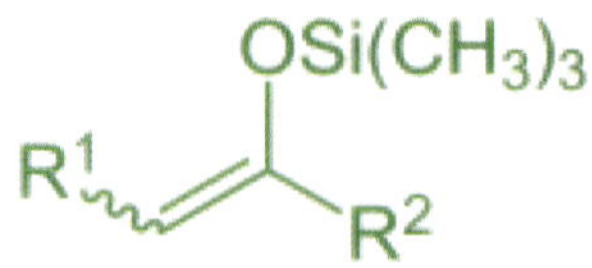 silyl enol ether (R1 = Alkyl, Aryl, H; R2 = Alkyl, Aryl, H, OR, SR)

The Mukaiyama aldol addition is an organic reaction and a type of aldol reaction between a silyl enol ether and an aldehyde or formate. The reaction was discovered by Teruaki Mukaiyama. His choice of reactants allows for a crossed aldol reaction between an aldehyde and a ketone or a different aldehyde without self-condensation of the aldehyde. For this reason the reaction is used extensively in organic synthesis.

General Reaction Scheme

The Mukaiyama aldol addition is a Lewis acid mediated addition of enol silanes to carbonyl compounds. In this reaction compounds with various organic groups can be used. A basic version (R_2=H) without the presence of chiral catalysts is shown below:

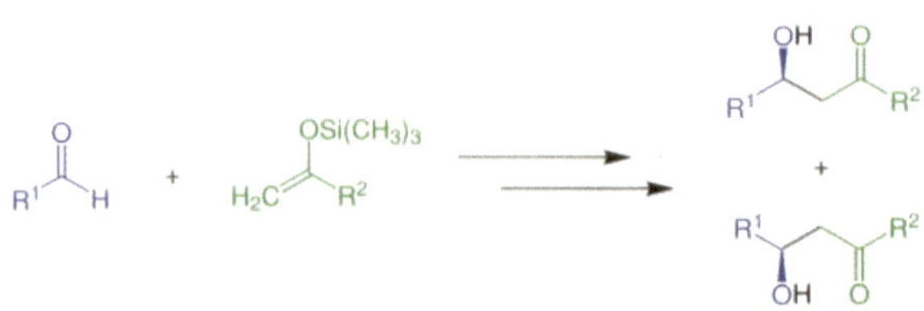

A racemic mix of enantiomers is built. If Z- or E-enol silanes are used in this reaction a mixture of four products occurs, yielding two race mates.

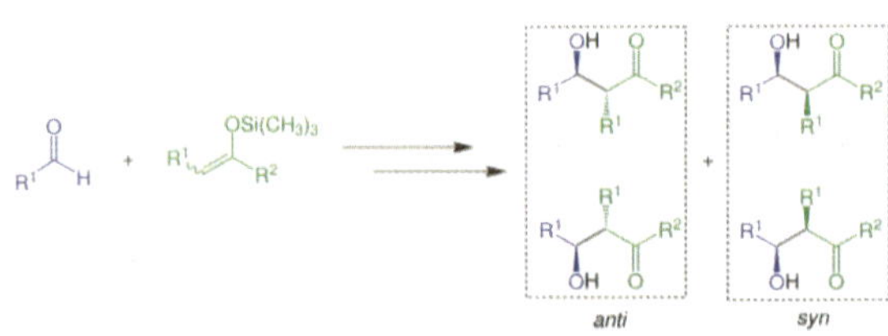

Whether the anti-diastereomer or the syn-diastereomer is built depends largely on reaction conditions, substrates and Lewis acids.

The archetypical reaction is that of the silyl enol ether of cyclohexanone with benzaldehyde. At room temperature it produces a diastereomeric mixture of threo (63%) and erythro (19%) β-hydroxyketone as well as 6% of the exocyclic enone condensation product. In its original scope the Lewis acid (titanium chloride) was used in stoichiometric amounts but truly catalytic systems exist as well. The reaction is also optimized for asymmetric synthesis.

Mechanism

Below, the reaction mechanism is shown with R2 = H:

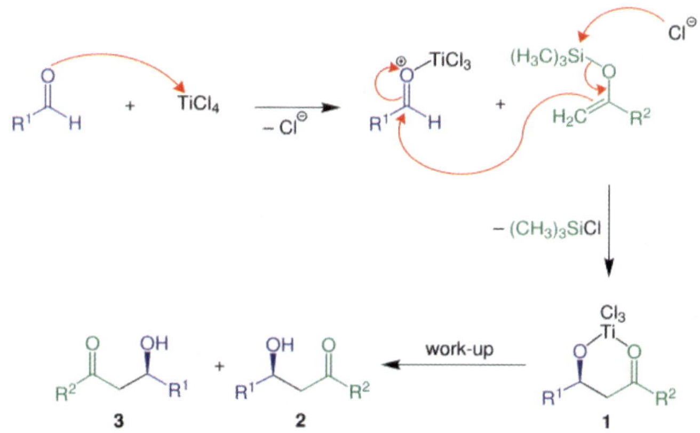

In the cited example the Lewis acid TiCl4 is used. First, the Lewis acid activates the aldehyde component followed by carbon-carbon bond formation between the enol silane and the activated aldehyde. With the loss of a chlorosilane the compound 1 is built. The desired product, a racemate of 2 and 3, is obtained by aqueous work-up.

Scope

A typical reaction involving two ketones is that between acetophenone as the enol and acetone:

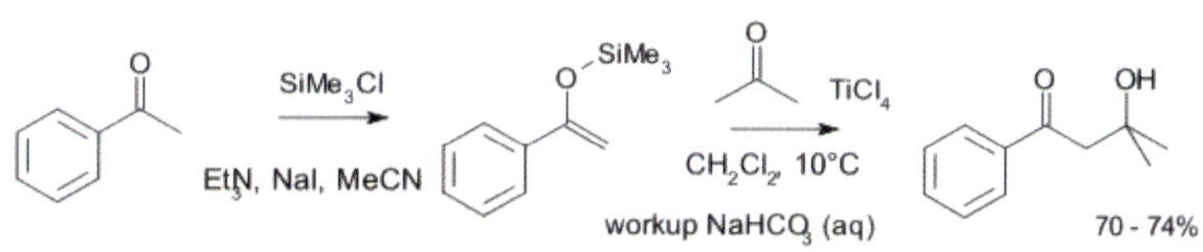

Ketone reactions of this type require higher reaction temperatures. For this work Mukaiyama was inspired by earlier work done by Georg Wittig in 1966 on crossed aldol reactions with lithiated imines.

Mukaiyama employed in his rendition of taxol total synthesis two aldol additions, one with an ketene silyl acetal and excess magnesium bromide:

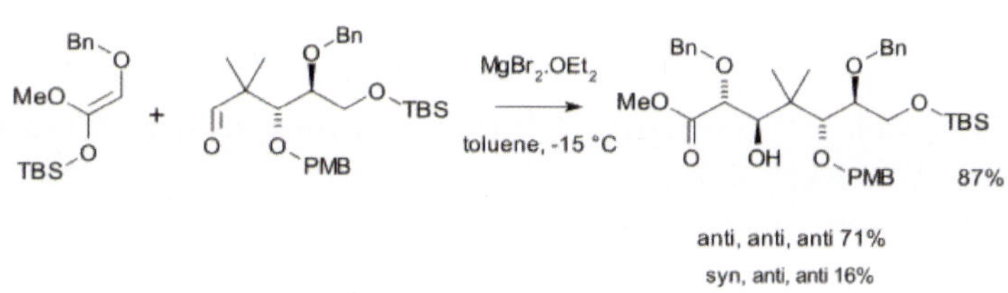

And a second one with an amine chiral ligand and a triflate salt catalyst:

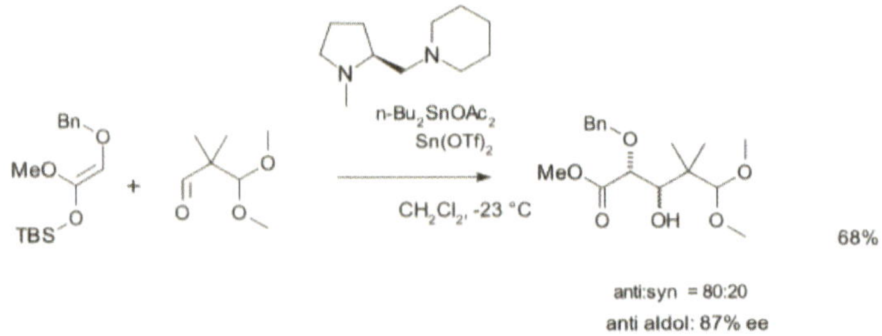

Utilization of chiral Lewis acid complexes and Lewis bases in asymmetric catalytic processes is the fastest-growing area in the usage of the Mukaiyama aldol reaction.

Electrophilic Addition

An electrophile can be defined as the species which attracts towards electron-rich regions of another molecule or ion. This is because electrophile either carries positive charge or an electron deficient molecule. Alkenes have pi-electrons so they easily attacked by electrophile and electrophile gets attach to double bonded carbon atom.

Such reactions are called as electrophilic addition reactions. Electrophilic addition reactions can be defined as the chemical reactions in which electrophile is added on an electron rich carbon atom to form an adduct product.

Electrophilic Addition Reaction Examples

Hydrogenation, hydrohalogenation, hydration are some common electrophilic addition reactions of alkenes. In general the molecule X-Y is added to double bonded carbon atoms of alkenes to form saturated compound.

So we can say that in an addition reaction two molecules are joined together to form a bigger one without any lose of other molecule. In electrophilic addition reactions, reaction starts with attack of electrophile on double bonded carbon atom. Let â€™s take an example of hydrohalogenation of ethene. In hydrohalogenation of alkene, H^+ acts as electrophile and reacts first with alkene. It results the formation of a carbocation as an intermediate. Later this intermediate is attacked by nucleophile (Br^-) to form haloalkane.

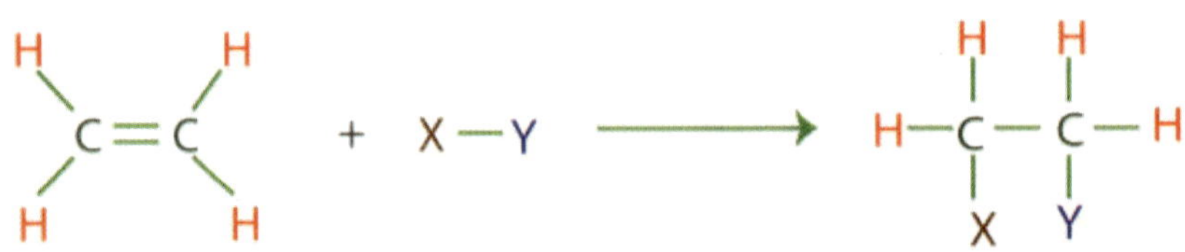

Similarly hydration of alkene is also an example of electrophilic addition reaction. Reaction of alkenes with aqueous acid like sulphuric acid results the formation of alcohols. Like hydrohalogenation, hydration of alkene also proceeds with the formation of carbocation as an intermediate. Reaction prefers the formation of stable carbocation which further reacts with nucleophile ($-OH^-$) to form alcohol. For example, reaction of ethene with aqueous acid forms ethanol:

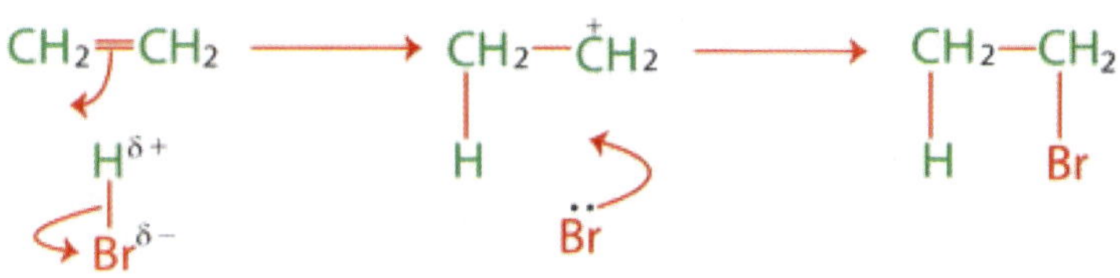

Electrophilic Addition Reaction Mechanism

Let â€™s discuss common mechanism for electrophilic addition reaction of alkene. Electrophile, E^+ which is an electron poor species reacts with electron rich alkene to form addition product. The general mechanism for electrophilic addition reaction can be shown as below:

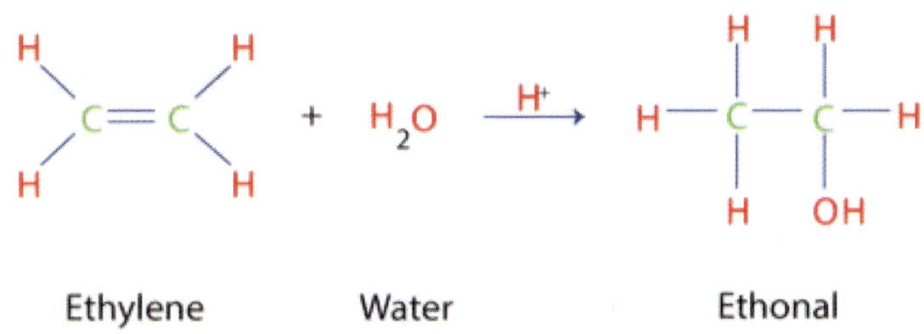

In the first step of reaction; the electrophile attacks on double bonded carbon atom of alkene. It forms a carbocation as an intermediate. If there is a possibility form a stable carbocation by rearrangement, it will form. The reaction follows the Markovnikov's rule which states that positive part will add on that carbon atom of alkene which has more H atoms. In the second step of mechanism, nucleophile reacts with carbocation to form adduct product. Since reactions proceed with the formation of planer carbocation so these reactions are not stereo selective but electrophile adds first to the alkene so they exhibit regioselectivity. Addition can occur in syn or anti direction. In syn-addition, both the groups are added at same side whereas in anti-addition they are added at opposite side of the double bond. For example addition of H_2 on double bond of alkene always occurs at syn position.

Conditions for Electrophilic Addition Reactions

Alkenes are unsaturated hydrocarbons which can easily give addition reactions due to presence of double covalent bond in the molecule. Some common examples of electrophilic addition reactions are dihalo addition reactions (halogenations), addition of H and X (hydrohalogenations), addition of H and OH (hydration), addition of H (hydrogenations), oxymercuration reactions with mercuric acetate and water, Hydroboration-oxidation reactions with diborane etc. Major conditions for electrophilic addition are:

- Presence of double covalent bond in reactant molecule;
- Presence of electrophile;
- Reactions conditions like solvent which can stabilize the poplar carbocation intermediate.

Electrophilic Addition to Conjugated Dienes

We have discussed about electrophilic addition reactions of alkenes. Dienes are also example alkenes which have two double bonds in the molecule. If the double bond is present in alternate manner, the dienes are called as conjugated dienes. Conjugated dienes are stable due to delocalization of pi-electrons. For example the resonance in 1, 3-butadiene is given below:

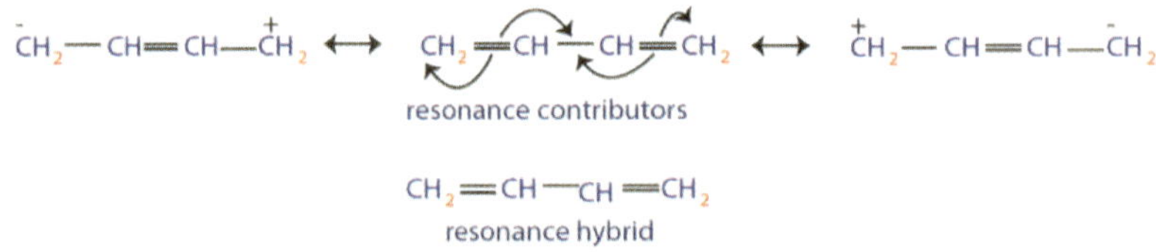

Since there are two pi-bonds in a molecule of 1, 3-bitadiene so during electrophilic addition of dienes, two modes of addition are possible. It can be either 1, 2 or 1, 4 addition. The product formed depends on the kinetic and thermodynamic conditions of reaction.

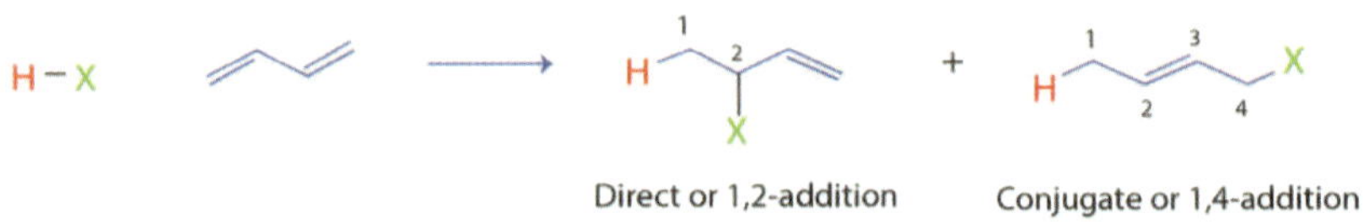

The reaction of HBr with 1, 3-butadiene. At low temperature (-80Â°C) the addition occurs with the formation of secondary carbocation whereas at high temperature, reaction occurs at both positions.

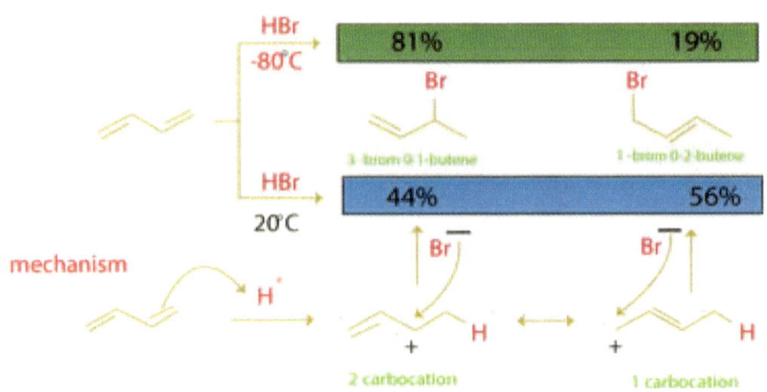

So we can say that addition at low temperature is a kinetically control reaction that form major product by secondary carbocation which forms more quickly in the reaction. On the contrary, at higher temperature reaction is thermodynamic control.

Michael Addition

The Michael reaction or Michael addition is the nucleophilic addition of a carbanion or another nucleophile to an α, β-unsaturated carbonyl compound. It belongs to the larger class of conjugate additions. This is one of the most useful methods for the mild formation of C–C bonds. Many asymmetric variants exist,

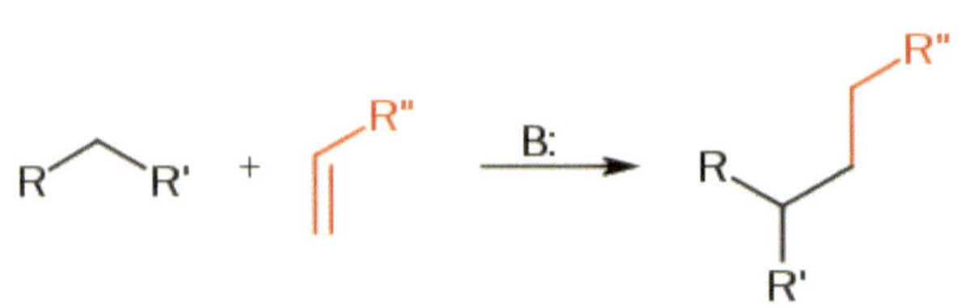

In this figure the R and R' substituents on the nucleophile (a Michael donor) are electron-withdrawing groups such as acyl and cyano making the methylene hydrogen acidic forming the carbanion

on reaction with base B. The substituent on the activated alkene, also called a Michael acceptor, is usually a ketone making it an enone, but it can also be a nitro group.

As originally defined by Arthur Michael, the reaction is the addition of an enolate of a ketone or aldehyde to an α, β-unsaturated carbonyl compound at the β carbon. A newer definition, proposed by Kohler, is the 1, 4-addition of a doubly stabilized carbon nucleophile to an α, β-unsaturated carbonyl compound. Some examples of nucleophiles include beta-ketoesters, malonates, and beta-cyanoesters. The resulting product contains a highly useful 1, 5-dioxygenated pattern.

Classical examples of the Michael reaction are the reaction between diethyl malonate (Michael donor) and diethyl fumarate (Michael acceptor), that of mesityl oxide and diethyl malonate, that of diethyl malonate and methyl crotonate, that of 2-nitropropane and methyl acrylate, that of ethyl phenylcyanoacetate and acrylonitrile and that of nitropropane and methyl vinyl ketone.

The Michael addition is an important atom-economical method for diastereoselective and enantioselective C–C bond formation. A classical tandem sequence of Michael and aldol additions is the Robinson annulation.

Mechanism

The reaction mechanism is 1 (with R an alkoxy group) as the nucleophile:

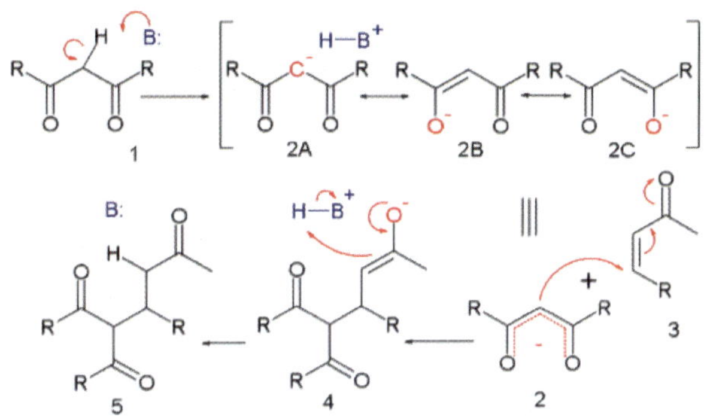

Deprotonation of 1 by base leads to carbanion 2 stabilized by its electron-withdrawing groups. Structures 2a to 2c are three resonance structures that can be drawn for this species, two of which have enolate ions. This nucleophile reacts with the electrophilic alkene 3 to form 4 in a conjugate addition reaction. Proton abstraction from protonated base (or solvent) by the enolate 4 to 5 is the final step.

The course of the reaction is dominated by orbital, rather than electrostatic, considerations. The HOMO of stabilized enolates has a large coefficient on the central carbon atom while the LUMO of many alpha, beta unsaturated carbonyl compounds has a large coefficient on the beta carbon. Thus, both reactants can be considered soft. These polarized frontier orbitals are of similar energy, and react efficiently to form a new carbon–carbon bond.

Like the aldol addition, the Michael reaction may proceed via an enol, silyl enol ether in the Mukaiyama-michael addition, or more usually, enolate nucleophile. In the latter case, the stabilized carbonyl compound is deprotonated with a strong base (hard enolization) or with a Lewis acid and a weak base (soft enolization). The resulting enolate attacks the activated olefin with 1,4-regioselectivity, forming a carbon–carbon bond. This also transfers the enolate to the electrophile. Since the electrophile is much less acidic than the nucleophile, rapid proton transfer usually transfers the enolate back to the nucleophile if the product is enolizable; however, one may take advantage of the new locus of nucleophilicity if a suitable electrophile is pendant. Depending on the relative acidities of the nucleophile and product, the reaction may be catalytic in base. In most cases, the reaction is irreversible at low temperature.

Asymmetric Michael Reaction

In recent years researchers have endeavoured to expand the scope of asymmetric Michael additions. The most common methods involve chiral phase transfer catalysis, such as asymmetric quaternary ammonium salts derived from the Cinchona alkaloids; or organocatalysis, which uses enamine or iminium activation with chiral secondary amines, usually derived from proline.

In the reaction between cyclohexanone and β-nitrostyrene sketched below, the base proline is derivatized and works in conjunction with a protic acid such as p-toluenesulfonic acid:

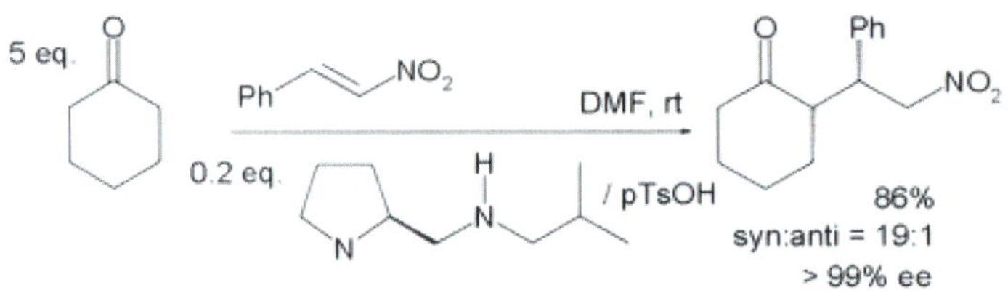

Syn addition is favored with 99% ee. In the transition state believed to be responsible for this selectivity, the enamine (formed between the proline nitrogen and the cycloketone) and β-nitrostyrene are co-facial with the nitro group hydrogen bonded to the protonated amine in the proline side group.

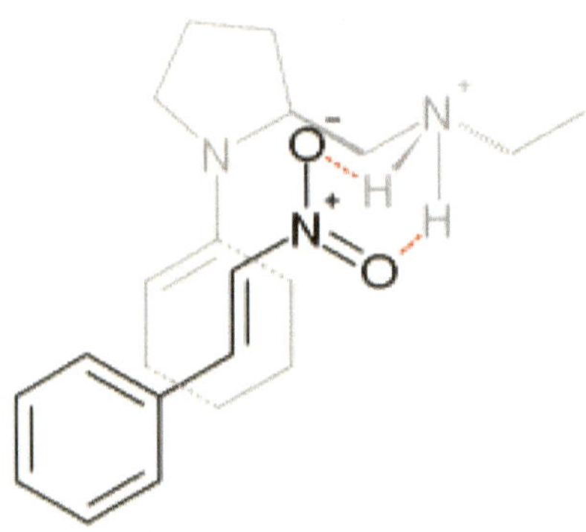

A well-known Michael reaction is the synthesis of warfarin from 4-hydroxycoumarin and benzylideneacetone:

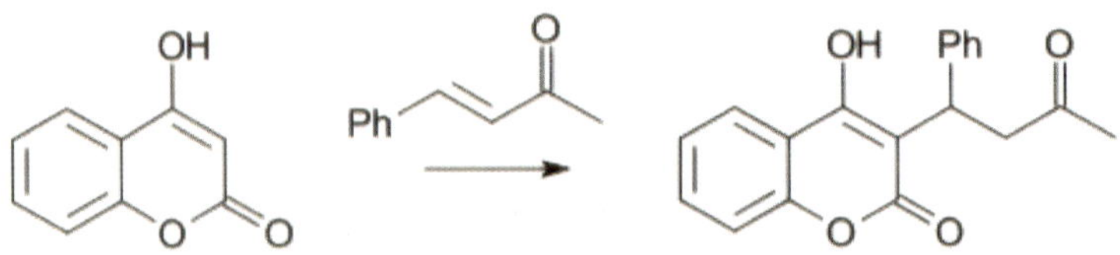

Several asymmetric versions of this reaction exist using chiral catalysts.

Mukaiyama-michael Addition

In the Mukaiyama-michael addition the nucleophile is a silyl ether and the catalyst usually titanium tetrachloride:

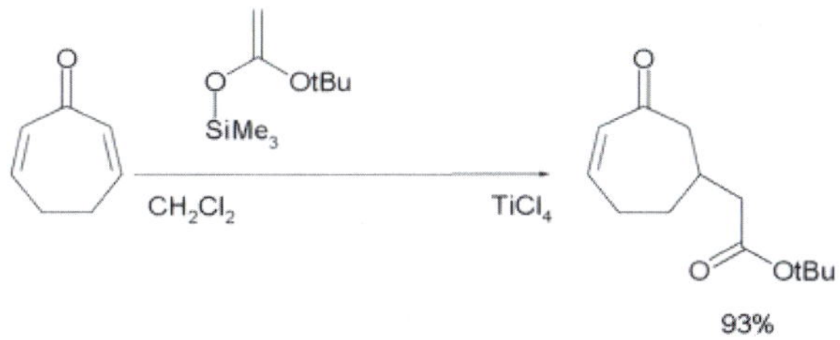

Nucleophilic addition,

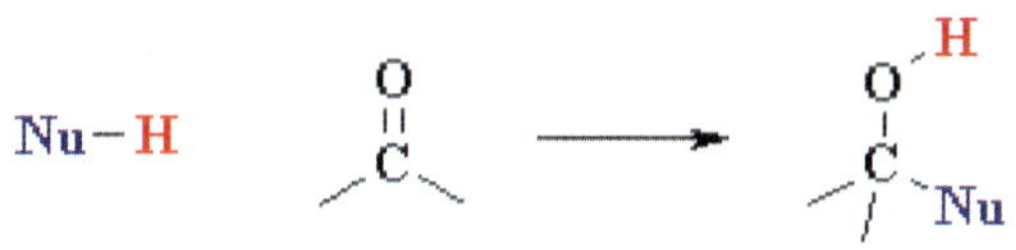

Nucleophilic addition reactions are an important class of reactions that allow the interconversion of C=O into a range of important functional groups.

Implication of the term "nucleophilic addition"

A nucleophile, Nu^-, is an electron rich species that will react with an electron poor species (here the C=O) An addition implies that two systems combine to a single entity.

There are three fundamental events in a nucleophilic addition reaction:

- Formation of the new s bond between the nucleophile, Nu, to the electrophilic C of the C=O group;
- Breaking of the p bond to the O resulting in the formation of an intermediate alkoxide;
- Protonation of the intermediate alkoxide to give an alcohol derivative.

Depending on the reactivity of the nucleophile, there are two possible general scenarios:

- Strong nucleophiles: (anionic) add directly to the C=O to form the intermediate alkoxide. The alkoxideis then protonated on work-up with dilute acid.

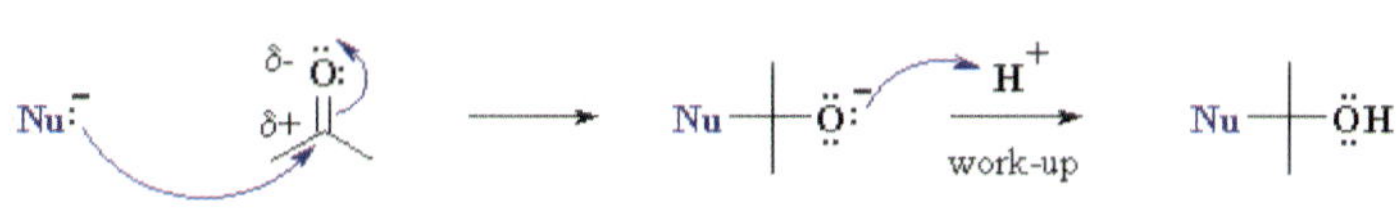

Nucleophilic Addition under **BASIC** conditions with stronger nucleophiles

Examples of such nucleophilic systems are: RMgX, RLi, RC≡CM, LiAlH4, NaBH4.

Weaker nucleophiles: (neutral) require that the C=O be activated prior to attack of the Nu.

This can be done using a acid catalyst which protonates on the Lewis basic O and makes the system more electrophilic.

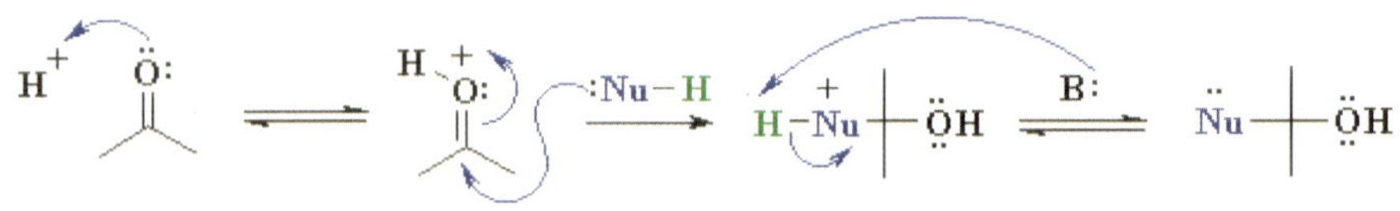

Nucleophilic Addition under **ACIDIC** condition with weaker nucleophiles

Examples of such nucleophilic systems are: H_2O, ROH, R-NH_2 .

The protonation of a carbonyl gives a structure that can be redrawn in another resonance form that reveals the electrophilic character of the C since it is a carbocation.

Elimination Reaction

Elimination reactions can be defined as the chemical reactions which involve elimination of leaving group to form unsaturated compound. A general elimination reaction can be written as given below:

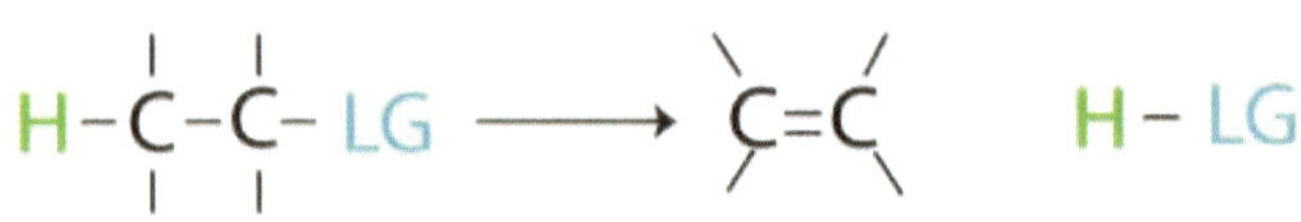

Dehydration $(-H_2O)(-H_2O)$ of alcohols or dehydrohalogenation (-HX) of haloalkanes are some typical examples of elimination reactions. In an elimination reaction, there are 3 main steps. First step is removal of proton that for a carbocation that form C-C pi bond in the molecule including breaking of the bond of the leaving group.

Elimination Reaction Mechanism

In first step, an elimination reaction loss of the leaving group results the formation of carbocation as an intermediate.

In the second step, the carbocation losses a proton in the presence of base to form pi-bond in the molecule.

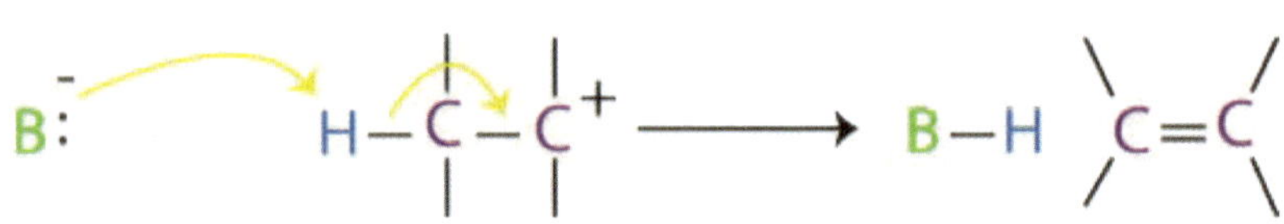

The reaction influences with the stability of carbocation formed during the reaction. The order of stability of carbocations formed during reaction is:

$$(CH_3)_3C^+ > (CH_3)_2CH^+ > CH_3CH_2^+ > CH_3^+$$

The mechanism of elimination reaction can be two types; E1 and E2. E1 mechanism involves the formation of carbocation as an intermediate during the reaction. Therefore the rate of reaction depends on the stability of carbocation formed during the reaction. The bond of carbon atom and leaving group breaks in rate determining step. So a better the leaving group will increase the rate of reaction. Since base is not the part of rate determining step so it does not affect the E1 mechanism but influences the E2 mechanism as it is part of rate determining step in that mechanism. So we can say that E1 mechanism needs good leaving group, a stable carbocation and weak bases.

Elimination Reaction Examples

One of the most common examples of elimination reaction is formation of alkene from haloalkane in the presence of base.

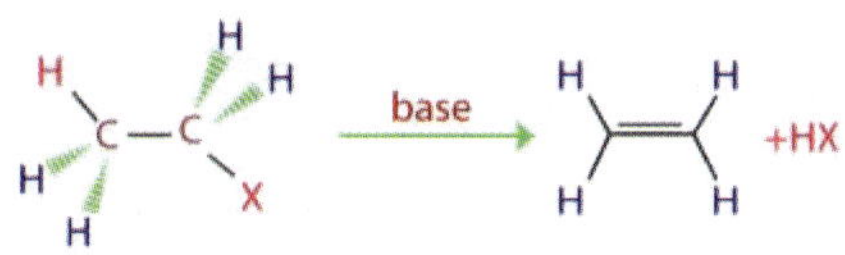

The elimination reaction proceeds with E_2 mechanism in the presence of base. The attack of base and elimination of leaving group occur simultaneously to form alkene. The hydrogen atom that is removed by the base must be at anti-position of the leaving group.

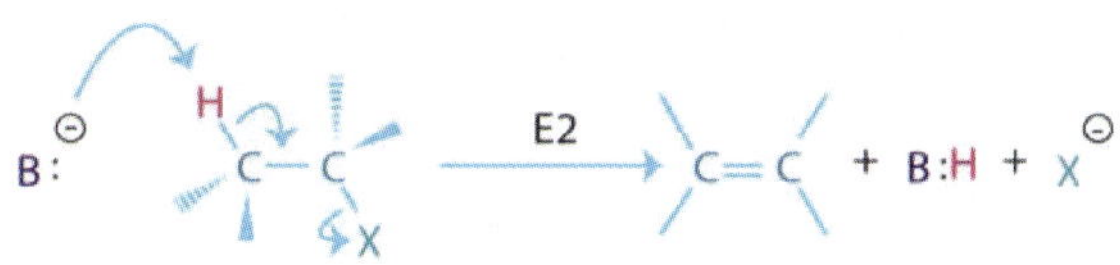

Dehydration of alcohol in the presence of base like ethoxide ion ($C_2H_5O^-$) is also an example of elimination reaction.

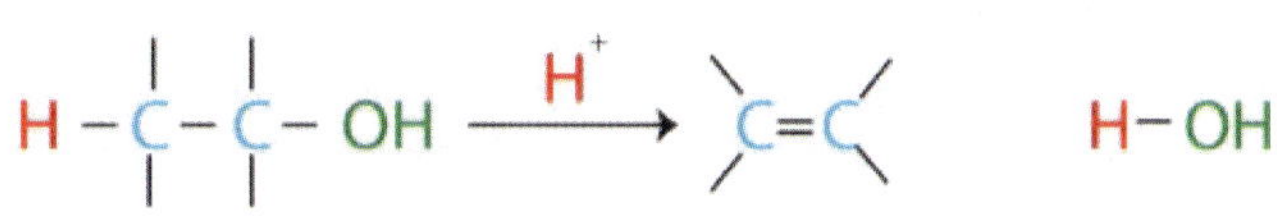

Beta Elimination Reaction

Beta elimination reaction can be defined as the chemical reaction in which the leaving group and H are placed at neighbor carbon atoms. This is called as beta elimination. For example; dehydration of alcohol is beta elimination. Heating of alcohol in the presence of strong acid like H_2SO_4, H_3PO_4 causes 1, 2 - elimination or beta elimination. It results the formation of an alkene and water. Since water is removed during the reaction so it is also called as dehydration reaction. The reaction proceeds with the formation of carbocation therefore the reactivity order of alcohol is tertiary > Secondary > Primary alcohol. The elimination reaction follows the Zaitsev's rule according to which the more highly substituted alkene forms as the major product. Out of cis and trans-isomers, trans-isomer is more stable therefore trans-isomer is always preferred over cis form. The carbocation that is formed during reaction can rearrange by 1, 2- hydride or 1, 2-alkyl shift to form the stable product. Since primary carbocations are not so stable so they follow the E2 mechanism and reaction proceeds with the formation of transition state as an intermediate.

Example of Beta Elimination Reaction

Elimination reaction removes certain groups from the reactant molecule and forms an unsaturated compound as a product. There are several examples of beta elimination reactions:

- Reaction of alcohol with mineral acid

  An alcohol reacts with mineral acid at high temperature to form alkene with elimination of water molecule.

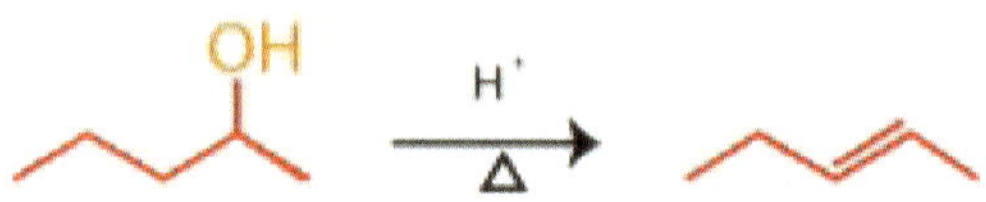

  The reaction can proceed with E1 or E2 mechanism. E1 mechanism occurs with formation of carbocation whereas E2 mechanism proceeds with a concerted path in single step with the formation of transition state.

- Another example of beta elimination is formation of alkene from alkyl halide in the presence of base.

  The rate determining step involves the formation of transition state due to reaction of alkyl halide with base like ethoxide ion.

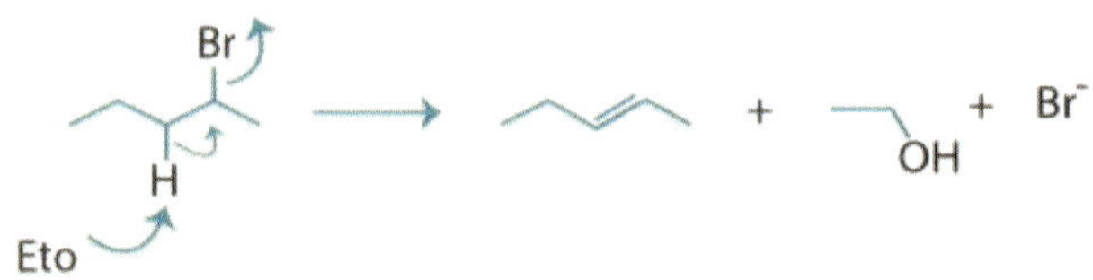

The presence of base eliminates beta hydrogen atom with leaving group that results the formation of alkene. In transition state, the bond between base and H forms whereas bond between C and leaving group and C-H is about to cleave. It results the formation of pi-bond between both carbon atoms. For beta elimination reaction beta hydrogen and the leaving group should be on the same plane. Because reaction involves:

- Formation of C-C pi-bond;
- Cleavage of C-H and C-X bond.

Since pi-bond is formed by the sideways overlap of two p-orbitals therefore for maximum overlapping both orbitals must be parallel to each other. Hence beta-H and leaving group should be on the same plane. Both beta-H and leaving group should preferably be anti-coplanar that results the anti-elimination. Reaction involves the formation of more stable more substituted alkene as major product.

Cope Reaction

The Cope Reaction of N-oxides, which can easily be prepared in situ from tertiary amines with an oxidant such as peracid, leads to alkenes via a thermally induced syn-elimination in aprotic solvents.

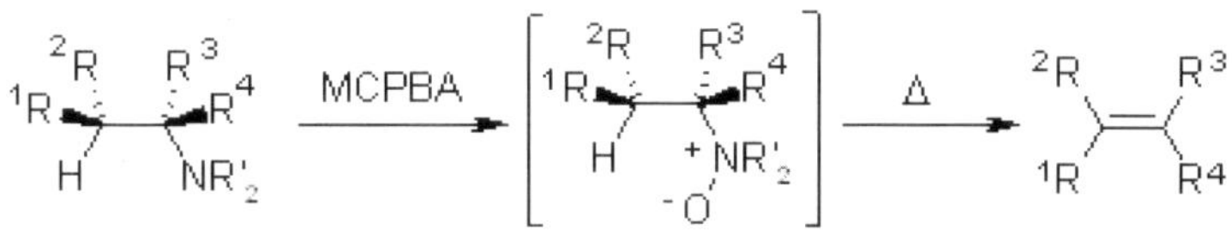

Mechanism of the Cope Elimination

The Cope Elimination is a syn periplanar elimination in which six electrons move in a five-membered ring according to a concerted, thermally-induced mechanism to yield an alkene and a hydroxylamine:

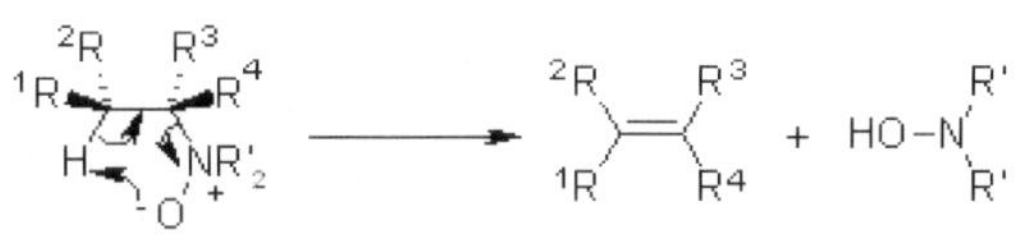

The sterically demanding amine oxide function reacts preferentially with the more easily accessible hydrogens, and often gives good selectivity favoring the less-substituted alkene. Thus, for simple alkenes, the reaction follows the Hofmann Rule.

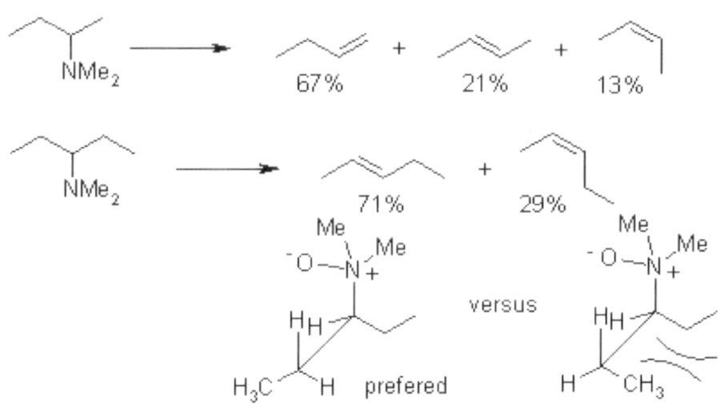

The following structures exemplify the stereochemical requirement for reaching a five-membered cyclic transition state and the influence of sterically demanding groups for some cyclic compounds:

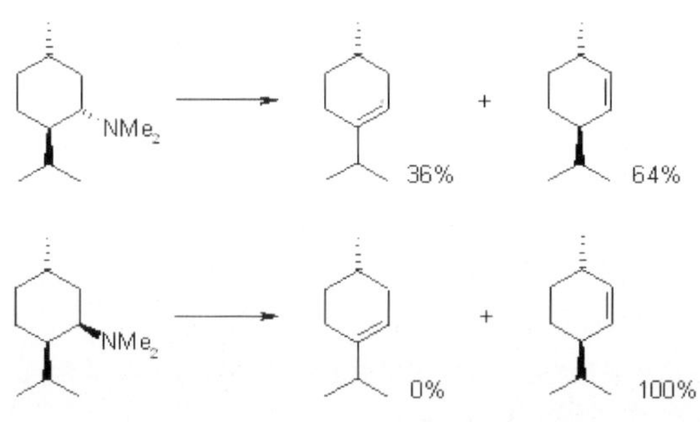

One synthetically useful exception to the general preference for more accessible hydrogens is the reaction of substrates that bear a β-phenyl group, or more generally speaking, an electron-withdrawing group in the β-position:

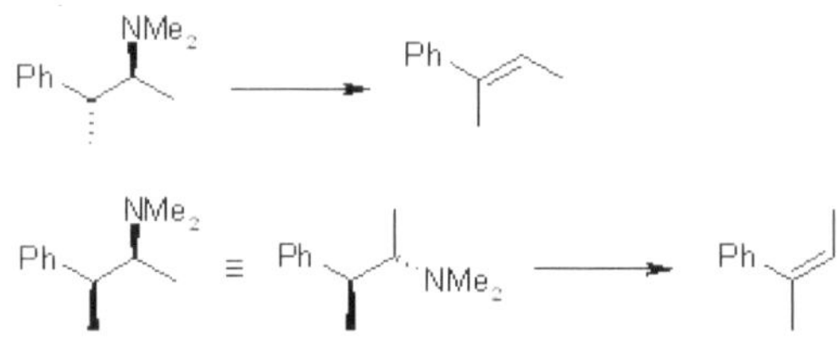

Compared with alkyl-substituted derivatives, a phenyl group provides a 100-fold rate increase. Phenyl and other electron-withdrawing groups lower the electron density of the carbon-hydrogen bond, making the hydrogen more acidic, and stabilizing the transition state. Computed minimal energy paths suggest that the Cope Elimination is slightly dissymmetric and nonsynchronous, with the H-transfer occurring slightly in advance of the other bond reorganizations.

Another interesting feature is the solvent dependence of the reaction rate. The Cope Elimination is extraordinarily sensitive to solvent effects, and a million-fold rate increase can be obtained going from protic to aprotic solvents. Within aprotic solvents, decreasing polarity significantly increases the reaction rate. The relative rate retardation for Cope Eliminations in protic solvents arises from hydrogen bonding between the amine oxide and the solvent. In addition, a fully solvated amine oxide in a protic solvent such as MeOH can even be relatively more favorable energetically than the less polar hydroxylamine, which can displace the equilibrium to favor the retro-Cope Elimination.

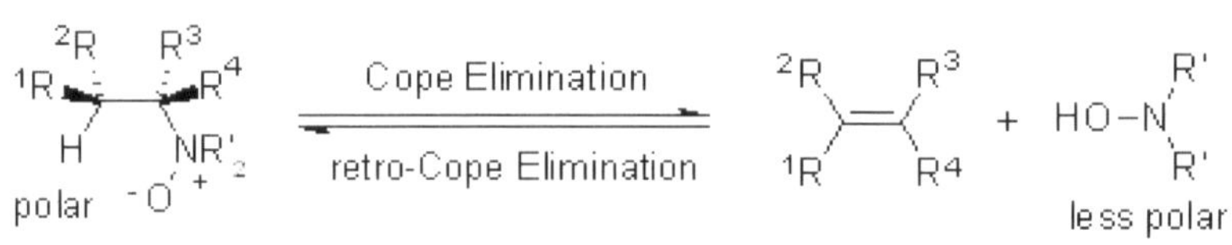

E1cB Elimination Reaction

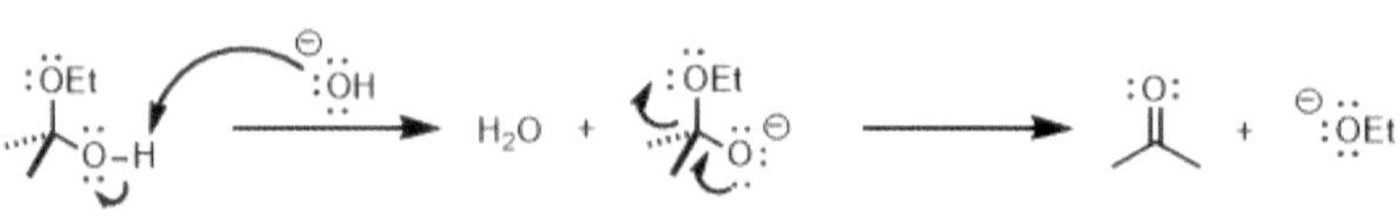

Example of the E1cB reaction mechanism in the degradation of a hemiacetal under basic conditions.

The E1cB elimination reaction is a type of elimination reactionwhich occurs under basic conditions, where a particularly poor leaving group (such as -OH or -OR) and an acidic hydrogen eliminate to form an additional bond. E1cB is a three-step process. First, a baseabstracts the most acidic proton to generate a stabilized anion. The lone pair of electrons on the anion then moves to the neighboring atom, thus expelling the leaving group and forming double or triple bond. The name of the mechanism - E1cB - stands for Elimination Unimolecular conjugate Base. Elimination refers to the fact that the mechanism is an elimination reaction and will lose two substituents. Unimolecular refers to the fact that the rate-determining step of this reaction only involves one molecular entity. Finally, conjugate base refers to the formation of the carbanion intermediate, which is the conjugate base of the starting material.

Mechanism

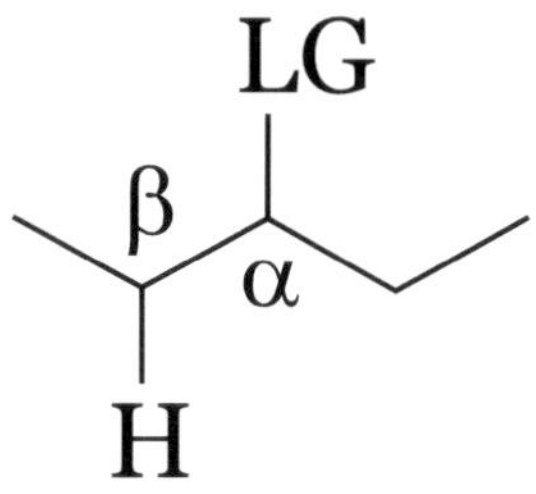

α and β assignments in a molecule with leaving group, LG.

There are two main requirements to have a reaction proceed down an E1cB mechanistic pathway. The compound must have an acidichydrogen on its β-carbon and a relatively poor leaving group on the α-carbon. The first step of an E1cB mechanism is the deprotonation of the β-carbon, resulting in the formation of an anionic transition state, such as a carbanion. The greater the stability of this transition state, the more the mechanism will favor an E1cB mechanism. This transition state can be stabilized through induction or delocalization of the electronlone pair through resonance. In general it can be claimed that an electron withdrawing group on the substrate, a strong base, a poor leaving group and a polar solvent triggers the E1cB mechanism. An example of an E1cB mechanism that has a stable transition state can be seen in the degradation of ethiofencarb a carbamate insecticide that has a relatively short half-life in earth's atmosphere. Upon deprotonation of the amine, the resulting amide is relatively stable because it is conjugated with the neighboring carbonyl. In addition to containing an acidic hydrogen on the β-carbon, a relatively poor leaving group is also necessary. A bad leaving group is necessary because a good leaving group will leave before the ionization of the molecule. As a result, the compound will likely proceed through an E2 pathway. Some examples of compounds that contain poor leaving groups and can undergo the E1cB mechanism are alcohols and fluoroalkanes. It has also been suggested that the E1cB mechanism is more common among alkenes eliminating to alkynes than from an alkane to alkene. One possible explanation for this is that the sp^2 hybridization creates slightly more acidic protons. Although it should be noted that this mechanism is not limited to carbon-based eliminations. It has been observed with other heteroatoms, such as nitrogen in the elimination of a phenol derivative from ethiofencarb.

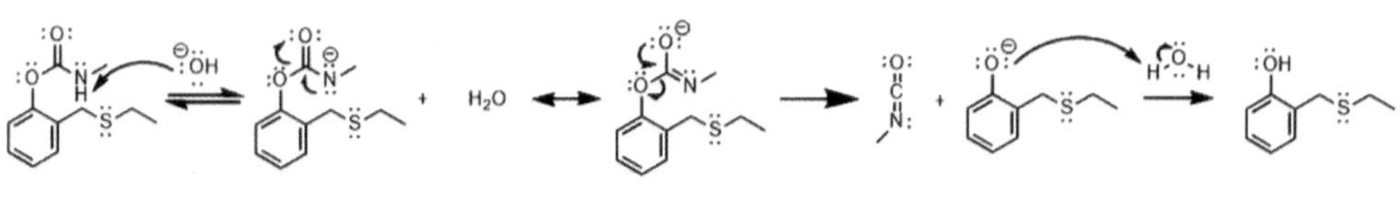

Degradation of ethiofencarb illustrating the presence of a stable anion due to resonance between the amide functional group and the carbonyl group.

Distinguishing E1cB-elimination Reactions from E1- and E2-elimination Reactions

All elimination reactions involve the removal of two substituents from a pair of adjacent atoms in a compound. Alkene, alkynes, or similar heteroatom variations (such as carbonyl and cyano) will form. The E1cB mechanism is just one of three types of elimination reaction. The other two elimination reactions are E1 and E2 reactions. Although the mechanisms are similar, they vary in the timing of the deprotonation of the α-carbon and the loss of the leaving group. E1 stands for unimolecular elimination, and E2 stands for bimolecular elimination. In an E1 mechanism, the molecule contains a good leaving group that departs before deprotonation of the α-carbon. This results in the formation of a carbocation intermediate. The carbocation is then deprotonated resulting in the formation of a new pi bond. The molecule involved must also have a very good leaving group such as bromine or chlorine, and it should have a relatively less acidic α-carbon.

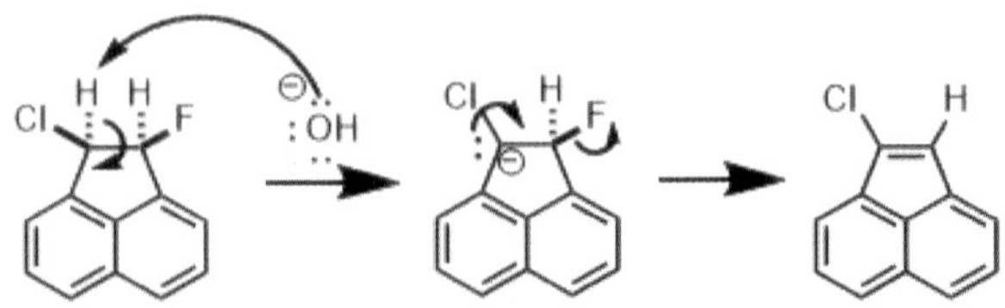

Example of the preferential elimination of fluorine in an E1cB-elimination reaction.

In an E2-elimination reaction, both the deprotonation of the α-carbon and the loss of the leaving group occur simultaneously in one concerted step. Molecules that undergo E2-elimination mechanisms have more acidic α-carbons than those that undergo E1 mechanisms, but their α-carbons are not as acidic as those of molecules that undergo E1cB mechanisms. The key difference between the E2 vs E1cb pathways is a distinct carbanion intermediate as opposed to one concerted mechanism. Studies have been shown that the pathways differ by using different halogen leaving groups. One example uses chlorine as a better stabilizing halogen for the anion than fluorine, which makes fluorine the leaving group even though chlorine is a much better leaving group. This provides evidence that the carbanion is formed because the products are not possible through the most stable concerted E2 mechanism. The following table summarizes the key differences between the three elimination reactions; however, the best way to identify which mechanism is playing a key role in a particular reaction involves the application of chemical kinetics.

E1	E2	E1cB
Stepwise reaction	Concerted reaction	Stepwise reaction
Carbocation intermediate	Simultaneous removal of proton, formation of double bond, and loss of leaving group	Carbanion intermediate
no kind of conclusion	No preference	no kind of conclusion
Good leaving groups	Leaving group	Poor leaving groups
Less acidic B-H	Acidic B-H	More acidic B-H

Chemical Kinetics of E1cB-elimination Mechanisms

When trying to determine whether or not a reaction follows the E1cB mechanism, chemical kinetics are essential. The best way to identify the E1cB mechanism involves the use of rate laws and the kinetic isotope effect. These techniques can also help further differentiate between E1cB, E1, and E2-elimination reactions.

Rate Law

When trying to experimentally determine whether or not a reaction follows the E1cB mechanism, chemical kinetics is essential. The best ways to identify the E1cB mechanism involves the use of rate laws and the kinetic isotope effect.

The rate law that governs E1cB mechanisms is relatively simple to determine. Consider the following reaction:

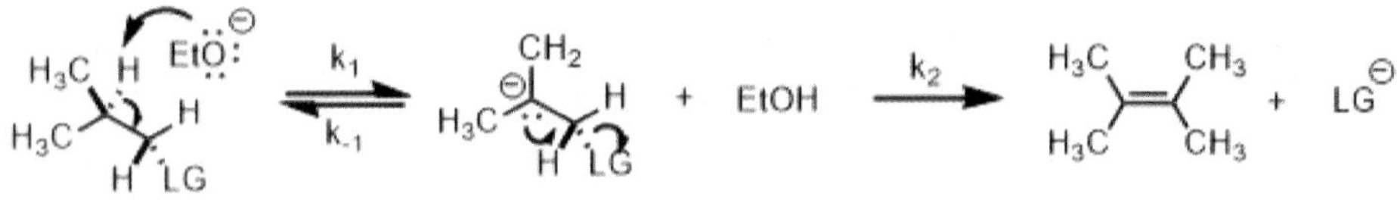

Example of an E1cB-elimination mechanism with a generic leaving group LG, and ethoxide as the base.

Assuming, that there is a steady-state carbanion concentration in the mechanism, the rate law for an E1cB mechanism.

$$\frac{d[P]}{dt} = \frac{k_1 k_2 [substrate][base]}{k_{-1}[conjugate\ acid] + k_2}$$

$$rate = k_{obs}[substrate][base]$$

From the above equation, it is clear the second order kinetics will be exhibited. E1cB mechanisms kinetics can vary slightly based on the rate of each step. As a result, the E1cB mechanism can be broken down into three categories:

- $E1cB_{anion}$ is when the carbanion is stable and a strong base is used in excess of the substrate, making deprotonation irreversible, followed by rate-determining loss of the leaving group (k_1[base] ≫ k_2).
- $E1cB_{rev}$ is when the first step is reversible but the formation of product is slower than reforming the starting material, this again results from a slow second step (k_{-1}[conjugate acid] ≫ k_2).
- $E1cB_{irr}$ is when the first step is slow, but once the anion is formed the product quickly follows (k_2 ≫ k_{-1}[conjugate acid]). This leads to an irreversible first step but unlike $E1cB_{anion}$, deprotonation is rate determining.

Kinetic Isotope Effect

Deuterium

Deuterium exchange and a deuterium kinetic isotope effect can help distinguish among $E1cB_{rev}$, $E1cB_{anion}$, and $E1cB_{irr}$. If the solvent is protic and contains deuterium in place of hydrogen (e.g., CH_3OD), then the exchange of protons into the starting material can be monitored. If the recovered starting material contains deuterium, then the reaction is most likely undergoing an E1cBrev type mechanism. Recall, in this mechanism protonation of the carbanion (either by the conjugate acid or by solvent) is faster than loss of the leaving group. This means after the carbanion is formed, it will quickly remove a proton from the solvent to form the starting material. If the reactant contains deuterium at the β position, a primary kinetic isotope effect indicates that deprotonation is rate determining. Of the three E1cB mechanisms, this result is only consistent with the $E1cB_{irr}$ mechanism, since the isotope is already removed in $E1cB_{anion}$ and leaving group departure is rate determining in $E1cB_{rev}$.

Fluorine-19 and Carbon-11

Another way that the kinetic isotope effect can help distinguish E1cB mechanisms involves the use of 19F. Fluorine is a relatively poor leaving group, and it is often employed in E1cB mechanisms. Fluorine kinetic isotope effects are also applied in the labeling of Radiopharmaceuticals and other compounds in medical research. This experiment is very useful in determining whether or not the loss of the leaving group is the rate-determining step in the mechanism and can help distinguish between E1cBirr and E2 mechanisms. 11C can also be used to probe the nature of the transition state structure. The use of 11C can be used to study the formation of the carbanion as well as study its lifetime which can not only show that the reaction is a two-step E1cB mechanism (as opposed to the concerted E2 mechanism), but it can also address the lifetime and stability of the transition state structure which can further distinguish between the three different types of E1cB mechanisms.

Aldol Reactions

The most well-known reaction that undergoes E1cB elimination is the aldol condensation reaction under basic conditions. This involves the deprotonation of a compound containing a carbonyl group that results in the formation of an enolate. The enolate is the very stable conjugate base of the starting material, and is one of the intermediates in the reaction. This enolate then acts as a nucleophile and can attack an electrophilic aldehyde. The Aldol product is then deprotonated forming another enolate followed by the elimination of water in an E1cB dehydration reaction. Aldol reactions are a key reaction in organic chemistry because they provide a means of forming carbon-carbon bonds, allowing for the synthesis of more complex molecules.

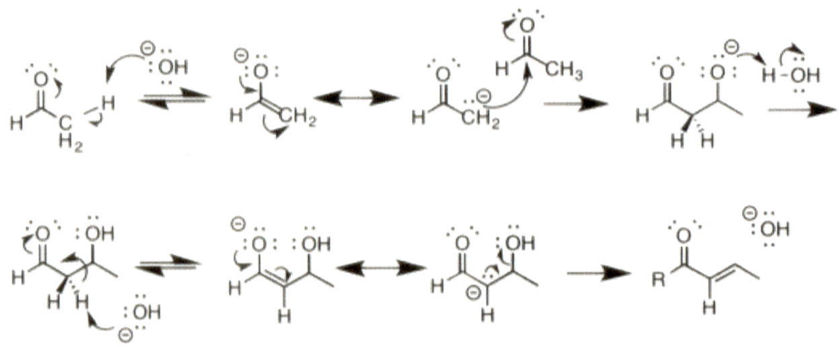

An Aldol condensation reaction is one of the most common examples of an E1cB mechanism.

Photo-induced E1cB

A photochemical version of E1cB has been reported by Lukeman et al. In this report, a photochemically induced decarboxylation reaction generates a carbanion intermediate, which subsequently eliminates the leaving group. The reaction is unique from other forms of E1cB since it does not require a base to generate the carbanion. The carbanion formation step is irreversible, and should thus be classified as $E1cB_{irr}$.

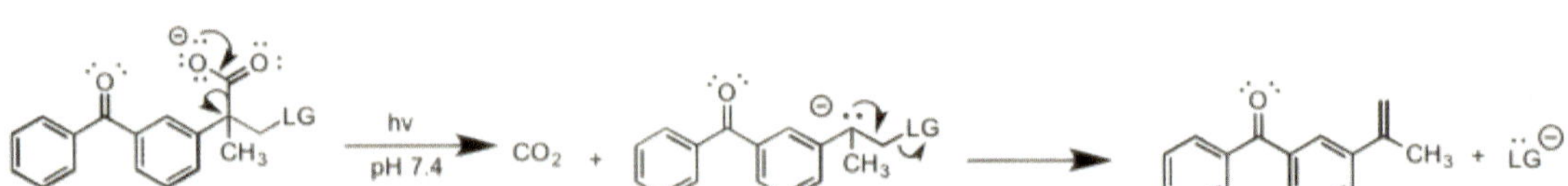

E1cB reaction mechanism through photo-induced decarboxylation.

Hofmann Elimination

When a primary amine bearing one or more beta hydrogens is treated with methyl iodide, followed by aqueous silver oxide, followed by heat, the primary amine is converted to an alkene. This reaction is known as Hofmann elimination,

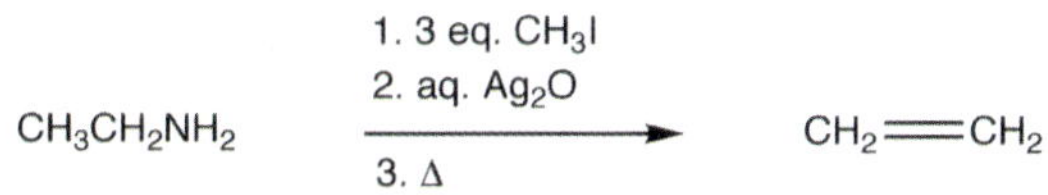

The net reaction is 1,2-elimination, hence the name Hofmann Elimination,

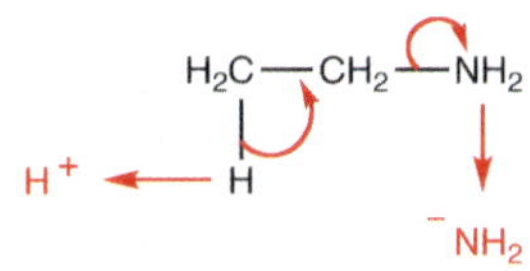

Hofmann elimination occurs in three stages,

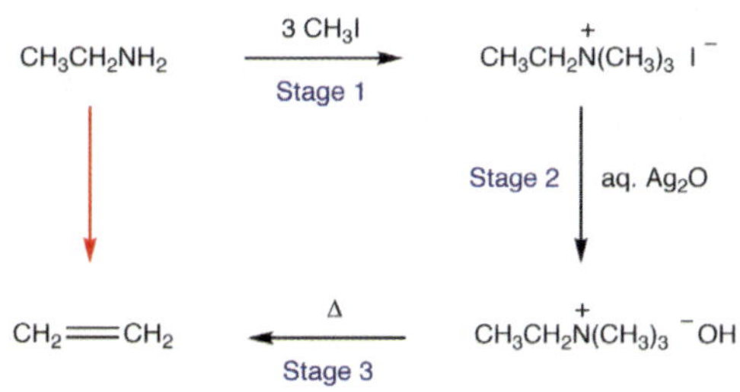

Mechanism

Stage 1:

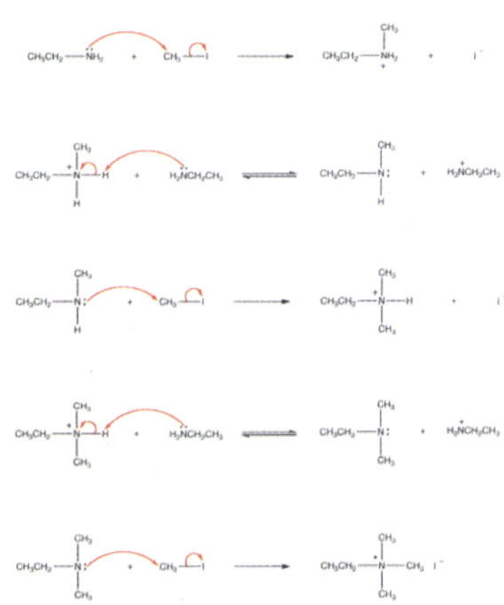

Stage 2:

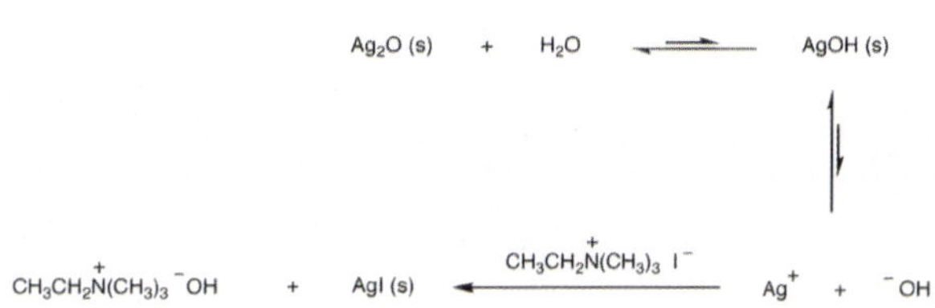

Stage 3:

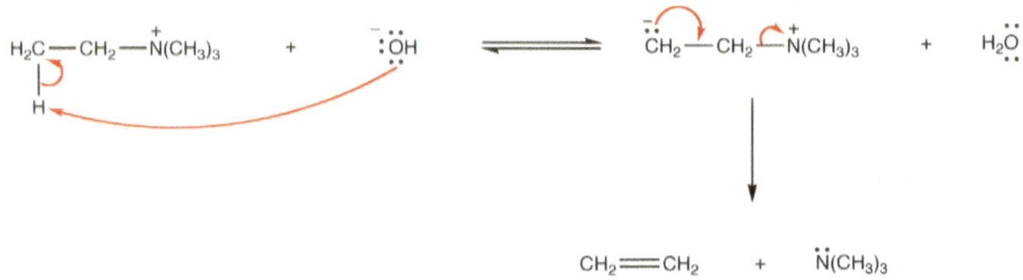

Stage 3 is a 1,2-elimination via E1cB mechanism.

Hofmann elimination is regioselective. Since the 1,2-elimination in Stage 3 occurs via E1cB mechanism, Hofmann rule is used to predict the major product.

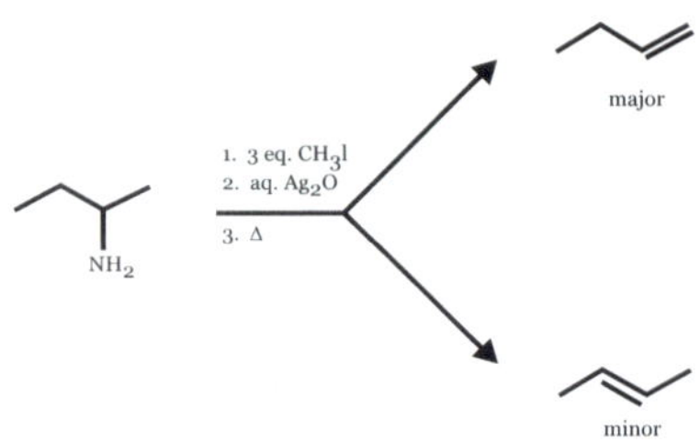

Redox Reaction

Oxidation-reduction reaction is also called redox reaction. It is a chemical reaction in which the oxidation number of a participating chemical species changes. The term covers a large and diverse body of processes. Many oxidation-reduction reactions are as common and familiar as fire, the rusting and dissolution of metals, the browning of fruit, and respiration and photosynthesis—basic life functions.

Major Classifications

Most oxidation-reduction (redox) processes involve the transfer of oxygen atoms, hydrogen atoms, or electrons, with all three processes sharing two important characteristics:

- They are coupled—i.e., in any oxidation reaction a reciprocal reduction occurs,
- They involve a characteristic net chemical change—i.e., an atom or electron goes from one unit of matter to another.

Both reciprocity and net change are illustrated below in examples of the three most common types of oxidation-reduction reactions.

Oxygen-atom Transfer

Carbon reacts with mercury(II) oxide (a compound in which mercury has a bonding capacity expressed as +2;) to produce carbon dioxide and mercury metal. This reaction can be written in equation form:

$$C + 2HgO \longrightarrow CO_2 + Hg$$

carbon mercury (II) oxide carbon dioxide mercury metal

Carbon, receiving oxygen, is oxidized; mercury(II) oxide, losing oxygen, undergoes the complementary reduction; and the net change is the transfer of two oxygen atoms from mercury(II) oxide units to a carbon atom.

Hydrogen-atom Transfer

Hydrogen atoms are transferred from hydrazine, a compound of nitrogen and hydrogen, to oxygen in the following reaction:

$$N_2H_4 + O_2 \longrightarrow N_2 + 2H_2O.$$

hydrazine oxygen nitrogen water

Hydrazine, losing hydrogen, is oxidized to molecular nitrogen, while oxygen, gaining hydrogen, is reduced to water.

Electron Transfer

Zinc metal and copper(II) ion react in water solution, producing copper metal and an aqueous (denoted by aq) zinc ion according to the equation:

$$Zn + Cu^{2+}(aq) \longrightarrow Zn^{2+}(aq) + Cu.$$

zinc metal aqueous copper(II) ion aqueous zinc(II) ion copper metal

With the transfer of two of its electrons, the zinc metal is oxidized, becoming an aqueous zinc ion, while the copper(II) ion, gaining electrons, is reduced to copper metal. Net change is the transfer of two electrons, lost by zinc and acquired by copper.

Because of their complementary nature, the oxidation and reduction processes together are referred to as redox reactions. The reactant that brings about the oxidation is called the oxidizing agent, and that reagent is itself reduced by the reducing agent. In the examples given above, mercury(II) oxide, oxygen, and the copper(II) ion are oxidizing agents, and carbon, hydrazine, and zinc are the reducing agents.

Oxidizing and Reducing agents

In redox processes, the reductant transfers electrons to the oxidant. Thus, in the reaction, the reductant or reducing agent loses electrons and is oxidized, and the oxidant or oxidizing agent gains electrons and is reduced. The pair of an oxidizing and reducing agent that are involved in a particular reaction is called a redox pair. A redox couple is a reducing species and its corresponding oxidizing form, e.g., Fe^{2+}/Fe^{3+}.

Oxidizers

Substances that have the ability to oxidize other substances (cause them to lose electrons) are said to be oxidative or oxidizing and are known as oxidizing agents, oxidants, or oxidizers. That is, the oxidant (oxidizing agent) removes electrons from another substance, and is thus itself reduced. And, because it "accepts" electrons, the oxidizing agent is also called an electron acceptor. Oxygen is the quintessential oxidizer.

Oxidants are usually chemical substances with elements in high oxidation states (e.g., H_2O_2, MnO_4^-, CrO_3, $Cr_2O_7^{2-}$, OsO_4), or else highly electronegative elements (O_2, F_2, Cl_2, Br_2) that can gain extra electrons by oxidizing another substance.

Reducers

Substances that have the ability to reduce other substances (cause them to gain electrons) are said to be reductive or reducing and are known as reducing agents, reductants, or reducers. The reductant (reducing agent) transfers electrons to another substance, and is thus itself oxidized. And, because it "donates" electrons, the reducing agent is also called an electron donor. Electron donors can also form charge transfer complexes with electron acceptors.

Reductants in chemistry are very diverse. Electropositive elemental metals, such as lithium, sodium, magnesium, iron, zinc, and aluminium, are good reducing agents. These metals donate or give away electrons readily. Hydride transfer reagents, such as $NaBH_4$ and $LiAlH_4$, are widely used in organic chemistry,primarily in the reduction of carbonyl compounds to alcohols. Another method of reduction involves the use of hydrogen gas (H_2) with a palladium, platinum, or nickel catalyst. These catalytic reductions are used primarily in the reduction of carbon-carbon double or triple bonds.

Standard Electrode Potentials (Reduction Potentials)

Each half-reaction has a standard electrode potential (E^o_{cell}), which is equal to the potential difference or voltage at equilibrium under standard conditions of an electrochemical cell in which the cathode reaction is the half-reaction considered, and the anode is a standard hydrogen electrode where hydrogen is oxidized:

$$\frac{1}{2} H_2 \rightarrow H^+ + e^-.$$

The electrode potential of each half-reaction is also known as its reduction potential Eored, or potential when the half-reaction takes place at a cathode. The reduction potential is a measure of the tendency of the oxidizing agent to be reduced. Its value is zero for $H^+ + e^- \rightarrow 1/2\ H_2$ by definition, positive for oxidizing agents stronger than H^+ (e.g., +2.866 V for F_2) and negative for oxidizing agents that are weaker than H^+ (e.g., −0.763 V for Zn^{2+}).

For a redox reaction that takes place in a cell, the potential difference is:

$$E^o_{cell} = E^o_{cathode} - E^o_{anode}$$

However, the potential of the reaction at the anode was sometimes expressed as an oxidation potential:

$$E^o_{ox} = -E^o_{red}.$$

The oxidation potential is a measure of the tendency of the reducing agent to be oxidized, but does not represent the physical potential at an electrode. With this notation, the cell voltage equation is written with a plus sign:

$$E^o_{cell} = E^o_{red(cathode)} + E^o_{ox(anode)}$$

Examples of Redox Reactions

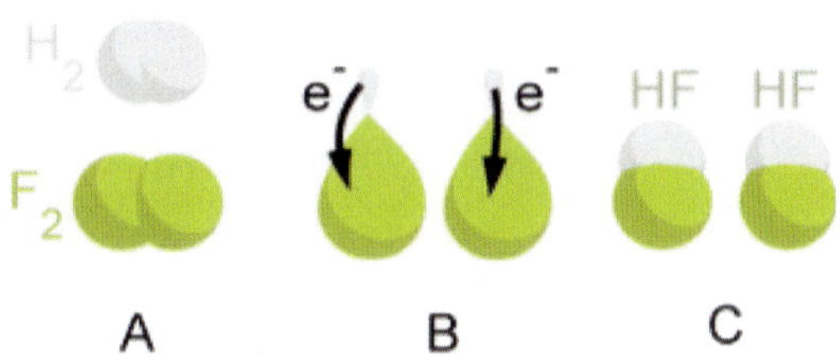

A redox reaction

A good example is the reaction between hydrogen and fluorine in which hydrogen is being oxidized and fluorine is being reduced,

$$H_2 + F_2 \rightarrow 2\ HF$$

We can write this overall reaction as two half-reactions,

The oxidation reaction,

$$H_2 \rightarrow 2\ H^+ + 2\ e^-$$

And the reduction reaction,

$$F_2 + 2\ e^- \rightarrow 2\ F^-$$

Analyzing each half-reaction in isolation can often make the overall chemical process clearer. Because there is no net change in charge during a redox reaction, the number of electrons in excess in the oxidation reaction must equal the number consumed by the reduction reaction.

Elements, even in molecular form, always have an oxidation state of zero. In the first half-reaction, hydrogen is oxidized from an oxidation state of zero to an oxidation state of +1. In the second half-reaction, fluorine is reduced from an oxidation state of zero to an oxidation state of −1.

When adding the reactions together the electrons are canceled:

$$\begin{array}{rcl} H_2 & \rightarrow & 2H^+ + 2e^- \\ F_2 + 2e^- & \rightarrow & 2F^- \\ \hline H_2 + F_2 & \rightarrow & 2H^+ + 2F^- \end{array}$$

And the ions combine to form hydrogen fluoride:

$$2\ H^+ + 2\ F^- \rightarrow 2\ HF$$

The overall reaction is:

$$H_2 + F_2 \rightarrow 2\ HF$$

Metal Displacement

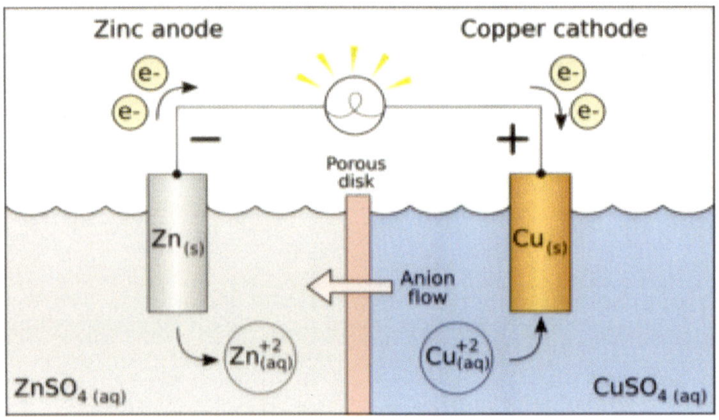

A redox reaction is the force behind an electrochemical cell like the Galvanic cell pictured. The battery is made out of a zinc electrode in a $ZnSO_4$ solution connected with a wire and a porous disk to a copper electrode in a $CuSO_4$ solution.

In this type of reaction, a metal atom in a compound (or in a solution) is replaced by an atom of another metal. For example, copper is deposited when zinc metal is placed in a copper(II) sulfatesolution:

$$Zn(s) + CuSO_4(aq) \rightarrow ZnSO_4(aq) + Cu(s)$$

In the above reaction, zinc metal displaces the copper(II) ion from copper sulfate solution and thus liberates free copper metal.

The ionic equation for this reaction is,

$$Zn + Cu^{2+} \rightarrow Zn^{2+} + Cu$$

As two half-reactions, it is seen that the zinc is oxidized,

$$Zn \rightarrow Zn^{2+} + 2\ e^-$$

And the copper is reduced,

$$Cu^{2+} + 2\ e^- \rightarrow Cu$$

Other Examples

- The reduction of nitrate to nitrogen in the presence of an acid (denitrification):

  $2\ NO_3^- + 10\ e^- + 12\ H^+ \rightarrow N_2 + 6\ H_2O$

- The combustion of hydrocarbons, such as in an internal combustion engine, which produces water, carbon dioxide, some partially oxidized forms such as carbon monoxide, and heat energy. Complete oxidation of materials containing carbon produces carbon dioxide.

- In organic chemistry, the stepwise oxidation of a hydrocarbon by oxygen produces water and, successively, an alcohol, an aldehyde or a ketone, a carboxylic acid, and then a peroxide.

Corrosion and Rusting

- The term corrosion refers to the electrochemical oxidation of metals in reaction with an oxidant such as oxygen. Rusting, the formation of iron oxides, is a well-known example of electrochemical corrosion; it forms as a result of the oxidation of iron metal. Common rust often refers to iron(III) oxide, formed in the following chemical reaction:

  $4Fe + 3O_2 \rightarrow 2Fe_2O_3$

Oxides, such as iron(III) oxide or rust, which consists of hydrated iron(III) oxides $Fe_2O_3 \cdot nH_2O$ and iron(III) oxide-hydroxide ($FeO(OH)$, $Fe(OH)_3$), form when oxygen combines with other elements.

- The oxidation of iron(II) to iron(III) by hydrogen peroxide in the presence of an acid:

  $Fe^{2+} \rightarrow Fe^{3+} + e^-$

  $H_2O_2 + 2e^- \rightarrow 2OH^-$

  Overall Equation:

  $2\ Fe^{2+} + H_2O_2 + 2H^+ \rightarrow 2Fe^{3+} + 2H_2O$

Iron rusting in pyrite cubes

Redox Reactions in Industry

Cathodic protection is a technique used to control the corrosion of a metal surface by making it the cathode of an electrochemical cell. A simple method of protection connects protected metal to a more easily corroded "sacrificial anode" to act as the anode. The sacrificial metal instead of the protected metal, then, corrodes. A common application of cathodic protection is in galvanized steel, in which a sacrificial coating of zinc on steel parts protects them from rust.

Oxidation is used in a wide variety of industries such as in the production of cleaning products and oxidizing ammonia to produce nitric acid, which is used in most fertilizers.

Redox reactions are the foundation of electrochemical cells, which can generate electrical energy or support electrosynthesis. Metal ores often contain metals in oxidized states such as oxides or sulfides, from which the pure metals are extracted by smelting at high temperature in the presence of a reducing agent. The process of electroplating uses redox reactions to coat objects with a thin layer of a material, as in chrome-plated automotive parts, silver platingcutlery, galvanization and gold-plated jewelry.

Redox Reactions in Biology

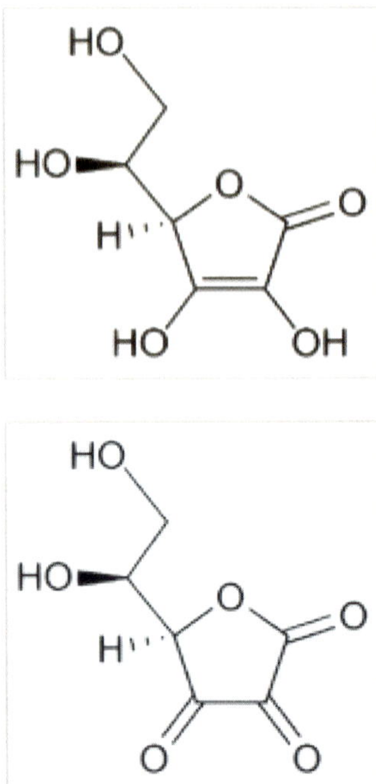

Top: ascorbic acid (reduced form of Vitamin C), Bottom: dehydroascorbic acid (oxidized form of Vitamin C).

Many important biological processes involve redox reactions.

Cellular respiration, for instance, is the oxidation of glucose ($C_6H_{12}O_6$) to CO_2 and the reduction of oxygen to water. The summary equation for cell respiration is:

$$C_6H_{12}O_6 + 6CO_2 \rightarrow 6CO_2 + 6H_2O$$

The process of cell respiration also depends heavily on the reduction of NAD^+ to NADH and the reverse reaction (the oxidation of NADH to NAD^+). Photosynthesis and cellular respiration are complementary, but photosynthesis is not the reverse of the redox reaction in cell respiration:

$$6CO_2 + 6H_2O + \text{light energy} \rightarrow C_6H_{12}O_6 + 6O_2$$

Biological energy is frequently stored and released by means of redox reactions. Photosynthesis involves the reduction of carbon dioxide into sugars and the oxidation of water into molecular oxygen. The reverse reaction, respiration, oxidizes sugars to produce carbon dioxide and water. As intermediate steps, the reduced carbon compounds are used to reduce nicotinamide adenine dinucleotide (NAD^+) to NADH, which then contributes to the creation of a proton gradient, which drives the synthesis of adenosine triphosphate (ATP) and is maintained by the reduction of oxygen. In animal cells, mitochondria perform similar functions.

Free radical reactions are redox reactions that occur as a part of homeostasis and killing microorganisms, where an electron detaches from a molecule and then reattaches almost instantaneously. Free radicals are a part of redox molecules and can become harmful to the human body if they do not reattach to the redox molecule or an antioxidant. Unsatisfied free radicals can spur the mutation of cells they encounter and are, thus, causes of cancer.

The term redox state is often used to describe the balance of GSH/GSSG, NAD^+/NADH and $NADP^+$/NADPH in a biological system such as a cell or organ. The redox state is reflected in the balance of several sets of metabolites (e.g., lactate and pyruvate, beta-hydroxybutyrate, and acetoacetate), whose inter conversion is dependent on these ratios. An abnormal redox state can develop in a variety of deleterious situations, such as hypoxia, shock, and sepsis. Redox mechanism also control some cellular processes. Redox proteins and their genes must be co-located for redox regulation according to the CoRR hypothesis for the function of DNA in mitochondria and chloroplasts.

Redox Cycling

A wide variety of aromatic compounds are enzymatically reduced to form free radicals that contain one more electron than their parent compounds. In general, the electron donor is any of a wide variety of flavoenzymes and their coenzymes. Once formed, these anion free radicals reduce molecular oxygen to superoxide, and regenerate the unchanged parent compound. The net reaction is the oxidation of the flavoenzyme's coenzymes and the reduction of molecular oxygen to form superoxide. This catalytic behavior has been described as a futile cycle or redox cycling.

Redox Reactions in Geology

In geology, redox is important to both the formation of minerals and the mobilization of minerals, and is also important in some depositional environments. In general, the redox state of most rocks can be seen in the color of the rock. The rock forms in oxidizing conditions, giving it a red color. It is then "bleached" to a green or sometimes white form when a reducing fluid passes through the rock. The reduced fluid can also carry uranium-bearing minerals. Famous examples of redox conditions affecting geological processes include uranium deposits and Moqui marbles.

Mi Vida uranium mine, near Moab, Utah. The alternating red and white/green bands of sandstone correspond to oxidized and reduced conditions in groundwater redox chemistry.

Balancing Redox Reactions

Describing the overall electrochemical reaction for a redox process requires a balancing of the component half-reactions for oxidation and reduction. In general, for reactions in aqueous solution, this involves adding H^+, OH^-, H_2O, and electrons to compensate for the oxidation changes.

Acidic Media

In acidic media, H^+ ions and water are added to half-reactions to balance the overall reaction.

For instance, when manganese(II) reacts with sodium bismuthate:

$$\textit{Unbalanced reaction}: \quad Mn^{2+}(aq) + NaBiO_3(s) \rightarrow Bi^{3+}(aq) + MnO^-_4(aq)$$

$$\textit{Oxidation}: \quad 4H_2O(l) + Mn^{2+}(aq) \rightarrow MnO^-_4(aq) + 8H^+(aq) + 5e^-$$

$$\textit{Reduction}: \quad 2e^- + 6H^+ + BiO^-_3(s) \rightarrow Bi^{3+}(aq) + 3H_2O(l)$$

The reaction is balanced by scaling the two half-cell reactions to involve the same number of electrons (multiplying the oxidation reaction by the number of electrons in the reduction step and vice versa):

- $8H_2O(l) + 2Mn^{2+}(aq) \rightarrow 2MnO_4^-(aq) + 16H^+(aq) + 10e^-$
- $10e^- + 30H^+ + 5BiO_3^-(s) \rightarrow 5Bi^{3+}(aq) + 15H_2O(l)$

Adding these two reactions eliminates the electrons terms and yields the balanced reaction:

$$14\,H^+(aq) + 2Mn^{2+}(aq) + 5NaBiO_3(s) \rightarrow 7H_2O(l) + 2MnO_4^-(aq) + 5Bi^{3+}(aq) + 5Na^+(aq)$$

Basic Media

In basic media, OH^- ions and water are added to half reactions to balance the overall reaction.

For example, in the reaction between potassium permanganate and sodium sulfite:

$$\textit{Unbalanced reaction:}\quad KMnO_4 + Na_2SO_3 + H_2O \rightarrow MnO_2 + Na_2SO_4 + KOH$$

$$\textit{Reduction:}\quad 3e^- + 2H_2O + MnO_4^- \rightarrow MnO_2 + 4OH^-$$

$$\textit{Oxidation:}\quad 2OH^- + SO_3^{2-} \rightarrow SO_4^{2-} + H_2O + 2e^-$$

Balancing the number of electrons in the two half-cell reactions gives:

$$6e^- + 4H_2O + 2MnO_4^- \rightarrow 2MnO_2 + 8OH^-$$

$$6OH^- + 3SO_3^{2-} \rightarrow 3SO_4^{2-} + 3H_2O + 6e^-$$

Adding these two half-cell reactions together gives the balanced equation:

$$2KMnO_4 + 3Na_2SO_3 + H_2O \rightarrow 2MnO_2 + 3Na_2SO_4 + 2KOH$$

Mnemonics

The key terms involved in redox are often confusing. For example, a reagent that is oxidized loses electrons; however, that reagent is referred to as the reducing agent. Likewise, a reagent that is reduced gains electrons and is referred to as the oxidizing agent. These mnemonics are commonly used by students to help memorise the terminology:

- "OIL RIG": Oxidation is loss of electrons, reduction is gain of electrons.
- "LEO the lion says GER": Loss of electrons is oxidation, gain of electrons is reduction.
- "LEORA says GEROA": Loss of electrons is oxidation (reducing agent), gain of electrons is reduction (oxidizing agent).
- "RED CAT" and "AN OX", or "AnOx RedCat" ("an ox-red cat"): Reduction occurs at the cathode and the anode is for oxidation.
- "RED CAT gains what AN OX loses": Reduction at the cathode gains (electrons) what anode oxidation loses (electrons).

Substitution Reaction

Substitution reaction is a class of chemical reactions in which an atom, ion, or group of atoms or ions in a molecule is replaced by another atom, ion, or group. An example is the reaction in which the chlorine atom in the chloromethane molecule is displaced by the hydroxide ion, forming methanol:

$$CH_3Cl + {}^-OH \rightarrow CH_3OH + Cl^-$$

If the chlorine atom is displaced by other groups—such as the cyanide ion $({}^-CN)$, the ethoxide ion ($C_2H_5O^-$), or the hydrosulfide ion (HS^-)—chloromethane is transformed, respectively, to acetonitrile (CH_3CN), methyl ethyl ether ($CH_3OC_2H_5$), or methanethiol (CH_3SH). Thus an organic compound such as an alkyl halide can give rise to numerous types of organic compounds by substitution reactions with suitable reagents.

Substitution Reaction Conditions

In order to substitution reaction to occur there are certain conditions that have to be used. They are:

- Maintaining low temperatures such as room temperature;
- The strong base such as NaOH has to be in dilute form. Suppose if the base is of higher; concentration, there are chances of dehydrohalogenation taking place;
- The solution needs to be in an aqueous state such as water.

Substitution Reactions- Types

Substitution Reactions are of two types naming nucleophilic reaction and electrophilic reactions. These two types of reactions mainly differ in the kind of atom which is attached to its original molecule. In the nucleophilic reactions the atom is said to be electron-rich species, whereas, in the electrophilic reaction, the atom is an electron-deficient species. A brief explanation of the two types of reactions are as given below:

- Nucleophilic Substitution Reaction

Firstly what are nucleophiles? Nucleophiles are those species in the form of ion or a molecule which are strongly attached to the region of a positive charge. These are said to be fully charged or have negative ions present on a molecule. The common examples of nucleophiles are cyanide ions, water, hydroxide ions, and ammonia.

A Nucleophilic substitution reaction in organic chemistry is a type of reaction where a nucleophile gets attached to the positive charged atoms or molecules of the other substance. A good example of a nucleophilic substitution reaction is the hydrolysis of alkyl bromide (R-Br), under the basic conditions, wherein the nucleophile is nothing but the base OH−, whereas the leaving group is the Br^-. The reaction for the following is as given below $\mathbf{R-Br + OH^- ------ R-OH + Br^-}$

Nucleophilic reactions are as important in the field of organic chemistry, and these reactions are broadly classified as to occur at the place of a carbon atom of a saturated aliphatic carbon compound.

- Electrophilic substitution reactions

The electrophilic substitution reaction involves the electrophiles. Electrophiles are those which donate a pair of electrons in the formation of a covalent bond. The Electrophilic reactions occur mostly with the aromatic compounds. These compounds have about an excess of electrons that can be shared throughout the system of reaction.

The Electrophilic substitution reactions are basically defined as those chemical reactions where the electrophile replaces the functional group in a compound but not the hydrogen atom. Some of the examples of species of electrophiles include hydronium ion (H_3O^+), halides of hydrogen such as HCl, HBr, HI, sulfur trioxide (SO_3), the nitronium ion (NO^{2+}), etc.

The other types of substitution reactions include radical reactions and organometallic substitution reaction.

References

- Heathcock, C. H. (1991). «The Aldol Reaction: Acid and General Base Catalysis". In Trost, B. M.; Fleming, I. Comprehensive Organic Synthesis. 2. Elsevier Science. pp. 133–179. doi:10.1016/B978-0-08-052349-1.00027-5. ISBN 978-0-08-052349-1
- Addition-reaction: britannica.com: Retrieved 15 March 2018
- Oxidation-reduction-reaction, science: britannica.com, Retrieved 25 May 2018
- Hajos, Zoltan G.; Parrish, David R. (1974). "Asymmetric synthesis of bicyclic intermediates of natural product chemistry". Journal of Organic Chemistry. 39 (12): 1615–1621. doi:10.1021/jo00925a003
- Aldol-addition-mechanism-reaction, academy: study.com, Retrieved 18 July 2018
- Substitution-reaction, science: britannica.com, Retrieved 11 July 2018
- Carreira, E. M.; Fettes, A.; Martl, C. (2006). "Catalytic Enantioselective Aldol Addition Reactions". Org. React. 67: 1–216. doi:10.1002/0471264180.or067.01. ISBN 0471264180
- Electrophilic-addition, organic-chemistry: tutorvista.com, Retrieved 31 March 2018
- Substitution-reaction, chemistry: byjus.com, Retrieved 22 June 2018

5

Organic Synthesis

The synthesis of organic compounds is under the domain of organic synthesis. In order to completely understand organic synthesis, it is necessary to understand the processes related to it, such as Achmatowicz reaction, carbohydrate synthesis, Wöhler synthesis and retrosynthesis. The following chapter elucidates these varied processes and mechanisms associated with this area of study.

Organic synthesis is the study of how we build molecules ranging from complex, biologically active natural products to new materials. Because synthesis allows a chemist to construct entirely new structures, it empowers chemists to probe the world around them in new, creative ways.

A multi-step synthesis of any organic compound requires the chemist to accomplish three related tasks:

1. Constructing the carbon framework or skeleton of the desired molecule.
2. Introducing, removing or transforming functional groups in a fashion that achieves the functionality of the desired compound.
3. Exercising selective stereo control at all stages in which centres of stereoisomerism are created or influenced.

   These are not discrete independent tasks to be attacked and solved in turn, but must be integrated and correlated in an overall plan. Thus, the assembly of the molecular framework will depend in part on the structure and functionality of available starting materials, the selectivity (region and stereo) of the various reactions that may be used to stitch them together, and the loss or relocation of functional groups in the intermediate compounds formed on the way to the final product.

Other factors must also be considered, always in context with those listed above:

4. Since a successful synthesis must produce the desired product in reasonable amount, it should be as short and efficient as possible. A two or three-step sequence is usually better than a six or seven step procedure, even if the individual step yields are better in the longer route. Most reactions do not proceed in 100% yield, and the losses are multiplied with each additional step. Furthermore, a long multi-step synthesis requires many hours of effort by the chemists conducting the reactions.
5. The format of a synthesis is important. Assuming a constant yield for each step, a linear sequence of reactions gives a poorer overall yield than the same number of convergent reactions, as shown in the following diagram.

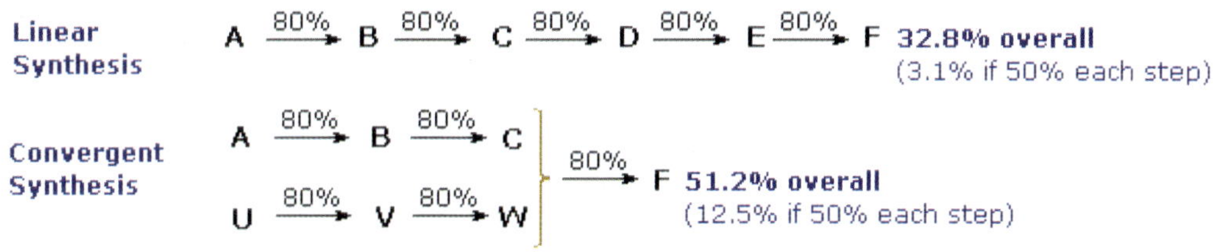

6. If numerous functional groups are present at intermediate stages, some of these may require protection from unwanted reaction. Since the use of a protective group requires its introduction and later removal, such operations can add many steps to a synthesis. A similar cost is attached to the use of blocking and activating groups.
7. The starting compounds and reagents for a synthesis must be purchased, so economic considerations are often significant. Recovery or recycling of expensive reagents and catalysts is often desirable. Also, if large amounts of useless by-products are formed in the synthesis, the expense of their disposal is a factor. The term atom efficiency has been coined to reflect the latter point.
8. For a given target molecule, several different reaction sequences may serve to accomplish its synthesis. Indeed, new synthetic routes to known compounds continue to be reported, particularly as new reactions are developed and applied to difficult transformations.

For all but the simplest of these, a majority of the reactions in the synthesis involve functional group modification, preceding or following a smaller number of carbon-carbon bond forming reactions. Because functional group chemistry consists of such a vast number of addition, elimination and substitution interconversions, it is not possible to identify a general pattern in their application to synthesis.

Total Synthesis

Total synthesis is the complete chemical synthesis of a complex molecule, often a natural product, from simple, commercially available precursors. It usually refers to a process not involving the aid of biological processes, which distinguishes it from semisynthesis. The target molecules can be natural products, medicinally important active ingredients, or organic compounds of theoretical interest. Often the aim is to discover new route of synthesis for a target molecule for which there already exist known routes. Sometimes no route exists and the chemist wishes to find a viable route for the first time. One important purpose of total synthesis is the discovery of new chemical reactions and new chemical reagents.

Scope

The moniker of total synthesis is less frequently, but nevertheless accurately applied to the synthesis of natural polypeptides and polynucleotides; for instance, the peptide hormones oxytocin and vasopressin were isolated.

Aims

Although untrue from a historical perspective total synthesis in the modern age has largely been an academic endeavour (in terms of manpower applied to problems), although industrial concerns may pick up particular avenues of total synthesis efforts, and expend considerable resources on particular natural product targets, especially in cases where semi-synthesis can be applied to complex, natural product-derived drugs.

Total synthesis projects often require a variety of reactions, and so efforts to achieve complex total syntheses serve to prepare chemists for pursuits in pharmaceutical discovery chemistry, in particular, as well as in process chemistry, in both cases, where comprehensive knowledge of chemical reactions and a strong and accurate chemical intuition are important qualifications.

Example

One classic in total synthesis is quinine total synthesis, which, before its total synthesis by Robert Burns Woodward and William von Eggers Doering in 1944, had a history of many partial syntheses that spanned 150 years and included disputes and frustration.

Achmatowicz Reaction

The Achmatowicz reaction converts furfuryl alcohols into substituted dihydropyranone acetals (also identified as pyranuloses), which are otherwise poorly accessible, through an oxidative rearrangement process. This reaction exploits furan-2-yl carbinols as substrates, which undergo oxidative ring expansion to smoothly provide 6-hydroxy-2*H*-pyran-2(3*H*)-ones). The reaction was initially employed in the conversion of dihydropyranones to methylglycosides and carbohydrates. Subsequently, this oxidative ring expansion strategy was further developed into a synthetically valuable transformation, particularly for the preparation of functionalized pyranones and substituted tetrahydropyrans. The Achmatowicz reaction can incorporate two chiral centers. Many asymmetric Achmatowicz reaction variants have also been developed. This has provided access to pyranones and tetrahydropyrans diastereoselectively and in optically active form. These heterocyclic templates have been extensively utilized in the synthesis of bioactive natural products, carbohydrates, and medicinal agents. Also, the Achmatowicz variant for the construction of piperidone derivatives starting from furan-2-yl amines, namely, the aza- Achmatowicz reaction was accomplished. The potential of Achmatowicz-type reactions, however, has not been fully exploited in synthesis.

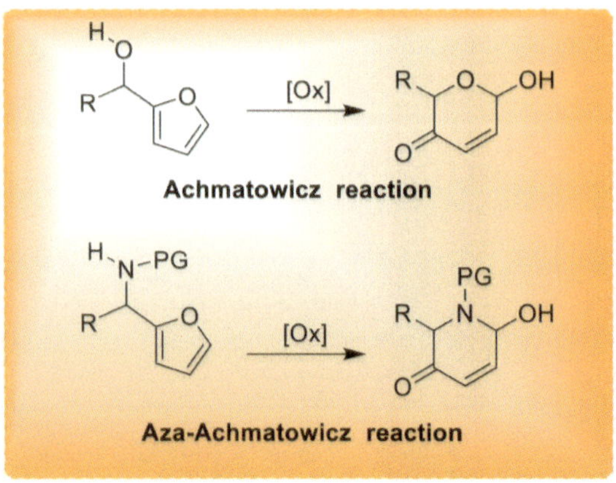

General outline of Achmatowicz and aza-Achmatowicz reactions.

Carbohydrate Synthesis

Carbohydrate synthesis may be a sub-field of chemical science involved specifically with the generation of natural and unnatural macromolecule structures. This may embody the synthesis of saccharide residues or structures containing over one saccharide, called oligosaccharides.

Chemical Synthesis of Oligosaccharides

In order to synthesize oligosaccharides with defined structures, iterative *f*--glycosylation reactions are required, where free OH-carrying sector 1 called glycosyl acceptor is coupled with glycosyl donor 2 to afford *O*-glycoside 3. Active investigations in the last two decades resulted in development of the socalled modern glycosylation technology that utilizes non-classical glycosyl donors 2 [X=F, OC(NH)CCl_3, SR, S(o)Ph, $OP(OR)_2$, etc.] which are relatively stable but can be activated into highly reactive glycosylating entities, under mild reaction conditions. For example, glycosyl fluoride, which used to attract little attention in synthetic carbohydrate chemistry, has been proved as a powerful glycosyl donor. In 1981, Mukaiyama and coworkers demonstrated that it can be activated by combined use of $AgClo_4$ and $SnCl_2$ to afford α-glycoside with a significant level of stereoselectivity. The utility of this system was quickly demonstrated in synthesis of glycolipids and natural products. Further refinement was achieved by Suzuki and coworkers, who introduced a combination of Cp_2MCl_2 (M=Ti, Zr, Hf)-Ag^+ as an activator and made glycosyl fluoride as one of the most strong glycosyl donors.

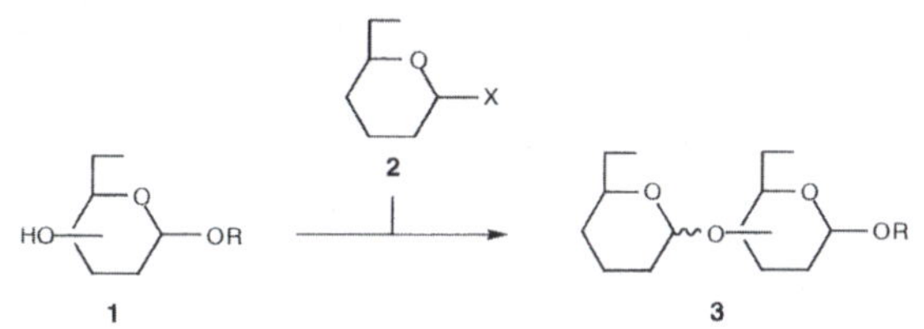

X	Promoter
Cl, Br	Hg(II), Ag(I), R_4NBr ..
F	Ag(I)-$SnCl_2$ (Cp_2HfCl_2), TMSOTf, BF_3, Tf_2O, Me_2GaX, $Ln(ClO_4)_3$ ..
OCOR OC(NH)CCl_3 OP(O)$(OR)_2$ $OP(OR)_2$	TMSOTf, BF_3, $Sn(OTf)_2$, $ZnCl_2$ ...
SR	MeOTf, $MeSSMe_2OTf$, NIS-TfOH, PhSeOTf, MeSOTf ..
S(O)Ph	Tf_2O

Meanwhile, hard Lewis acids are also demonstrated to be effective to promote the activation of glycosyl fluorides. Likewise, thioglycoside and trichloroacetimidate are now recognized to be highly effective glycosyl donors in oligosaccharide synthesis. Combination of these methodologies allows us to design synthetic routes to even more highly complex oligosaccharide structures. This aspect was clearly exemplified in our synthesis of blanched polylactosamine type oligosaccharide which consists of as many as 25 sugar residues.

Although such advancement has dramatically improved the overall efficiency of oligosaccharide synthesis, these modern methodologies are still not free of stereochemical problems inherent in o-glycosylation reactions. Among o-glycosides found in naturally occurring glycoconjugates aglycoside of sialic acid and 13-glycoside of mannose are considered to be "difficult sequences", because neither stereoelectronic control nor neighboring effect is available to control the stereochemistry in these particular cases.

Stereoselective Synthesis of NeuAc Glycoside

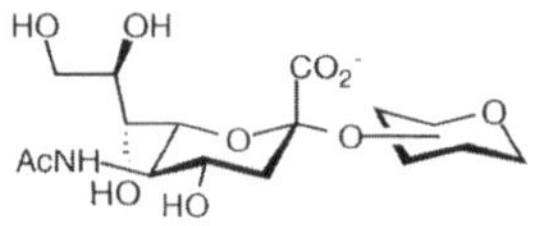

4

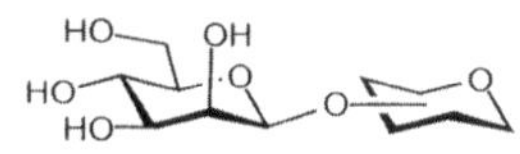

5

At the outset of our investigation on this subject, we assumed proper functionalization of the NeuAc donor is required. As shown in figure below, NeuAc donors having a PhSe or PhS group at the C-3 position were designed and synthesized. Having the neighboring participation active functional group as an auxiliary, glycosylation should proceed via the episelenonium or episulfonium ion to afford the desired a-glycoside. Requisite stereochemical control at the C-3 position can be easily obtained by having hemiketal 6 as an intermediate, because this position can be epimerized into thermodynamically more stable (and desired) 7, irrespective of the stereochemical outcome of the preceding addition reaction. In this sense, our synthetic scheme can be viewed as translating thermodynamically disfavored equatorial glycoside into thermodynamically favored C-3 equatorial stereochemistry. NeuAc donors 8a,b afforded a-glycosides in a highly stereoselective manner, and in an unprecedentedly high yield in the case that 8a was utilized. In particular, stereoselective synthesis of the NeuAcα2→48NeuAc sequence was realized for the first time and the first chemical synthesis of disialoganglioside GD3 was achieved. Subsequently, a similar strategy was reported from other groups, and it was demonstrated by Magnusson and ourselves that incorporation of thioglycosiclde functionality as a leaving group further improves the glycosylation efficiency.

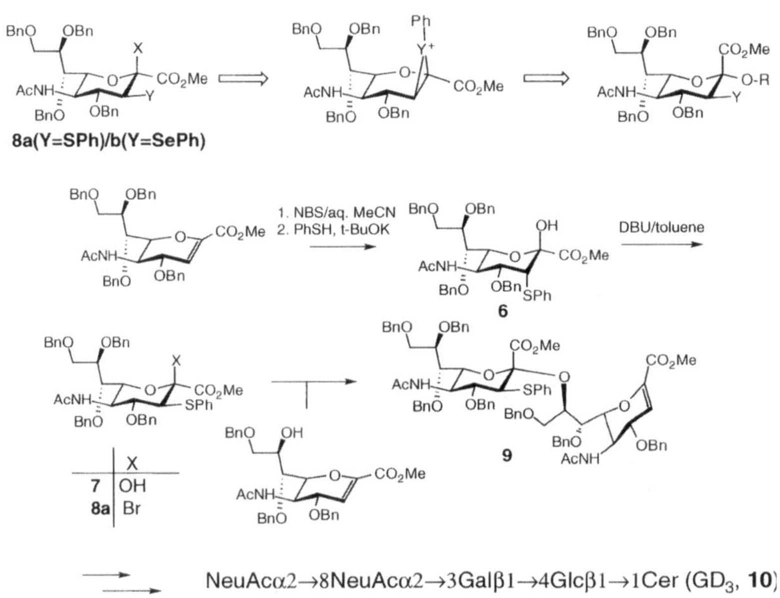

NeuAcα2→8NeuAcα2→3Galβ1→4Glcβ1→1Cer (GD_3, **10**)

Stereoselective Synthesis of NeuAc Glycoside

Meanwhile, a research team lead by Hasegawa and Kiso first demonstrated that glycosylation with NeuAc thioglycoside 12 gives α-glycoside with substantial stereo selectivity when it was conducted in acetonitrile. They have extensively applied this protocol to systematic synthetic studies of various gangliosides.

11 12

Stereoselective Formation of 13-manno Glycoside

β-manno glycoside exists as a central core unit of all types of asparagine (Asn)-linked glycoprotein oligosaccharides. Therefore, stereoselective synthesis of this particular type of glycoside is of fundamental significance aiming at chemical synthesis of glycoprotein-related molecules. Our approach depicted in figure below is based upon the concept called intramolecular aglycon delivery (IAD) which was first developed by Baressi and Hindsgaul and later by Stork and coworkers. In our version of IAD, mannosyl donor 13 that has a p-methoxybenzyl group (PMB) was utilized. The p-methoxybenzyl group has been extensively investigated as a valuable OH protecting group which is removable under nearly neutral conditions by the action of a certain oxidant, most typically DDQ. Since deprotection is believed to proceed through a hydrolytic quench of the quinonemethide-like intermediate 14, it was assumed treatment of 13 with alcohol (glycosyl acceptor) would give us mixed acetal 16 which serves as a tethered intermediate for IAD.

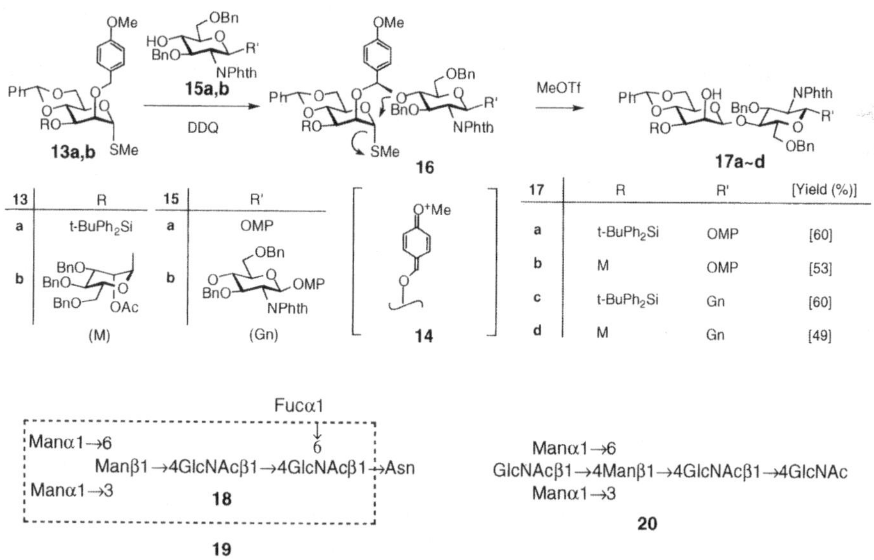

β-manno glycosylation via mixed Acetal

According to this expectation, PMB-carrying mannosyl donors 13a,b were prepared and subjected to a two-step sequence (1. DDQ, molecular sieves 4A/CH_2C_{12}; 2. MeoTf, DBMP/Cl(CH2)2Cl) using 15a,b as glycosyl acceptors. In all cases, the desired p-glycoside products 17a-d were obtained as single isomers. Even p-mannosylations of disaccharide with a trisaccharide donor was successful. These products were further converted into the core pentasaccharide structure 18 commonto

Asn-linked glycans. Likewise, fucose-containing hexasaccharidic asparagine 19, as well as"bisecting" GlcNAc-containing hexasaccharide 20, both of which stand for the most significant structural modifications found in Asn-linked glycans, were achieved.

More recently, the efficiency of β -mannosylation was further improved by using a 4,6-o- cyclohexylidene carrying mannosyl donor 21, which gives β -manno glycoside in >80% yield

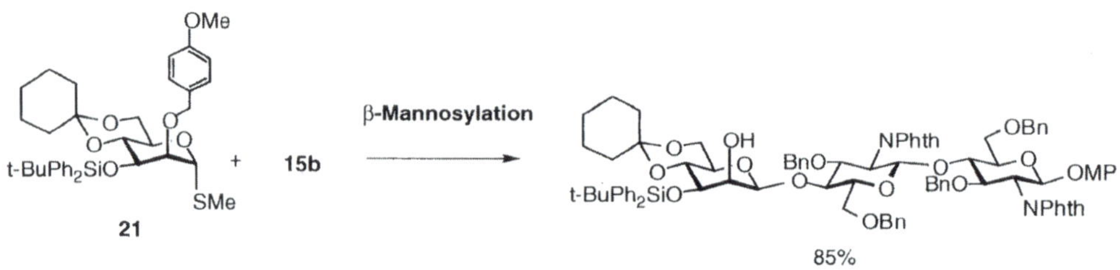

Wöhler Synthesis

In chemistry, we often separate the study of organic and inorganic compounds. This is because the types of reactions and how they react tend to be really different between the two fields of chemistry. They are so different, in fact, that prior to the 1820s scientists believed that there was something fundamentally different between organic and inorganic compounds. They believed that the only way that something could become organic was through a 'vital force' which was found in living creatures. This entire belief was completely dispelled when Friedrich Wohler created urea from non-organic compounds.

Wohler was not attempting to dispel the vital force theory during this experiment, he was simply trying to make ammonium cyanate. When he ended up making urea it dispelled the vital force theory. This was an important experiment which led to the start of organic chemistry as we know it today.

Urea is a great fertilizer, and comes from the urine of animals. It contains a carbon, doubled bonded to an oxygen, and bound to two nitrogen atoms, the nitrogen atoms each have two hydrogen atoms.

The Mechanism

The first step in this reaction takes lead cyanate (silver or potassium cyanate can also be used) and reacts it with ammonia and water. Wohler expected the ammonia to form a salt with the cyanate, forming ammonia cyanate. The first step of this reaction occurred just as he expected it, with the salts rearranging to form ammonia cyanate:

$$NH_4(OCN) \rightarrow NH_3 + HOCN \rightleftharpoons (NH_2)_2CO$$

One such reaction used lead cyanate and ammonia in aqueous solution:

$$Pb(OCN)_2 + 2NH_3 + 2H_2O \rightarrow Pb(OH)_2 + 2NH_4(OCN)$$

Retrosynthesis

Retrosynthetic analysis (retrosynthesis) is a technique for planning a synthesis, especially of complex organic molecules, whereby the complex target molecule (TM) is reduced into a sequence of progressively simpler structures along a pathway which ultimately leads to the identification of a simple or commercially available starting material (SM) from which a chemical synthesis can then be developed. Retrosynthetic analysis is based on known reactions (e.g the Wittig reaction, oxidation, reduction etc). The synthetic plan generated from the retrosynthetic analysis will be the roadmap to guide the synthesis of the target molecule.

Synthetic Planning

Synthesis is a construction process that involves converting simple or commercially available molecules into complex molecules using specific reagents associated with known reactions in the retrosynthetic scheme.

Retrosynthetic Scheme

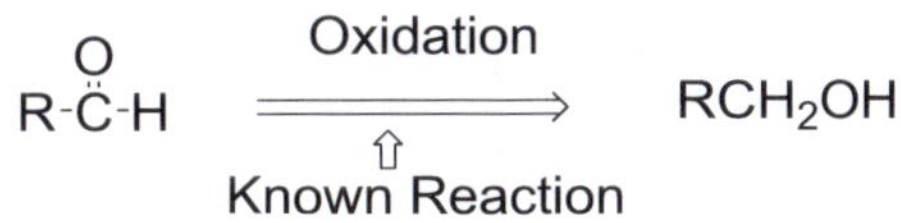

Synthetic Scheme

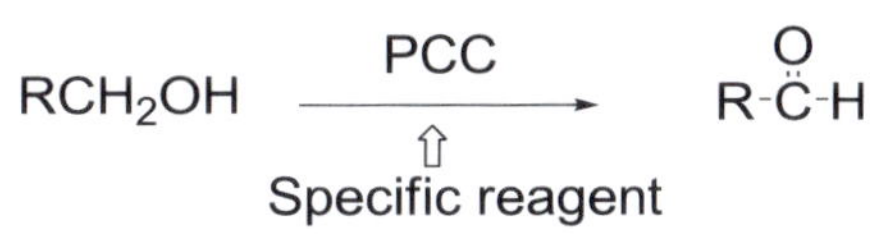

Syntheses can be grouped into two broad categories:

i) Linear syntheses

ii) Convergent syntheses

Linear Synthesis

In linear synthesis, the target molecule is synthesized through a series of linear transformations.

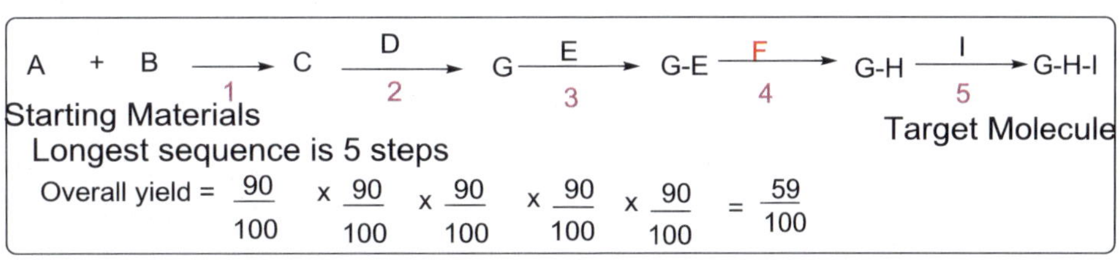

Since the overall yield of the synthesis is based on the single longest route to the target molecule, by being long, a linear synthesis suffers a lower overall yield.

The linear synthesis is fraught with failure for its lack of flexibility leading to potential large losses in the material already invested in the synthesis at the time of failure.

Convergent Synthesis

In convergent synthesis, key fragments of the target molecule are synthesized separately or independently and then brought together at a later stage in the synthesis to make the target molecule.

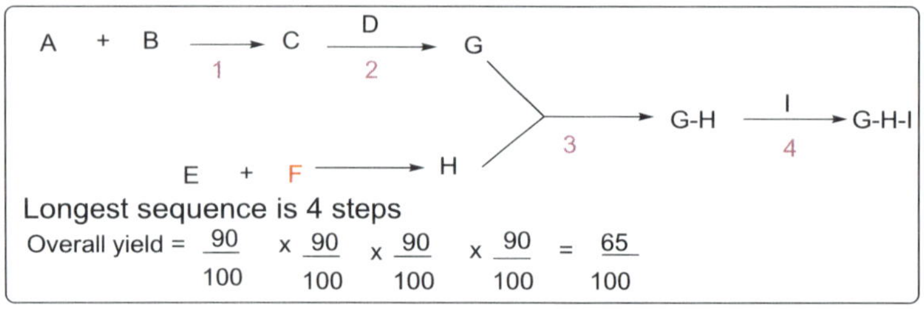

A convergent synthesis is shorter and more efficient than a linear synthesis leading to a higher overall yield.

It is flexible and easier to execute due to the independent synthesis of the fragments of the target molecule.

Retrosynthetic Analysis

Consider the retrosynthetic analysis of Muscalure, the pheromone of the domestic housefly, to illustrate the concept of retrosynthetic analysis:

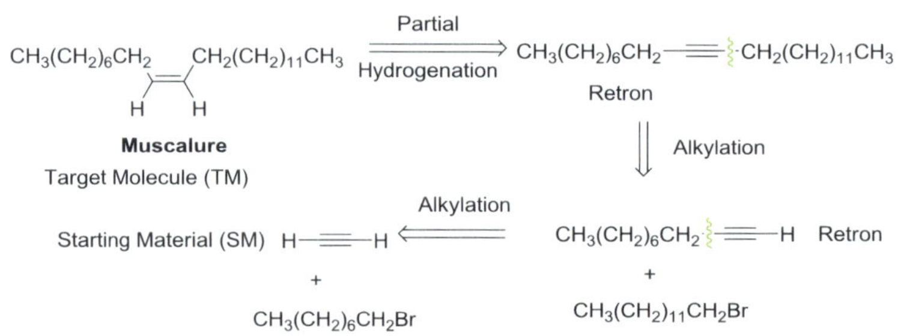

The retrosynthetic analysis of any target molecule must be based on known chemical reactions for it to stand a realistic chance of being translated to a chemical synthesis.

Muscalure can also be deconstructed based on the Wittig reaction.

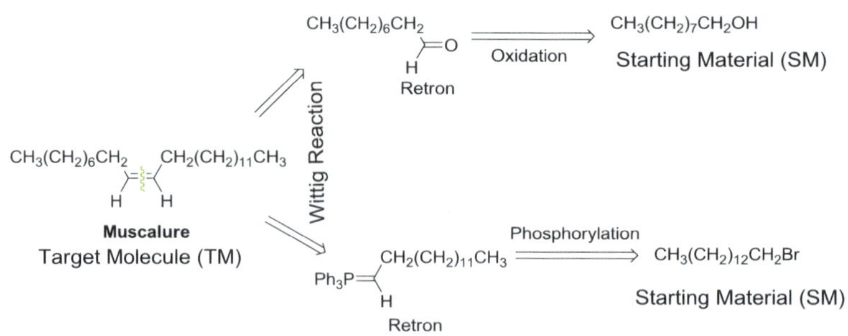

A thorough evaluation of the alternative retrosynthetic pathways should be made to identify the pathway that best stands a realistic chance of being translated to an efficient chemical synthesis.

When evaluating the various retrosynthetic schemes of the target molecule, it is important to recognize the strategic features of the molecule that the synthesis must address. Consider the chemical syntheses of Muscalure to appreciate how the two retrosynthetic plans fit in, but most importantly, note the use of specific reagents that transform the intermediates in the retrosynthetic scheme eventually to the target molecule.

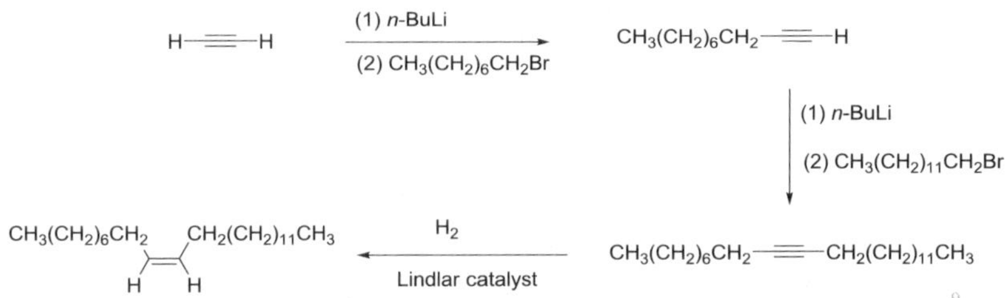

Synthesis of Muscalure via the wittig reaction:

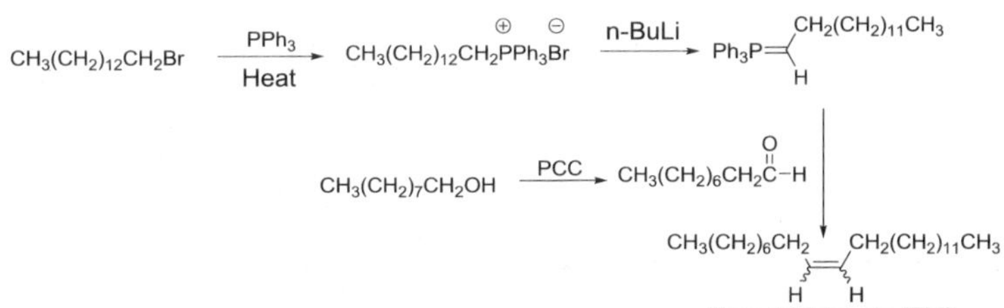

Target Molecule (TM)
(Mixture of geometric isomers)

Comparing the two synthetic pathways shows that while the wittig route is convergent, the stereochemistry of the double bond is not stereospecifically controlled.

The linear pathway of the stereospecific partial reduction of a terminal alkyne is the preferred route to Muscalure.

Retrosynthetic analysis and synthetic planning requires training (knowledge of chemistry) and experience (practical application of the chemistry).

The wider someone's knowledgebase is in organic chemistry, the more the options available to one to develop a variety of synthetic routes to a target molecule. One of these retrosynthetic pathways may turn out to be more practical and executable than the others.

A good synthetic plan should consider taking into account the advantage of a convergent synthesis, if possible, over a linear synthesis.

Terminology of Retrosynthetic Analysis

Disconnection

During retrosynthetic analysis the target molecule is systematically broken down by a combination of disconnection and functional group interconversion (FGI). The term disconnection relates to breaking a carbon-carbon bond of a molecule to generate shorter or simpler fragments. A good disconnection must achieve the greatest simplification of the target molecule. For a complex molecule, this basic disconnection process is repeated until the target is reduced to simple starting materials.

The complete set of disconnections and functional group interconversions for a specified target molecule is what constitutes a retrosynthetic pathway or plan.

Symbols of Retrosynthetic Analysis

- A disconnection is represented by a wavy (⌇) line through the bond being disconnected.
- A retrosynthetic arrow (⇒): This open arrow represents going from the target molecule "backwards" to simpler molecules (retrons).
- A synthetic arrow (→): This closed arrow represents going in the forward direction.

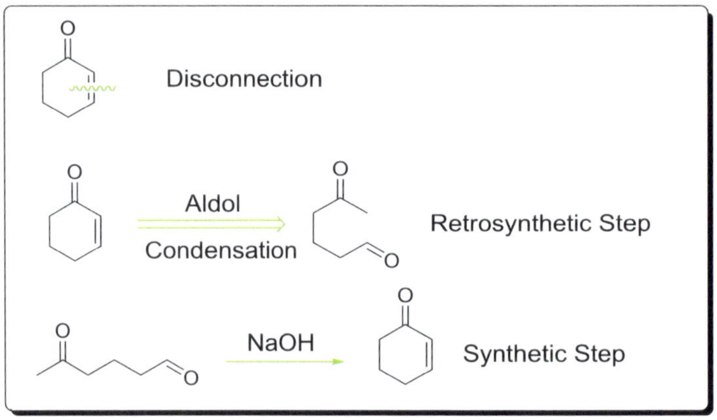

Functional Group Interconversion

Functional group interconversion (FGI) describes a process of converting one functional group to another e.g. an alcohol to an aldehyde, alkyne to alkene etc. Although FGI doesn't offer much gain to a synthesis, it sets the stage for subsequent disconnection of the intermediate. Revisit the retrosynthetic analysis of Muscalure to identify the disconnections and Functional group interconversions.

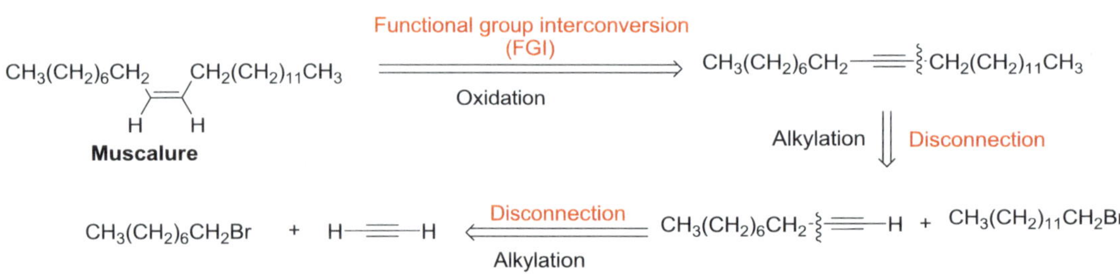

Retrosynthetic Strategy

The concept of bond polarity within functional groups is of prime importance in disconnections. The disconnection of a bond based on this innate polarity may lead to two pairs of idealized (imaginary) fragments called synthons from which a functional group may be generated.

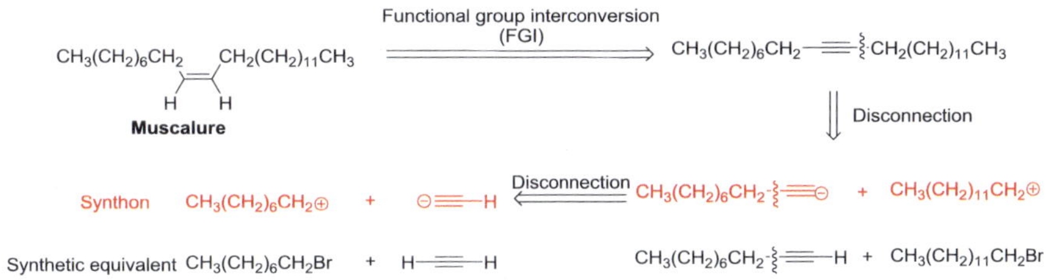

A synthon is an idealized fragment or species (e.g. $\left(CH_3^+ \text{ or } CH_3^-\right)$)generated from a bond disconnection during retrosynthetic analysis. It may not necessarily correspond to a real molecule. A synthetic equivalent is a real molecule or reagent (e.g. CH_3Br or CH_3MgBr) that can be ascribed to a synthon and can be employed in a synthetic step.

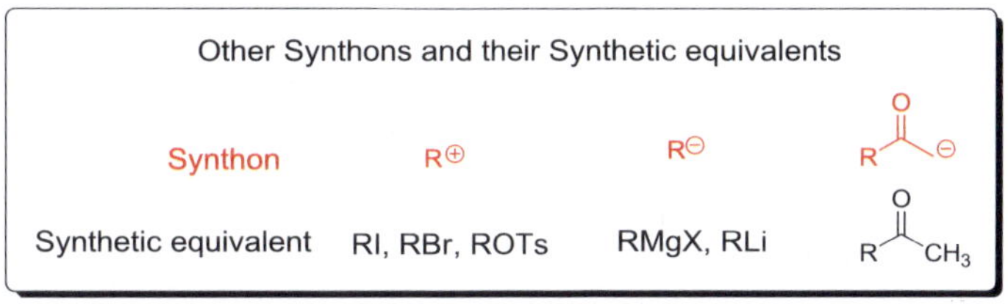

Strategies in Synthetic Planning

Strive for Success and Good Cost Management

In planning a synthesis generate a large number of retrosynthetic pathways to the target molecule: Examine these retrosynthetic pathways to identify among them an optimal synthetic route for which reagents are readily available and inexpensive.

Convergent vs Linear Synthesis

When considering a disconnection in the retrosynthetic analysis of a complex target molecule, try (if possible) to divide the molecule into halves at convenient bonds. This will make possible the formulation of a convergent synthesis with several mini-syntheses leading to the target molecule.

Disconnections that Lead to the Greatest Simplification of the Target Molecule

Given a choice of possible disconnections, those located at branch points or on rings are more strategic as they usually give straight chain fragments which are more likely to be commercially available or simply prepared.

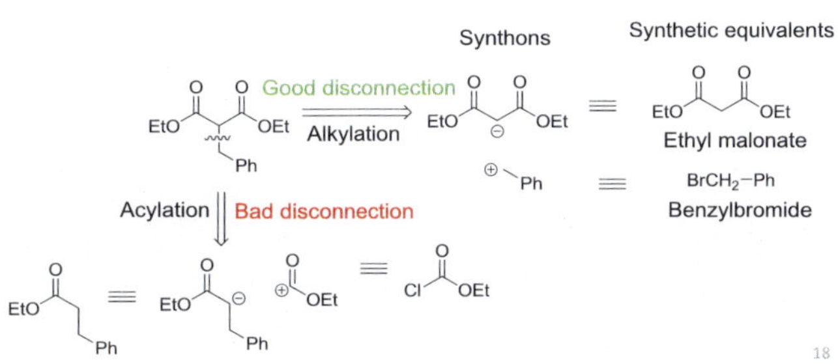

Identify and Exploit any Inherent Symmetry in a Target Molecule

Exploiting any symmetry in a TM or its intermediate can dramatically simplify its retrosynthesis. This may also provide an opportunity to identify a convergent pathway in the synthesis.

Retrosynthetic Plan

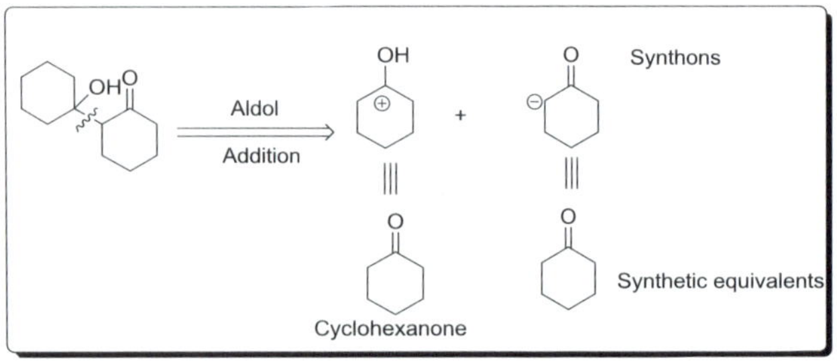

Synthetic Plan

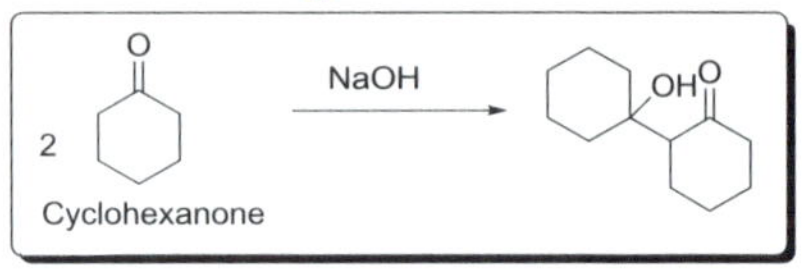

Self-Aldol Addition

Introduce Reactive Functional Groups at a Late Stage in the Synthesis

It is often difficult to selectively react at a less reactive functional group when a more reactive functionality is present within the same molecule. Such reactive functional groups are usually among the first to be disconnected during retrosynthetic analysis. The retrosynthetic analysis of 2, 4-dichlorophenoxyacetic acid (2, 4-D), a common herbicide for the control of broad leafed weeds, is shown below:

Retrosynthetic Plan

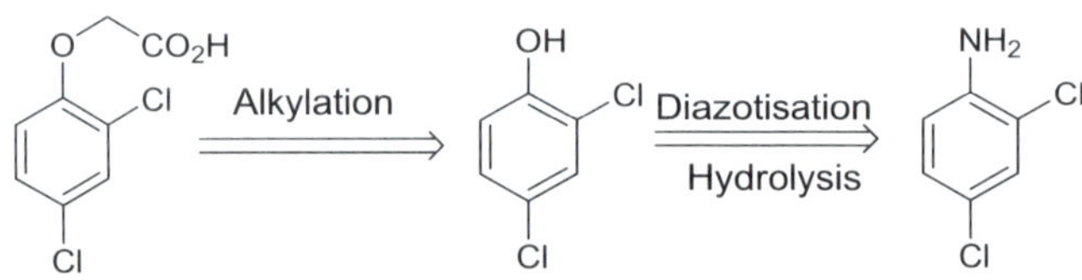

2,4-D

Strategies in Synthetic Planning Synthesis of the Weed-Killer, 2, 4-D

Based on the preceding retrosynthetic plan, 2, 4- dichlorophenoxyacetic acid (2, 4-D) can be synthesized as shown below:

Synthetic Plan

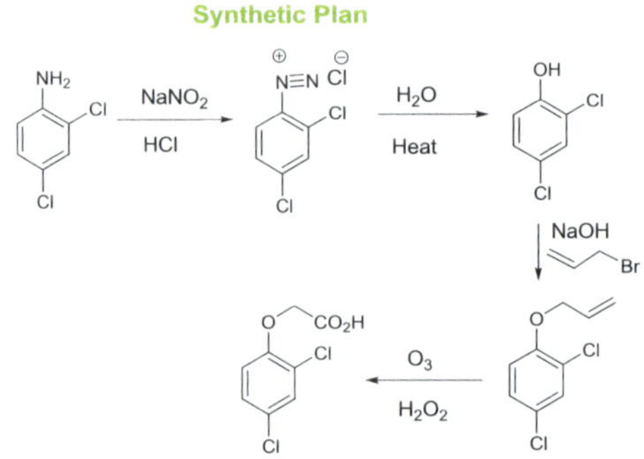

The synthesis of 2, 4-D can also be approached based on the alternative retrosynthetic and synthetic plan highlighted below:

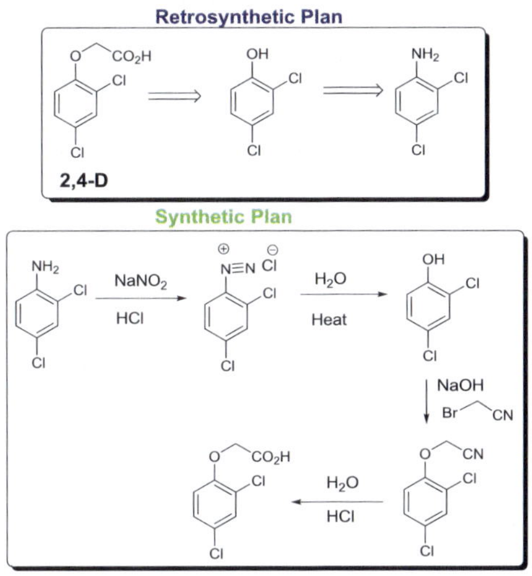

- During retrosynthetic analysis introduce additional functional groups, if necessary, to facilitate further disconnection: Functional group addition (FGA) strategy.
- The functional group addition strategy in retrosynthetic analysis involves introducing additional functional groups at strategic locations in a retron, if necessary, to guide further disconnections based on known powerful bond making reactions.
- Addition of functional groups e.g. double bonds or carbonyl groups can serve to direct reactivity to specific sites of a molecule significantly simplifying a synthesis.

Functional Group Addition Strategy

For example, one may introduce a carbonyl group in a substituted cyclohexane target molecule which may help guide introduction of a substituent through enolate alkylation.

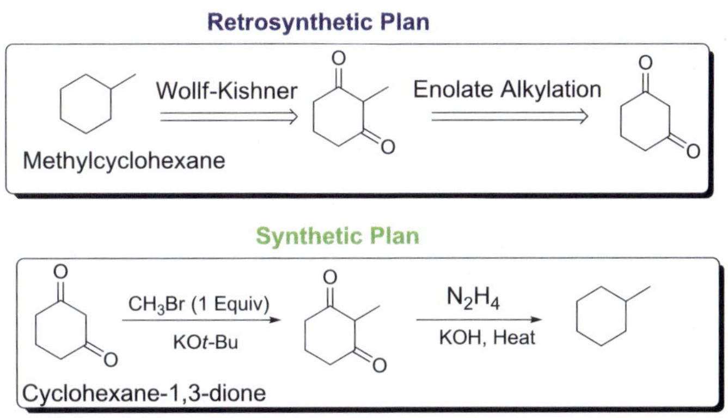

Use Protecting Groups if Inevitable

Given that the use of protecting groups adds to the number of steps of a synthesis, use them only when it is absolutely necessary.

Sample Retro Syntheses and Syntheses

Retrosynthetic Analysis of Methyl Cinnamate

Methyl cinnamate is found naturally in fruits, like strawberry. It is used in the flavor and perfume industries.

Retrosynthetic Plan

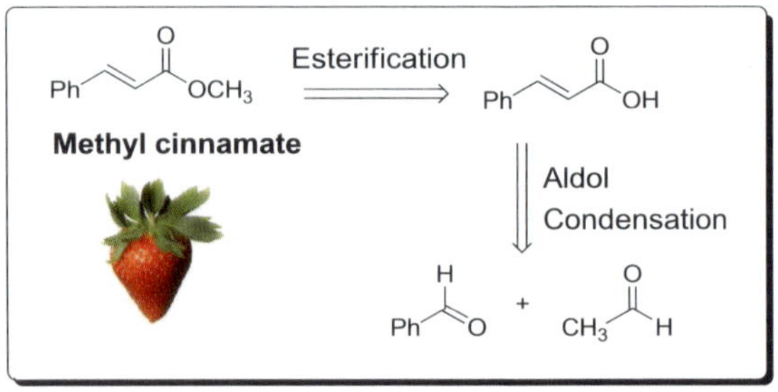

Synthetic Plan for Methyl cinnamte

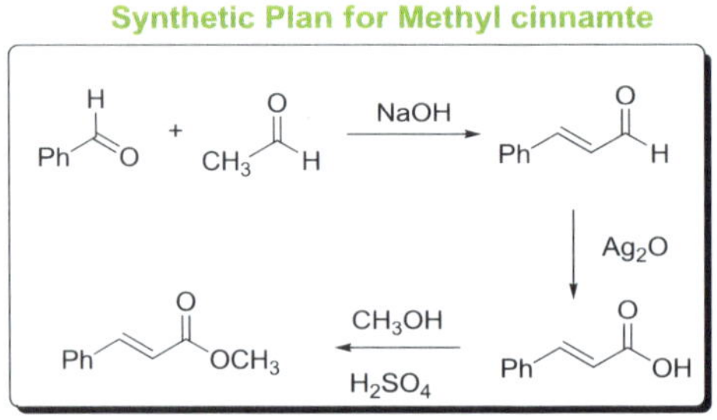

References

- Discovery of Novel Synthetic Methodologies and Reagents during Natural Product Synthesis in the Post-Palytoxin Era Ahlam M. Armaly, Yvonne C. DePorre, Emilia J. Groso, Paul S. Riehl, and Corinna S. Schindler Chem. Rev., Article ASAP doi:10.1021/acs.chemrev.5b00034
- Friedrich-wohlers-synthesis-of-urea-mechanism-experiment: study.com, Retrieved 22 June 2018
- Nicolaou, K. C. & Sorensen, E. J. 1996, Classics in Total Synthesis: Targets, Strategies, Methods, New York:-John Wiley & Sons, ISBN 978-3-527-29231-8
- Wohler-Synthesis-of-Urea-4270: mun.ca, Retrieved 29 May 2018
- du Vigneaud V, Ressler C, Swan JM, Roberts CW, Katsoyannis PG (1954). “The Synthesis of Oxytocin”. Journal of the American Chemical Society. 76 (12): 3115–3121. doi:10.1021/ja01641a004

6

Fundamental Concepts of Organic Chemistry

A complete understanding of organic chemistry requires a comprehensive study of aromaticity, covalent bonding, functional groups, heterocyclic compounds, the IUPAC nomenclature of organic chemistry, etc. which have been elaborately discussed in this chapter.

Aromaticity

Aromaticity is a property of conjugated cycloalkenes in which the stabilization of the molecule is enhanced due to the ability of the electrons in the π orbitals to delocalize. This act as framework to create a planar molecule.

Without aromatic compounds, we would not only be lacking many material necessities, our bodies would also not be able to function. Aromatic compounds are essential in industry; about 35 million tons of aromatic compounds are produced in the world every year to produce important chemicals and polymers, such as polyester and nylon. Aromatic compounds are also vital to the biochemistry of all living things. Three of the twenty amino acids used to form proteins ("the building blocks of life") are aromatic compounds and all five of the nucleotides that make up DNA and RNA sequences are all aromatic compounds. Needless to say, aromatic compounds are vital to us in many aspects.

The three general requirements for a compound to be aromatic are:

- The compound must be cyclic;
- Each element within the ring must have a p-orbital that is *perpendicular* to the ring, hence the molecule is planar;
- The compound must follow Hückel's Rule (the ring has to contain 4n+2 p-orbital electrons).

Among the many distinctive features of benzene, its aromaticity is the major contributor to why it is so unreactive. This topic will try to clarify the theory of aromaticity and why aromaticity gives unique qualities that make these conjugated alkenes inert to compounds such as Br_2 and even hydrochloric acid. It will also go into detail about the unusually large resonance energy due to the six conjugated carbons of benzene.

The delocalization of the p-orbital carbons on the sp^2 hybridized carbons is what gives the aromatic qualities of benzene.

Cyclic

Cycle or cyclic implies a ring. A cyclic molecule is one in which the first and last carbon in the chain are bound to each other.

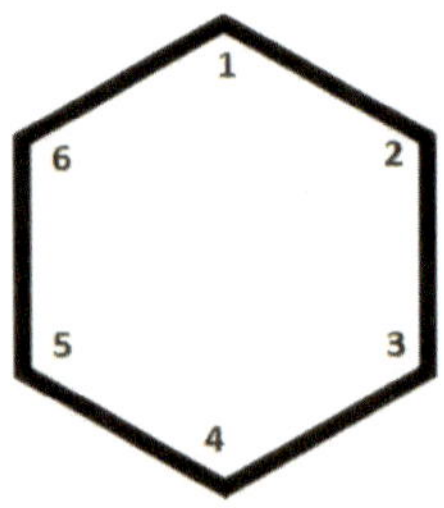

cyclohexane

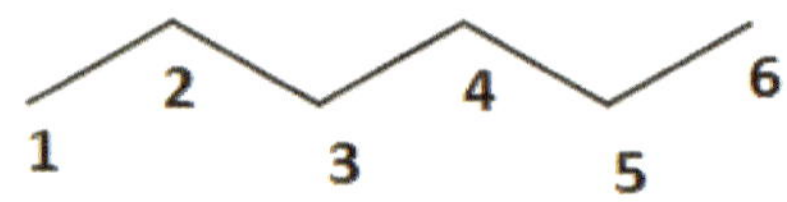

hexane

Carbon-1 is attached to carbon-2 but also attached to C-6; In hexane C-1 and C-6 are not attached to each other making it a linear compound.

Planar

Since aromaticity relies on the ability of orbitals to overlap, the ring must have its atoms in the same plane.

In short, the molecule must be flat.

Huckel's Rule: 4n+2 = Number of Resonating Electrons

$$4n+2 = \text{Number of electrons}(e^-)$$

Resonating electrons include both pi electrons and lone pairs.

When solving for 'n', n must equal to a whole number. If we get a fraction then the molecule does not obey Huckel's rule.

Benzene is the most common aromatic molecule. A benzene ring has 3 pi bonds thus 6 resonating electrons, $4n+2 = 6$

Solve for n:

$$4n+2=6$$

$$4n=4$$

$$n=1$$

n = 1 which is a whole number.Benzene follows Huckel's rule.

Evidence of Aromaticity: Heats of Hydrogenation

One of the ways to test the relative amounts of resonance energy in a molecule is to compare the heats of hydrogenation between similar compounds. For instance, if we compare cyclohexene, 1,3-cyclohexadiene, and benzene, we would expect that their heats of hydrogenation will increase since the number of double bonds increases respectively. However, experimental evidence suggests that the actual heat of hydrogenation for benzene is actually 49.3 kcal/mole, making it even more stable than the 1,3-cyclohexadiene even though it has two double bonds, compared to benzene's three double bonds. This characteristic can be attributed to the aromaticity of benzene which delocalizes the electrons of the six pi orbitals.

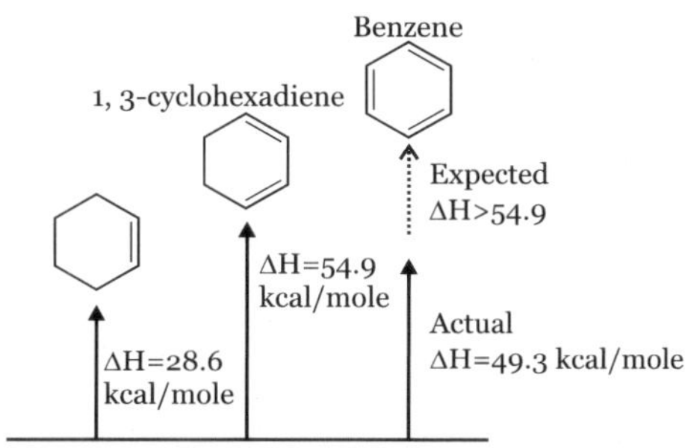

Aromatic compounds, originally named because of their fragrant properties, are unsaturated hydrocarbon ring structures that exhibit special properties, including unusual stability, due to their aromaticity. They are often represented as resonance structures containing single and double bonds. However, the bonding is stronger than expected for a conjugated structure, and it is more accurately depicted as delocalized electron density shared between all the atoms in the ring.

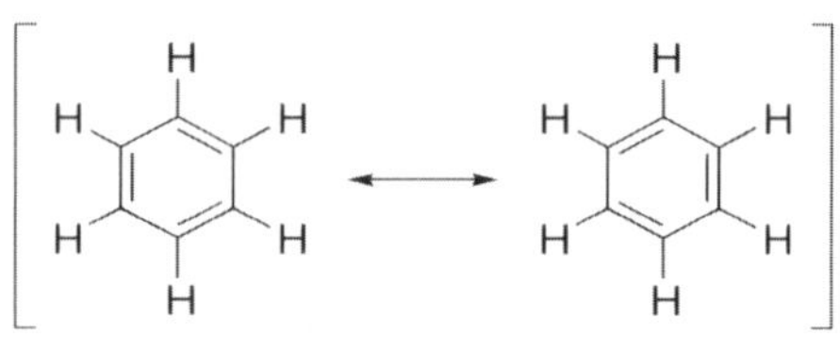

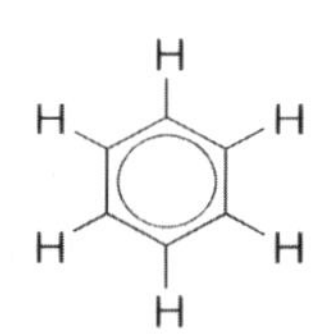

Benzene can only be fully depicted with all of its resonance structures, which show how its pi-electrons are delocalized throughout its six-carbon ring. This delocalization leads to a lower overall energy for the molecule, giving it greater stability.

Benzene

Benzene is symmetrical, planar, hexagonal strcture that boils at 80 C and melts at 5 C.

Benzene is the simplest of the aromatics (n=1 for Hückel's rule). Although you may see benzene written like that, benzene actually exists as two different compounds.

Kekulé structures for benzene

The benzene ring is constantly changing its form because of resonance. You can see the two structures that are possible for benzene.

Benzene Substituents

Two substituents that involve the benzene ring are phenyl- and benzyl- groups.

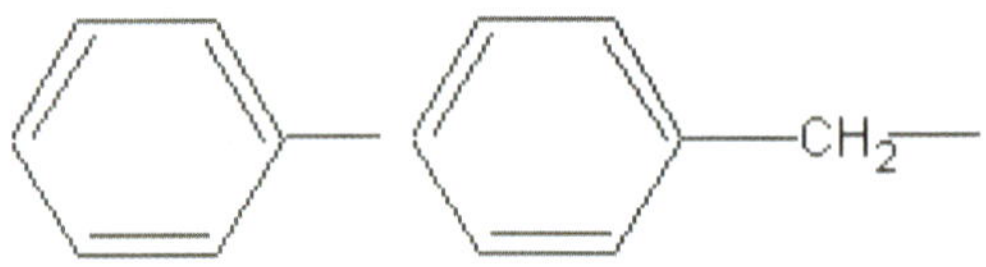

The molecule on the left is the phenyl- group and the one on the right is the benzyl- group. Aromatic groups are sometimes written Ar- (which is short for aryl) while R- represents an alkyl group. So if you saw someting written as Ar-CH_3, that would be a methylbenzene (or phenylmethane).

Importance of Aromatic Compounds

Aromatic compounds play key roles in the biochemistry of all living things. The four aromatic amino acids histidine, phenylalanine, tryptophan, and tyrosine each serve as one of the 20 basic building-blocks of proteins. Further, all 5 nucleotides (adenine, thymine, cytosine, guanine, and uracil) that make up the sequence of the genetic code in DNA and RNA are aromatic purines or pyrimidines. The molecule heme contains an aromatic system with 22 π-electrons. Chlorophyll also has a similar aromatic system.

Aromatic compounds are important in industry. Key aromatic hydrocarbons of commercial interest are benzene, toluene, *ortho*-xylene and *para*-xylene. About 35 million tonnes are produced worldwide every year. They are extracted from complex mixtures obtained by the refining of oil or by distillation of coal tar, and are used to produce a range of important chemicals and polymers, including styrene, phenol, aniline, polyester and nylon.

Types of Aromatic Compounds

The overwhelming majority of aromatic compounds are compounds of carbon, but they need not be hydrocarbons.

Neutral Homocyclics

Benzene, as well as most other annulenes (cyclodecapentaene excepted) with the formula C_nH_n where n is an even number, such as cyclotetradecaheptaene.

Heterocyclics

In heterocyclic aromatics (heteroaromatics), one or more of the atoms in the aromatic ring is of an element other than carbon. This can lessen the ring's aromaticity, and thus (as in the case of furan) increase its reactivity. Other examples include pyridine, pyrazine, pyrrole, imidazole, pyrazole, oxazole, thiophene, and their benzannulated analogs. In all these examples, the number of π-electrons is 6, due to the π-electrons from the double bonds as well as the two electrons from any lone pair that is in the p-orbital that is in the plane of the aromatic π system. For example, in pyridine, the five sp^2-hybridized carbons each have a p-orbital that is perpendicular to the plane of the ring, and each of these p-orbitals contains one π-electron. Additionally, the nitrogen atom is also sp^2-hybridized and has one electron in a p-orbital, which adds up to 6 p-electrons, thus making pyridine aromatic. The lone pair on the nitrogen is not part of the aromatic π system. Pyrrole and imidazole are both five membered aromatic rings that contain heteroatoms. In pyrrole, each of the four sp^2-hybridized carbons contributes one π-electron, and the nitrogen atom is also sp^2-hybridized and contributes two π-electrons from its lone pair, which occupies a p-orbital. In imidazole, both nitrogens are sp^2-hybridized; the one in the double bond contributes one electron and the one which is not in the double bond and is in a lone pair contributes two electrons to the π system.

Fused Aromatics and Polycyclics

Polycyclic aromatic hydrocarbons are molecules containing two or more simple aromatic rings fused together by sharing two neighboring carbon atoms. Examples are naphthalene, anthracene, and phenanthrene. In fused aromatics, not all carbon–carbon bonds are necessarily equivalent, as the electrons are not delocalized over the entire molecule. The aromaticity of these molecules can be explained using their orbital picture. Like benzene and other monocyclic aromatic molecules, polycyclics have a cyclic conjugated pi system with p-orbital overlap above and below the plane of the ring.

Substituted Aromatics

Many chemical compounds are aromatic rings with other functional groups attached. Examples include trinitrotoluene (TNT), acetylsalicylic acid (aspirin), paracetamol, and the nucleotides of DNA.

Aromatic Ions

Aromatic molecules need not be neutral molecules. Ions that satisfy Huckel's rule of $4n + 2$ π-electrons in a planar, cyclic, conjugated molecule are considered to be aromatic ions. For example, the cyclopentadienyl anion and the cycloheptatrienylium cation are both considered to be aromatic ions, and the azulene molecule can be approximated as a combination of both.

In order to convert the atom from sp^3 to sp^2, a carbocation, carbanion, or carbon radical must be formed. These leave sp^2-hybridized carbons that can partake in the π system of an aromatic molecule. Like neutral aromatic compounds, these compounds are stable and form easily. The cyclopentadienyl anion is formed very easily and thus 1, 3-cyclopentadiene is a very acidic hydrocarbon with a pK_a of 16. Other examples of aromatic ions include the cyclopropenium cation (2 π-electrons) and cyclooctatetraenyl dianion (10 pi electrons).

Atypical Aromatic Compounds

Aromaticity also occurs in compounds that are not carbocyclic or heterocyclic; inorganic six-membered-ring compounds analogous to benzene have been synthesized. For example, borazine is a six-membered ring composed of alternating boron and nitrogen atoms, each with one hydrogen attached. It has a delocalized π system and undergoes electrophilic substitution reactions appropriate to aromatic rings rather than reactions expected of non-aromatic molecules.

Quite recently, the aromaticity of planar Si_5^{6-} rings occurring in the Zintl phase $Li_{12}Si_7$ was experimentally evinced by Li solid-state NMR. Metal aromaticity is believed to exist in certain clusters of aluminium.

Homoaromaticity is the term used to describe systems where conjugation is interrupted by a single sp^3 hybridized carbon atom.

Möbius aromaticity occurs when a cyclic system of molecular orbitals, formed from p_π atomic orbitals and populated in a closed shell by $4n$ (n is an integer) electrons, is given a single half-twist to form a Möbius strip. A π system with $4n$ electrons in a flat (non-twisted) ring would be antiaromatic, and therefore highly unstable, due to the symmetry of the combinations of p atomic orbitals. By twisting the ring, the symmetry of the system changes and becomes allowed, because the twist can be left-handed or right-handed, the resulting Möbius aromatics are *dissymmetric* or chiral. But as of 2012, no Möbius aromatic molecules had been synthesized. Aromatics with two half-twists corresponding to the paradromic topologies were first suggested by Johann Listing. In carbo-benzene the ring bonds are extended with alkyne and allene groups.

Y-aromaticity is a concept developed to explain the extraordinary stability and high basicity of the guanidinium cation. Guanidinium is not a ring molecule, and is cross-conjugated rather than a linear π system, but is reported to have its six π-electrons delocalized over the whole molecule. The concept is controversial and some authors emphasize different effects.

σ-aromaticity refers to stabilization arising from the delocalization of sigma bonds. It is often invoked in cluster chemistry and is closely related to Wade's Rule.

Covalent Bonding

In Chemistry, Covalent bond is the interatomic linkage that results from the sharing of an electron pair between two atoms.

The binding arises from the electrostatic attraction of their nuclei for the same electrons. A covalent bond forms when the bonded atoms have a lower total energy than that of widely separated atoms.

Molecules that have covalent linkages include the inorganic substances hydrogen, nitrogen, chlorine, water, and ammonia (H_2, N_2, Cl_2, H_2O, NH_3) together with all organic compounds. In structural representations of molecules, covalent bonds are indicated by solid lines connecting pairs of atoms:

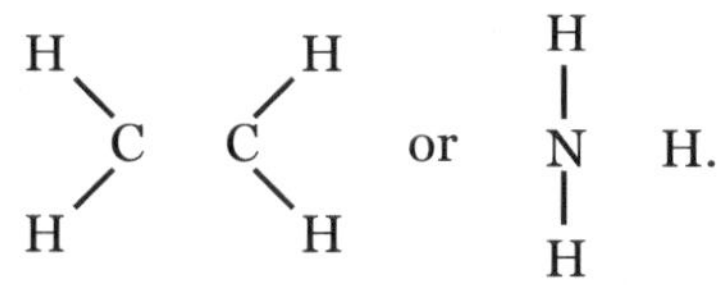

A single line indicates a bond between two atoms (*i.e.,* involving one electron pair), double lines (=) indicate a double bond between two atoms (*i.e.,* involving two electron pairs), and triple lines (≡) represent a triple bond, as found, for example, in carbon monoxide (C≡O). Single bonds consist of one sigma (σ) bond, double bonds have one σ and one pi (π) bond, and triple bonds have one σ and two π bonds.

The idea that two electrons can be shared between two atoms and serve as the link between them was first introduced in 1916 by the American chemist G.N. Lewis, who described the formation of such bonds as resulting from the tendencies of certain atoms to combine with one another in order for both to have the electronic structure of a corresponding noble-gas atom.

Covalent bonds are directional, meaning that atoms so bonded prefer specific orientations relative to one another; this in turn gives molecules definite shapes, as in the angular (bent) structure of the H_2O molecule. Covalent bonds between identical atoms (as in H_2) are nonpolar—*i.e.,* electrically uniform—while those between unlike atoms are polar—*i.e.,* one atom is slightly negatively charged and the other is slightly positively charged. This partial ionic character of covalent bonds increases with the difference in the electro negativities of the two atoms.

Covalent Structures

There are several types of structures for covalent substances, including individual molecules, molecular structures, macromolecular structures and giant covalent structures. Individual molecules have strong bonds that hold the atoms together, but there are negligible forces of attraction between molecules. Such covalent substances are usually gases, for example, HCl, SO_2, CO_2, and CH_4. In molecular structures, there are weak forces of attraction. Such covalent substances are low-boiling-temperature liquids (such as ethanol), and low-melting-temperature solids (such as iodine and solid CO_2). Macromolecular structures have large numbers of atoms linked by covalent bonds in chains, including synthetic polymers such as polyethylene and nylon, and biopolymers such as proteins and starch. Network covalent structures (or giant covalent structures) contain large numbers of atoms linked in sheets (such as graphite), or 3-dimensional structures (such as diamond and quartz). These substances have high melting and boiling points, are frequently brittle, and tend to have high electrical resistivity. Elements that have high electronegativity, and the ability to form three or four electron pair bonds, often form such large macromolecular structures.

One-and Three-electron Bonds

Bonds with one or three electrons can be found in radical species, which have an odd number of electrons. The simplest example of a 1-electron bond is found in the dihydrogen cation H_2^+ One-electron bonds often have about half the bond energy of a 2-electron bond, and are therefore called "half bonds". However, there are exceptions: in the case of dilithium, the bond is actually stronger for the 1-electron Li_2^+ than for the 2-electron Li_2. This exception can be explained in terms of hybridization and inner-shell effects.

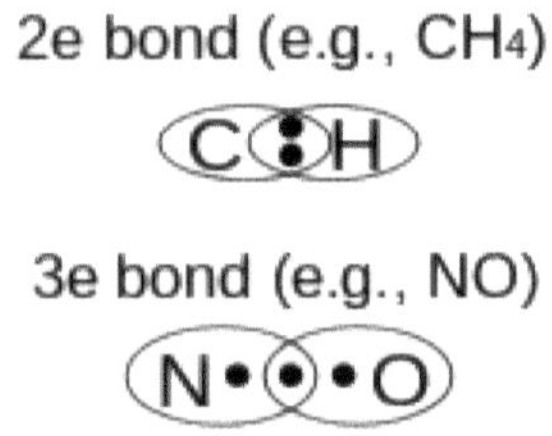

Comparison of the electronic structure of the three-electron bond to the conventional covalent bond.

The simplest example of three-electron bonding can be found in the helium dimer cation He_2^+. It is considered a "half bond" because it consists of only one shared electron (rather than two); in molecular orbital terms, the third electron is in an anti-bonding orbital which cancels out half of the bond formed by the other two electrons. Another example of a molecule containing a 3-electron bond, in addition to two 2-electron bonds, is nitric oxide, NO. The oxygen molecule, O_2 can also be regarded as having two 3-electron bonds and one 2-electron bond, which accounts for its paramagnetism and its formal bond order of 2. Chlorine dioxide and its heavier analogues bromine dioxide and iodine dioxide also contain three-electron bonds.

Molecules with odd-electron bonds are usually highly reactive. These types of bond are only stable between atoms with similar electro negativities.

Resonance

There are situations whereby a single Lewis structure is insufficient to explain the electron configuration in a molecule, hence a superposition of structures are needed. The same two atoms in such molecules can be bonded differently in different structures (a single bond in one, a double bond in another, or even none at all), resulting in a non-integer bond order. The nitrate ion is one such example with three equivalent structures. The bond between the nitrogen and each oxygen is a double bond in one structure and a single bond in the other two, so that the average bond order for each N–O interaction is: $(2+1+1)/3 = 4/3$.

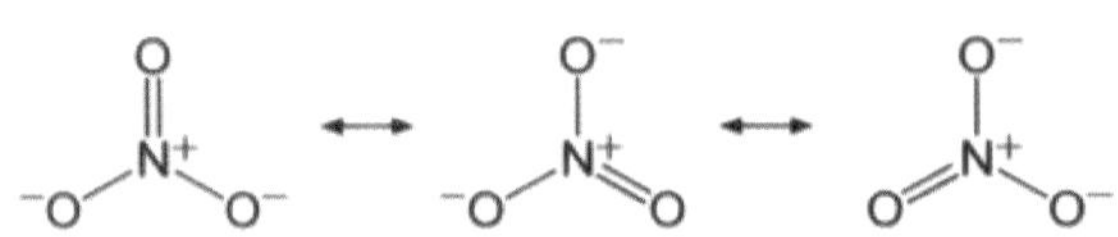

Aromaticity

In organic chemistry, when a molecule with a planar ring obeys Hückel's rule, where the number of π electrons fit the formula $4n + 2$ (where n is an integer), it attains extra stability and symmetry. In benzene, the prototypical aromatic compound, there are 6 π bonding electrons ($n = 1$, $4n + 2 = 6$). These occupy three delocalized π molecular orbitals (molecular orbital theory) or form conjugate π bonds in two resonance structures that linearly combine (valence bond theory), creating a regular hexagon exhibiting a greater stabilization than the hypothetical 1, 3, 5-cyclohexatriene.

In the case of heterocyclic aromatics and substituted benzenes, the electronegativity differences

between different parts of the ring may dominate the chemical behavior of aromatic ring bonds, which otherwise are equivalent.

Hypervalence

Certain molecules such as xenon difluoride and sulfur hexafluoride have higher co-ordination numbers than would be possible due to strictly covalent bonding according to the octet rule. This is explained by the three-center four-electron bond ("3c–4e") model which interprets the molecular wave function in terms of non-bonding highest occupied molecular orbitals in molecular orbital theory and ionic-covalent resonance in valence bond theory.

Electron-deficiency

In three-center two-electron bonds ("3c–2e") three atoms share two electrons in bonding. This type of bonding occurs in electron deficient compounds like diborane. Each such bond (2 per molecule in diborane) contains a pair of electrons which connect the boron atoms to each other in a banana shape, with a proton (nucleus of a hydrogen atom) in the middle of the bond, sharing electrons with both boron atoms. In certain cluster compounds, so-called four-center two-electron bonds also have been postulated.

Quantum Mechanical Description

After the development of quantum mechanics, two basic theories were proposed to provide a quantum description of chemical bonding: valence bond (VB) theory and molecular orbital (MO) theory. A more recent quantum description is given in terms of atomic contributions to the electronic density of states.

Covalency from Atomic Contribution to the Electronic Density of States

In COOP, COHP and BCOOP, evaluation of bond covalency is dependent on the basis set. To overcome this issue, an alternative formulation of the bond covalency can be provided in this way.

The center mass $cm(n,l,m_l,m_s)$ of an atomic orbital $|n,l,m_l,m_s\rangle$, with quantum numbers n, l, m_l, m_s, for atom A is defined as:

$$cm^{A}(n,l,m_l,m_s)=\frac{\int_{E_0}^{E_1} E g^{A}_{|n,l,m_l,m_s\rangle}(E)dE}{\int_{E_0}^{E_1} g^{A}_{|n,l,m_l,m_s\rangle}(E)dE}$$

Where, $g^{A}_{|n,l,m_l,m_s\rangle}(E)$ E is the contribution of the atomic orbital $|n,l,m_l,m_s\rangle$ of the atom A to the total electronic density of states $g(E)$ of the solid:

$$g(E)=\sum_{A}\sum_{n,l}\sum_{m_l,m_s} g^{A}_{|n,l,m_l,m_s\rangle}(E)$$

Where, the outer sum runs over all atoms A of the unit cell. The energy window $[E_0, E_1]$ is chosen in such a way that it encompasses all relevant bands participating in the bond. If the range to select is unclear, it can be identified in practice by examining the molecular orbitals that describe the electron density along the considered bond.

The relative position $C_{n_A l_A, n_B l_B}$ of the center mass of $|n_A, l_A\rangle$ levels of atom A with respect to the center mass of $|n_B, l_B\rangle$ levels of atom B is given as:

$$C_{n_A l_A, n_B l_B} = -\left|cm^A(n_A, l_A) - cm^B(n_B, l_B)\right|$$

Where, the contributions of the magnetic and spin quantum numbers are summed. According to this definition, the relative position of the A levels with respect to the B levels is,

$$C_{A,B} = -\left|cm^A - cm^B\right|$$

Where, for simplicity, we may omit the dependence from the principal quantum number n in the notation referring to $C_{n_A l_A, n_B l_B}$

In this formalism, the greater the value of $C_{A,B}$, the higher the overlap of the selected atomic bands, and thus the electron density described by those orbitals gives a more covalent A–B bond. The quantity $C_{A,B}$ is denoted as the *covalency* of the A–B bond, which is specified in the same units of the energy E.

Single Bonds

A single bond is when two electrons one pair of electrons are shared between two atoms. It is depicted by a single line between the two atoms. Although this form of bond is weaker and has a smaller density than a double bond and a triple bond, it is the most stable because it has a lower level of reactivity meaning less vulnerability in losing electrons to atoms that want to steal electrons.

Example: HCL

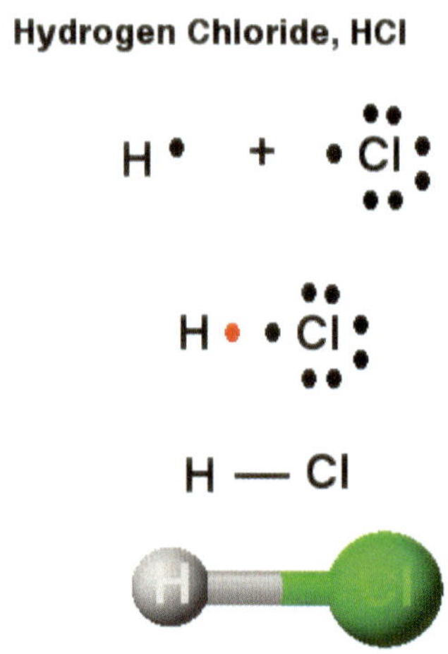

Double Bonds

A Double bond is when two atoms share two pairs of electrons with each other. It is depicted by two horizontal lines between two atoms in a molecule. This type of bond is much stronger than a single bond, but less stable; this is due to its greater amount of reactivity compared to a single bond.

Example: CO_2

Below is a Lewis dot structure of Carbon dioxide demonstrating a double bond. As you can see from the picture below, Carbon dioxide has a total of 1 Carbon atom and 2 Oxygen atoms. Each Oxygen atom has 6 valence electrons whereas the Carbon atom only has 4 valence electrons. To satisfy the Octet Rule, Carbon needs 4 more valence electrons. Since each Oxygen atom has 3 lone pairs of electrons, they can each share 1 pair of electrons with Carbon; as a result, filling Carbon's outer valence shell (Satisfying the Octet Rule).

Triple Bond

A Triple bond is when three pairs of electrons are shared between two atoms in a molecule. It is the least stable out of the three general types of covalent bonds. It is very vulnerable to electron thieves.

Example: Acetylene

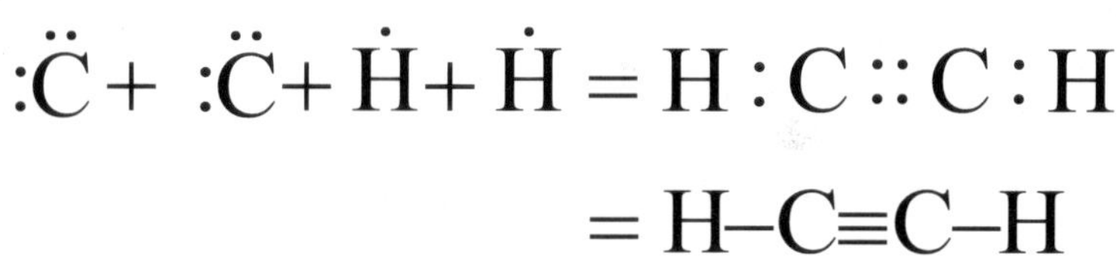

Below is a Lewis dot structure of Acetylene demonstrating a triple bond. As you can see from the picture below, Acetylene has a total of 2 Carbon atoms and 2 Hydrogen atoms. Each Hydrogen atom has 1 valence electron whereas each Carbon atom has 4 valence electrons. Each Carbon needs 4 more electrons and each Hydrogen needs 1 more electron. Hydrogen shares its only electron with Carbon to get a full valence shell. Now Carbon has 5 electrons. Because each Carbon atom has 5 electrons -1 single bond and 3 unpaired electrons the two Carbons can share their unpaired electrons, forming a triple bond. Now all the atoms are happy with their full outer valence shell.

Polar Covalent Bond

A Polar Covalent Bond is created when the shared electrons between atoms are not equally shared. This occurs when one atom has a higher electronegativity than the atom it is sharing with. The atom with the higher electronegativity will have a stronger pull for electrons (Similiar to a Tug-O-War game, whoever is stronger usually wins). As a result, the shared electrons will be closer to the atom with the higher electronegativity, making it unequally shared. A polar covalent bond will result in the molecule having a slightly positive side (the side containing the atom with a lower electronegativity) and a slightly negative side (containing the atom with the higher electronegativity) because the shared electrons will be displaced toward the atom with the higher electronegativity. As a result of polar covalent bonds, the covalent compound that forms will have an electrostatic potential. This potential will make the resulting molecule slightly polar, allowing it to form weak bonds with other polar molecules. One example of molecules forming weak bonds with each other as a result of an unbalanced electrostatic potential is hydrogen bonding, where a hydrogen atom will interact with an electronegative hydrogen, fluorine, or oxygen atom from another molecule or chemical group.

Example: Water, Sulfide, Ozone, etc.

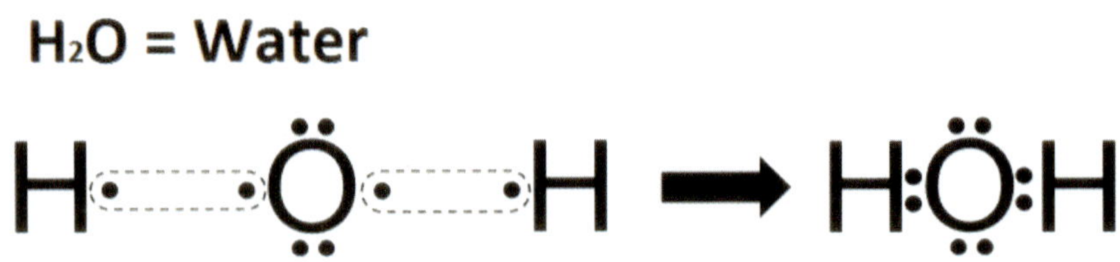

Water is a molecule which has the molecular formula of H2O. Each hydrogen atom bonds with a central oxygen atom by sharing a pair of electrons through a covalent bond. The result is a bent molecule with a mean H-O-H angle of 106°.

Nonpolar Covalent Bond

A Nonpolar Covalent Bond is created when atoms share their electrons equally. This usually occurs when two atoms have similar or the same electron affinity. The closer the values of their electron affinity, the stronger the attraction. This occurs in gas molecules; also known as diatomic elements. Nonpolar covalent bonds have a similar concept as polar covalent bonds; the atom with the higher electronegativity will draw away the electron from the weaker one. Since this statement is true if we apply this to our diatomic molecules all the atoms will have the same electronegativity since they are the same kind of element; thus, the electro negativities will cancel each other out and will have a charge of 0(A.K.A. Nonpolar covalent bond).

Examples of gas molecules that have a nonpolar covalent bond: Hydrogen gas atom, Nitrogen gas atoms, etc.

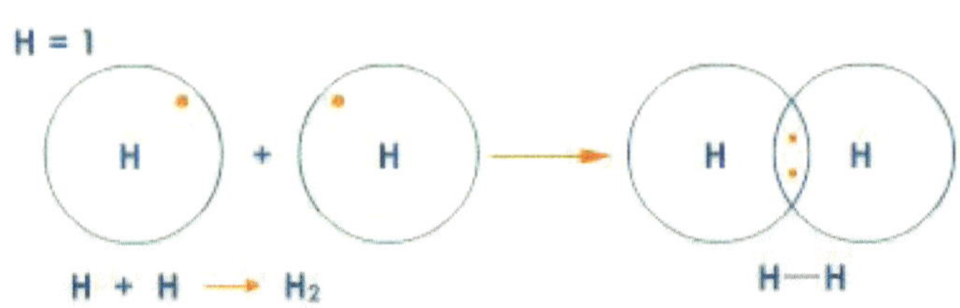

As you can see from the picture above, Hydrogen gas has a total of 2 Hydrogen atoms. Each Hydrogen atom has 1 valence electron. Since Hydrogen can only fit a max of 2 valence electrons in its orbital, each Hydrogen atom only needs 1 electron. Each atom has 1 valence electron, so they can just share, giving each atom two electrons each.

Sigma (σ) and Pi bonds

Sigma (σ) Bond

This type of covalent bond is formed by the axial overlapping of half-filled atomic orbitals. The atomic orbitals overlap along the inter-nuclear axis and involve end to end or head' on overlap. The electron cloud formed as a result of axial overlap is cylindrically symmetrical about inter-nuclear axis. The electrons constituting sigma bond are called sigma electrons. There can be three types of axial overlap among s and p-orbitals as discussed below:

- s-s overlap: It involves mutual overlap of half-filled s-orbitals of the atoms approaching to form a bond. The bond formed is called s-s σ bond.
- s-p overlap: It involves mutual overlap of half-filled s-orbital of the one atom with half-filled p-orbital of the other. The bond so formed is called s-p σ bond.
- p-p overlap: It involves mutual overlap of half-filled p-orbitals of the two atoms. The bond so formed is called p-p σ bond.

The s -s, s-p and p-p overlaps have been shown diagrammatically in Figure below:

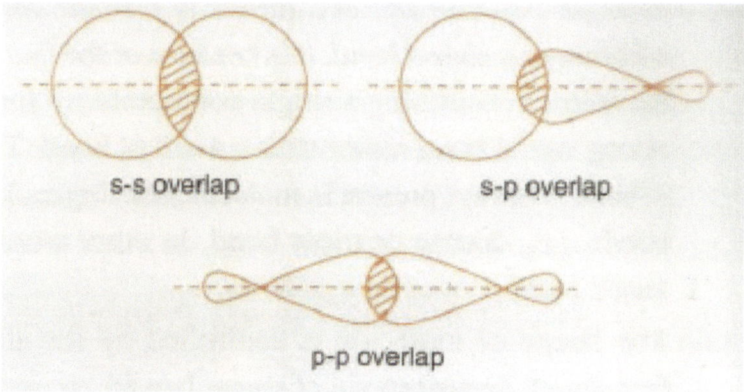

Figure: s-s, s-p and p-p overlap of atomic orbitals.

Strength of three types of sigma bonds, the strength of three types of sigma bonds varies as follows:

p-p > p-s > s-s

It is because of the fact, that p-orbitals allow overlap to greater extent as compared top-s which is larger as compared to s-s overlap.

Pi Bond

This type of covalent bond is formed by the lateral or sidewise overlap of the atomic orbitals. The orbital overlap takes place in such a way that their axes are parallel to each other but perpendicular to the inter-nuclear axis. The pi bond consists of two charge clouds above and below the plane of the atoms involved in the bond formation. The electrons involved in the n-bond formation are called n-electrons.

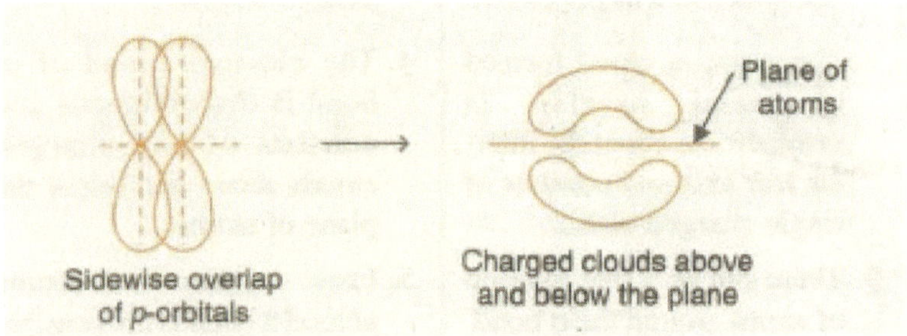

Figure: Sidewise overlap of p-orbitals.

It may be noted that:

- Sigma bond is stronger than pi bond, It is because of the fact that overlapping of atomic orbitals can take place to a greater extent during the formation of sigma bond whereas overlapping of orbitals occurs to a smaller extent during the formation of pi bond.
- Pi bond between the two atoms is formed only in addition to a sigma bond. It is because of the fact that the atoms constituting a single bond prefer to form a strong sigma bond rather than a weak pi bond. Thus, pi bond is always present in molecules having multiple bonds, i.e., double or triple bond. In other words, a single bond cannot be a pi bond.
- The shape of molecule is controlled by the sigma framework (orientations of sigma bonds) around the central atom. Pi bonds are superimposed on sigma bonds hence they simply modify the dimensions of the molecule.

Comparison Between Sigma and Pi Bonds

The various points of distinction between sigma and pi bonds are given in table.

Sigma. Bond	Pi Bond
1. This bond is formed by the axial overlap of atomic orbitals.	1. The bond is formed by the sidewise overlap of atomic orbitals.
2. This bond can be formed by overlap of s-s, s-p or p-p orbitals.	2. It involves of overlap of p-orbitals only.
3. The bond is stronger because overlapping can take place to a larger extent.	3. The bond is weaker because the overlapping occurs to a smaller extent.
4. The electron cloud formed by axial overlap is symmetrical about the inter-nuclear axis and consists of single charged cloud.	4. The electron cloud of pi bond is discontinuous and consists of two charged clouds above and below the plane of atoms.
5. There can be a free rotation of atoms around the σ bond.	5. Free rotation of atoms around 1t bond is not possible because it involves breaking of 1t bond.
6. The bond may be present between the two atoms either alone or along with 1t bond.	6. The bond is always present between the two atoms along with the sigma bond, i.e., it is always superimposed on sigma bond.
7. The shape of molecule is determined by the sigma framework around the central atom.	7. The 1t bonds do not contribute to the shape, however, they modify the dimensions of the molecule.

Formation of sigma and pi bonds in F_2, HF and O_2 molecules are given in Figure below:

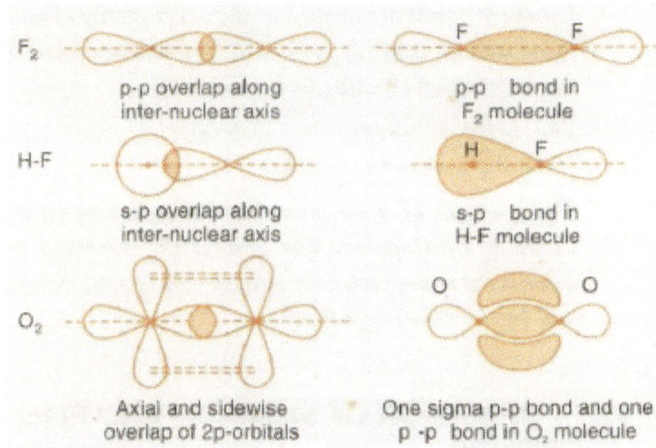

Fig: Formation of F_2, HF and o_2 molecules.

Coordinate Covalent Bond

A coordinate bond is a type of covalent bond where both of the electrons that form the bond originate from the same atom (more generally, a "dative" covalent bond).

Coordinate bonds form between a central electrophile (low electron density, such as metal cations) and one or more nucleophiles (high electron density, such as the hydroxide anion) oriented around the former. Nucleophiles acts as "ligands" by supplying two electrons per coordinate bond to the electrophile to satisfy the octet rule. A "polydentate" ligand molecule may form multiple coordinate bonds.

As with any covalent bond, the atoms in a coordinate bond redistribute electrons in order to satisfy the octet rule, which states that each atom will lose, gain, or share electrons in order to have a full valence of eight electrons in its outer shell. Hydrogen is a common exception to the octet rule, and instead follows the duet rule because it only has one 1s orbital and its outer valence is full when it has two electrons.

It is important to note that coordinate bonding between oppositely charged ionic species is of a covalent character, where the bonding electrons are shared, as opposed to ionic compounds.

Coordinate Covalent Bond(Dative Bond) Formation

One of the two atoms must have at least one lone pair of electrons, e.g., ammonia (NH_3), water(H_2O). A pair of electrons which is not shared with any other atom is known as the lone pair of electrons but it is provided to the other atom for the formation of the coordinate Covalent bond.

Another atom should be short of at least a lone pair of electrons, e.g., Hydrogen ion (H^+).

Examples

Ammonium Ion Coordinate Bond(NH_3)

The Nitrogen atom has five valence electrons and hydrogen atom has one valence electron. In ammonia, each of the three hydrogen atoms is bonded to the nitrogen atom by a pair of shared electrons. (One electron from the hydrogen atom and one from the nitrogen).

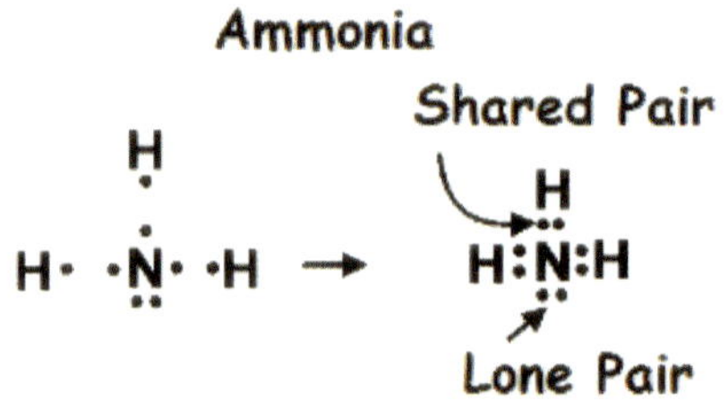

Thus, the nitrogen atom in ammonia is left alone pair of electrons.

This ammonia molecule combines with a hydrogen ion H^+, which has no electrons in its outermost orbit in such a way that the lone pair of electrons on the nitrogen atom is shared by the hydrogen ion. The latter thus acquires the helium configuration. This bond which is formed between the nitrogen atom in ammonia and the hydrogen ion is a coordinate bond.

Lewis Structure of Ammonium Ion($NH4^+$)

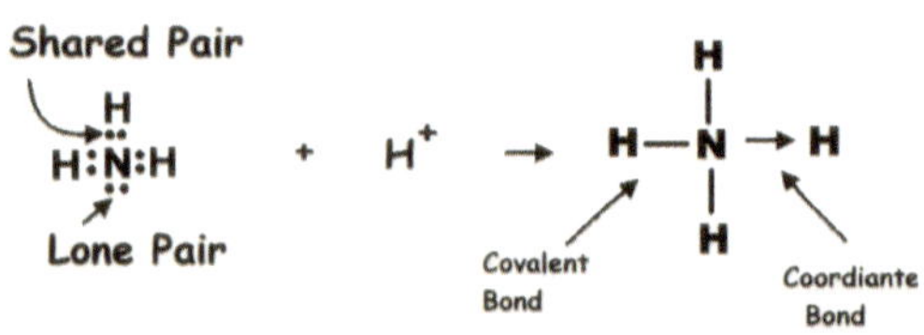

The nitrogen atom in an ammonia molecule contains a lone pair of electrons. It is a donor. The hydrogen ion (H^+) accepts that lone pair of electrons (Acceptor) and forms a coordinate bond. Once this bond is formed, all four N-H bonds become identical. The linkage forms an ammonium having a single positive charge.

When ammonium chloride NH_4Cl is formed cation $NH4^+$ (having 3 covalent and one coordinate bond) and anion Cl^- are attracted towards each other due to electrical charge existing between them ionic bond is formed. Thus ammonium chloride is a good example of a compound having all the three types of bonds i.e., covalent, coordinate and ionic bond.

Hydronium Ion Coordinate Covalent Bond(H_3O^+)

Water contains two hydrogen atoms and one oxygen atom. The hydrogen atom has one electron in its valence shell while the oxygen atom has six electrons in its Valence shell.

Lewis Structure of H_2O

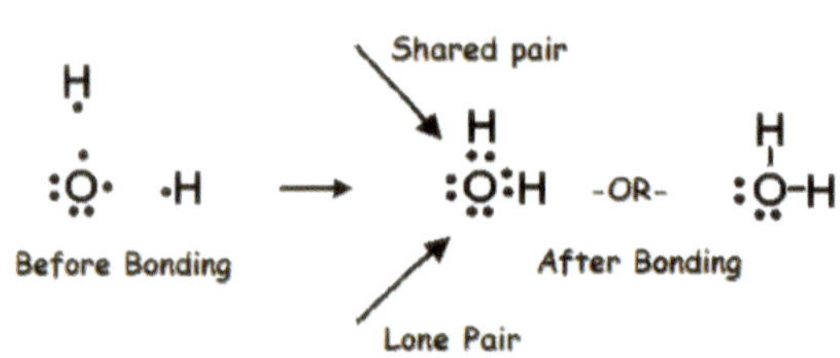

Two hydrogen atoms share one electron each with the oxygen atom to form a water molecule. The two unshared pairs of electrons known as lone pairs of electrons, do not participate in the bond formation in the water molecule.

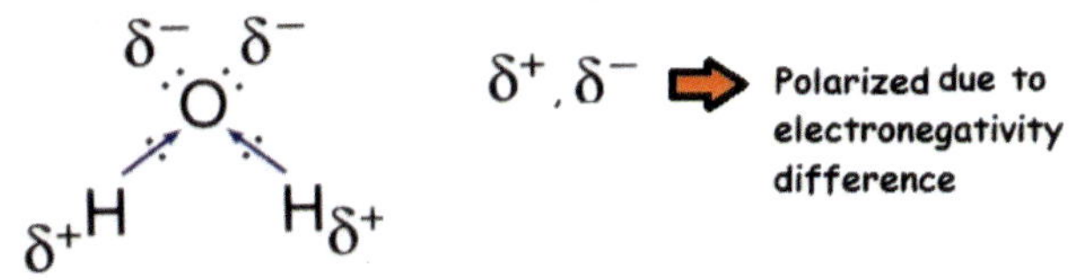

The O-H bonds in a water molecule are polarised due to the large difference in the electronegativity values of the oxygen and the hydrogen atoms. (The oxygen atom has an electronegativity value of 3.5, while that of the hydrogen atom is 2.1). Oxygen has higher electronegativity so it attained slightly negative charge while hydrogen atoms attain slight positive charges. Therefore, a molecule of water due to these charges is a polar molecule and is represented as shown alongside.

When an acid is added to water, strong dipole interaction (attraction between slightly positive and slight negative charge) occurs between the polar water molecules and the polar acid molecules.

As a result of such a dipole interaction, the hydrogen ion gets released from the polar bonds of the acid molecules:

$$HCl + H_2O \rightarrow H^+ + Cl^-$$

The hydrogen ion spontaneously adds on to one of the two lone pairs of electrons of the oxygen atom of the water molecule.

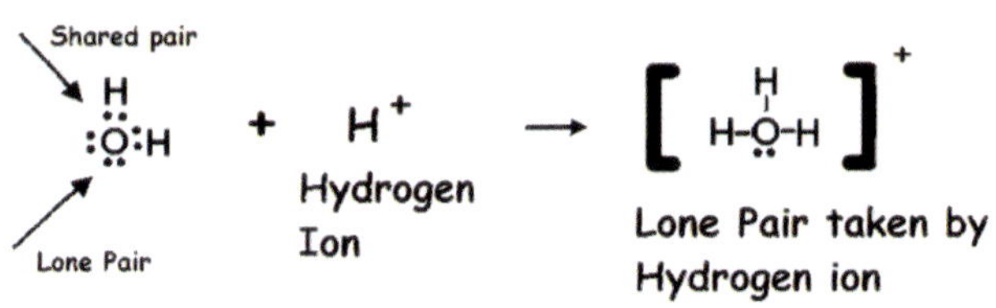

An addition of released H^+ ion to a lone pair of electrons of the oxygen atom of the polar water molecule leads to the formation of a hydronium ion.

Proton +Water →Hydronium ion

$H^+ + H_2O \rightarrow H_3O^+$

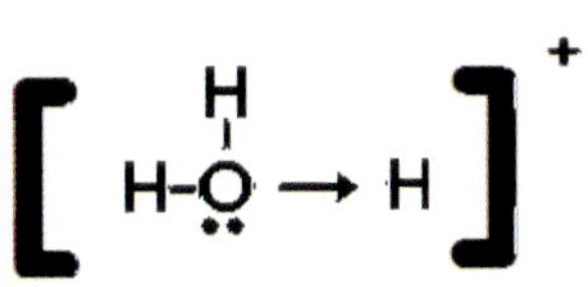

Hydroxyl ion coordinate covalent bond OH^-.

The hydroxyl ion or hydroxide ion is OH^-. It is formed when one hydrogen ion (H^+) is removed from the water molecule.

$H_2O \rightarrow H^+ + OH^-$

When H^+ is removed from the water molecule, the shared pair of electron remains with oxygen.

As, oxygen is more electronegative and thus hydroxyl ion has the negative charge.

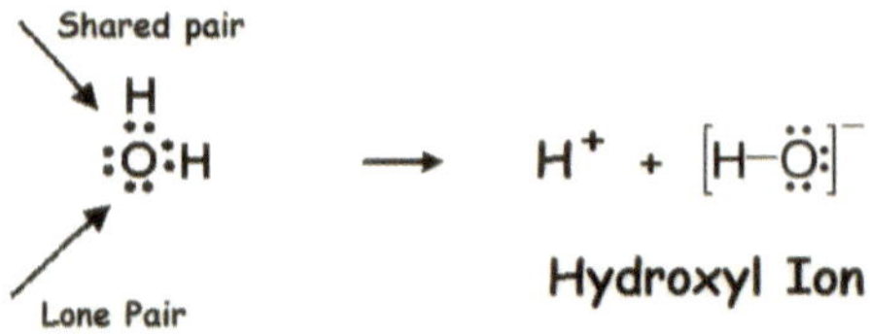

This is Coordinate Covalent Bond.

Applications

Understanding how coordinate bonds form can be instrumental in designing complex organic molecules.

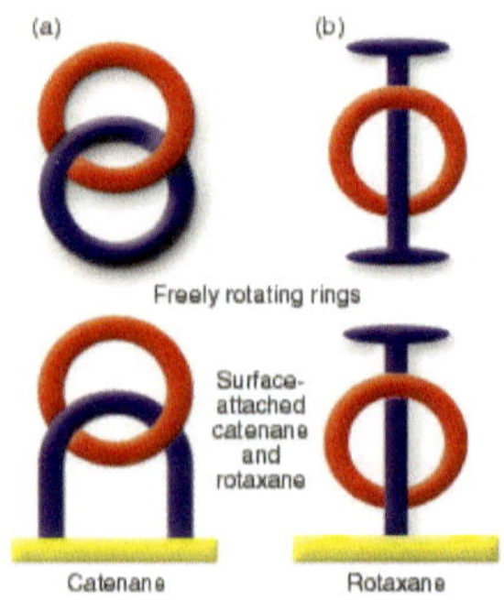

(a) Catenanes are formed from two or more molecular rings interlocked like links in a chain.
(b) rotaxanes are formed from a ring molecule threaded on an axle molecule capped by two molecular stoppers.

Rotaxanes, shown in the righthand column, are one of many mechanically interlocked molecular shapes.

Rotaxanes are an example of a mechanically interlocked molecule, a complex organic molecule that is physically joined. These molecules are interesting to synthetic chemists as "potential building blocks for future molecular-scale devices—motors, sensors, and machines on the nanometer scale".

The image below shows an organic molecule that can self-assemble to form a rotaxane because the molecule was synthesized with coordinate bonds in targeted places. Those bonds are then broken and re-formed to give the molecule its desired shape.

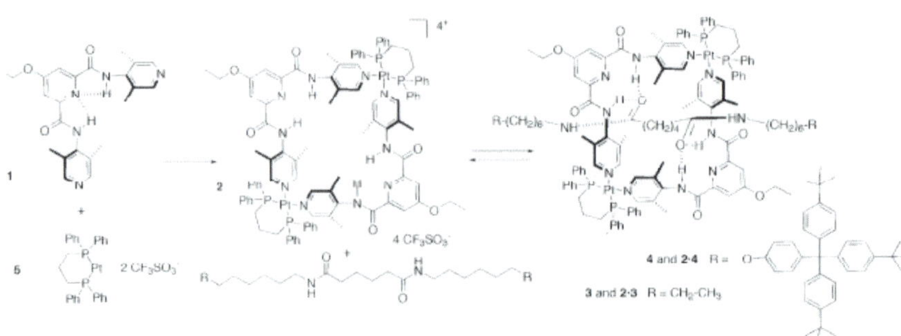

Precursor molecules to a rotaxane shape.

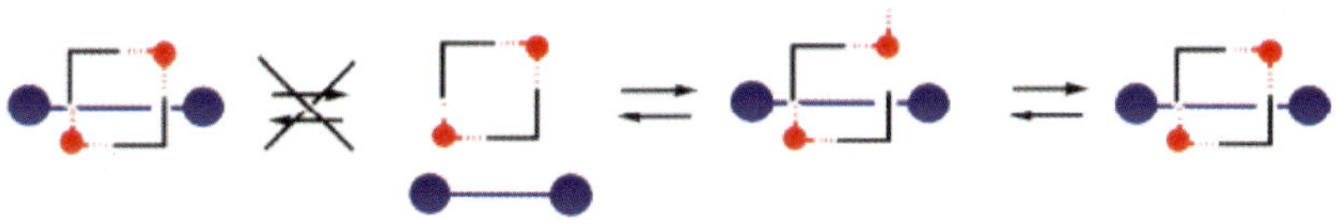

Schematic showing how coordinate bond dissolution allows the rotaxane to self-assemble.

Covalent Bond Classification (LXZ Notation)

The covalent bond classification (CBC) method is also referred to as the LXZ notation. It was published by M. L. H. Green in the mid-1990s as a solution for the need to describe covalent compounds such as organometallic complexes in a way that is not prone to limitations resulting from the definition of oxidation state. Instead of simply assigning a charge to an atom in the molecule (i.e. the oxidation state), the covalent bond classification method analyzes the nature of the ligands surrounding the atom of interest, which is often a transition metal. According to this method, there are three basic types of interactions that allow for coordination of the ligand. The three types of interaction are classified according to whether the ligating group donates two, one, or zero electrons. These three classes of ligands are respectively given the symbols L, X and Z.

Types of Ligands

X-type ligands are those that donate one electron to the metal and accept one electron from the metal when using the neutral ligand method of electron counting, or donate two electrons to the metal when using the donor pair method of electron counting. Regardless of whether it is considered neutral or anionic, these ligands yield normal covalent bonds. A few examples of this type of ligand are H, halogens (Cl, Br, F, etc.), OH, CN, CH_3, and NO.

L-type ligands are neutral ligands that donate two electrons to the metal center regardless of the electron counting method being used. These electrons can come from lone pairs, pi or sigma donors. The bonds formed between these ligands and the metal are dative covalent bonds, which are also known as coordinate bonds. Examples of this type of ligand include CO, PR_3, NH_3, H_2O, carbenes (=CRR'), and alkenes.

Z-type ligands are those that accept two electrons from the metal center as opposed to the donation occurring with the other two types of ligands. However, these ligands also form dative covalent bonds like the L-type. This type of ligand is not usually used, because in certain situations it can be written in terms of L and X. For example, if a Z ligand is accompanied by an L type, it can be written as X_2. Examples of these ligands are Lewis acids, such as BR_3.

Uses of the Notation

When given a metal complex and the trends for the ligand types, the complex can be written in a more simplified manner with the form $[ML_lX_xZ_z]^{Q\pm}$. The subscripts represent the numbers of each ligand type present in that complex, M is the metal center and Q is the overall charge on the complex. Some examples of this overall notation are as follows:

Condensed formula	LXZ notation
$[Mn(CO)_6]^+$	$[ML_6]^+$
$[Ir(CO)(PPh_3)_2(Cl)(NO)]^{2+}$	$[ML_3X_2]^{2+}$
$[Fe(CO)_2(CN)_4]^{2-}$	$[ML_2X_4]^{2-}$

Also from this general form, the values for electron count, oxidation state, coordination number, number of d-electrons, valence number and the ligand bond number can be calculated.

Electron Count = $N + x + 2l - Q$

Where, N is the group number of the metal.

Oxidation State (OS) = $x + Q$

Coordination Number (CN) = $x + l$

Number of d-electrons (dn) = $N - OS$

= $N - (x + Q)$

Valence Number (VN) = $x + 2z$ S

Ligand Bond Number (LBN) = $l + x + z$

Other Uses

This template of writing a metal complex also allows for better comparison of molecules with different charges. This can happen when the assignment is reduced to its "equivalent neutral class". The equivalent neutral class is the classification of the complex if the charge was localized on the ligand as opposed to the metal center. In other words, the equivalent neutral class is the representation of the complex as though there was no charge.

Atomic Radius

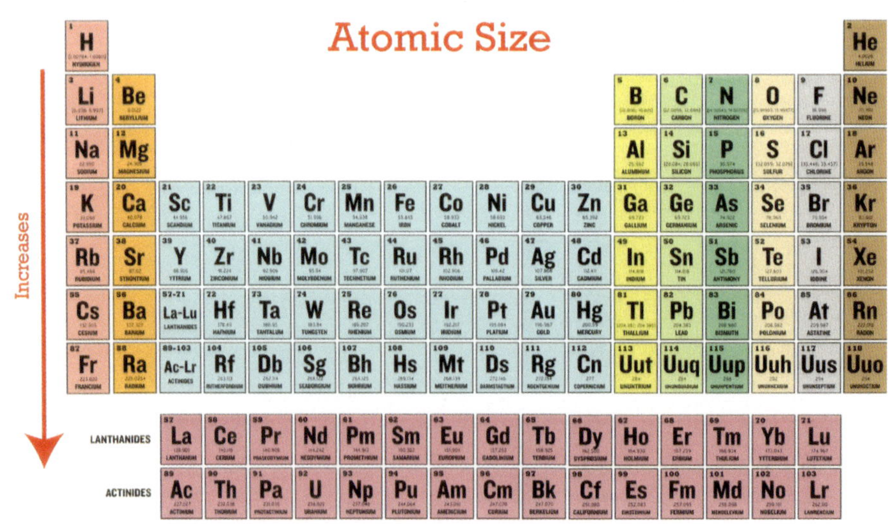

Atomic radius is the distance from the center of the nucleus to the outermost shell containing electrons. In other words, it is the distance from the center of the nucleus to the point up to which the density of the electron cloud is maximum.

Types of Atomic Radii

Atomic radii are divided into three types:

- Covalent radius
- Van der Waals radius
- Metallic radius

Therefore, we will study these three types of radius because they are vital for the better understanding of the subject.

Covalent Radius

Covalent radius is one half the distance between the nuclei of two covalently bonded atoms of the same element in a molecule. Therefore, $r_{covalent}$ = ½ (inter-nuclear distance between two bonded atoms). The inter-nuclear distance between two bonded atoms is called the bond length.

Therefore,

$$r_{covalent} = \frac{1}{2}(\text{bond length})$$

Van der Waals Radius

It is one half the distance between the nuclei of two identical non-bonded isolated atoms or two adjacent identical atoms belonging to two neighboring molecules of an element in the solid state. The magnitude of the Van der Waals radius is dependent on the packing of the atoms when the element is in the solid state.

For example, the inter-nuclear distance between two adjacent chlorine atoms of the two neighboring molecules in the solid state is 360 pm. Therefore, the Van der Waals radius of the chlorine atom is 180 pm.

Metallic Radius

A metal lattice or crystal consists of positive kernels or metal ions arranged in a definite pattern in a sea of mobile valence electrons. Each kernel is simultaneously attracted by a number of mobile electrons and each mobile electron is attracted by a number of metal ions.

Force of attraction between the mobile electrons and the positive kernels is called the metallic bond. It is one half the inter-nuclear distances between the two adjacent metal ions in the metallic lattice. In a metallic lattice, the valence electrons are mobile, therefore, they are only weakly attracted by the metal ions or kernels.

In a covalent bond, a pair of electrons is strongly attracted by the nuclei of two atoms. Thus, a metallic radius is always longer than its covalent radius. For example, the metallic radius of sodium is 186 pm whereas its covalent radius as determined by its vapor which exists as Na_2 is 154 pm. The metallic radius of Potassium is 231 pm while its covalent radius is 203 pm.

Variation of Atomic Radii in the Periodic Table

Variation within a Period

- The Covalent and Van der Waals radii decrease with increase in atomic number as we move from left to right in a period. The alkali metals at the extreme left of the periodic table have the largest size in a period. The halogens at the extreme right of the periodic table have the smallest size. The atomic size of nitrogen is the smallest. After nitrogen, atomic size increases for Oxygen and then decreases for fluorine. The size of atoms of inert gases is larger than those of the preceding halogens.
- As we move from left to right in a period, nuclear charge increases by 1 unit in each succeeding element while the number of shells remains the same. This enhanced nuclear charge pulls the electrons of all the shells closer to the nucleus. This makes each individual shells smaller and smaller. This result in a decrease of the atomic radius as we move from left to right in a period.
- The atomic radius abruptly increases as we move from halogens to the inert gas. This is because inert gases have completely filled orbitals. Hence, the inter-electronic are maximum. We express the atomic size in terms of Van der Waals radius since they do not form covalent bonds. Van der Waals radius is larger than the covalent radius. Therefore, the atomic size of an inert gas in a period is much higher than that of preceding halogen.

Variation Within a Group

The atomic radii of elements increase with an increase in atomic number from top to bottom in a group. As we move down the group, the principal quantum number increases. A new energy shell is added at each succeeding element. The valence electrons lie farther and farther away from the nucleus. As a result, the attraction of the nucleus for the electron decreases. Hence, the atomic radius increases.

Functional Group

Functional groups in organic chemistry are those which are responsible for the changes in the physical and chemical properties of an organic compound. There are various such groups in organic chemistry. It can be defined as:

> "An atom or group of atoms which show a characteristic unique reactivity when treated with some reagents and also which impart some unique chemical, as well as physical properties to the primary molecule, is defined as a functional group."

Molecules are divided into two types in organic chemistry:

- Organic molecules
- Inorganic molecules

The molecules which consist of carbon atoms along with hydrogen atoms are called as organic molecules like hydrocarbons, and the functional groups are mainly dealt with the organic molecules since they exhibit non-polarity kind of nature and also due to the high electro-negativity. These groups also decide the properties and the functionality of a chemical compound. Some of the examples of functional groups are carboxylic acid (RCOO-) and ester (R-COO-R). This comprises of an acyl group (RCO-) and an alkoxy group (-OR').

Types of Functional Groups

A variety of functional groups exist in organic chemistry. Most of them are carbon-based compounds. Some groups form a single bond with most of the hetero-atoms such as oxygen, iodine, nitrogen, sulfur, etc. There is also some other kind of groups having multiple bonds with the hetero-atoms such as oxygen, nitrogen atoms, etc.

The lists of some of the common functional groups are contained in the table below:

Group formula	Class name	Example	Common name
C=C	Alkene	$CH_2=CH_2$	Ethylene
Arene	C_6H_6	Benzene	
-OH	Alcohol	CH_3OH	Wood alcohol
-Cl	Chloro	CH_3Cl	Methyl chloride
C-X	Halide	CH_3-I	Methyl iodide
C-O-C	Ether	CH_3-CH_2-O-CH_2-CH_3	Ether
C-N	Amine	CH_3-NH_2	Methyl amine
Aldehyde	CH_3-CHO	Acetaldehyde	
Ketone	CH_3-CO-CH_3	Acetone	
Carboxylic acid	CH_3-COOH	Acetic acid	
Ester	CH_3-CO_2-CH_2-CH_3	Ethyl acetate	

The 'default' in organic chemistry (essentially, the *lack* of any functional groups) is given the term alkane, characterized by single bonds between carbon and carbon, or between carbon and hydrogen. Methane, CH_4, is the natural gas you may burn in your furnace. Octane, C_8H_{18}, is a component of gasoline.

Alkanes

Alkenes (sometimes called olefins) have carbon-carbon double bonds, and alkynes have carbon-carbon triple bonds. Ethene, the simplest alkene example, is a gas that serves as a cellular signal in fruits to stimulate ripening. (If you want bananas to ripen quickly, put them in a paper bag along with an apple - the apple emits ethene gas, setting off the ripening process in the bananas). Ethyne, commonly called acetylene, is used as a fuel in welding blow torches.

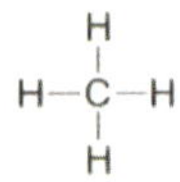

methane

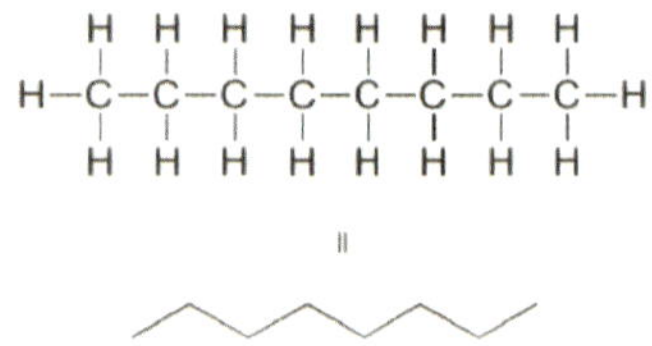

octane

Alkenes and Alkynes

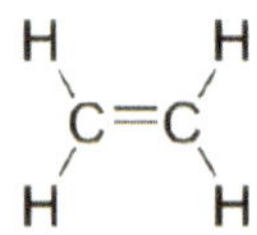

ethene
(an alkene)

ethyne
(an alkyne)

Furthermore, many alkenes can take two geometric forms: cis or trans, The cis and trans forms of a given alkene are different molecules with different physical properties because, there is a very high energy barrier to rotation about a double bond. In the example below, the difference between cis and trans alkenes is readily apparent.

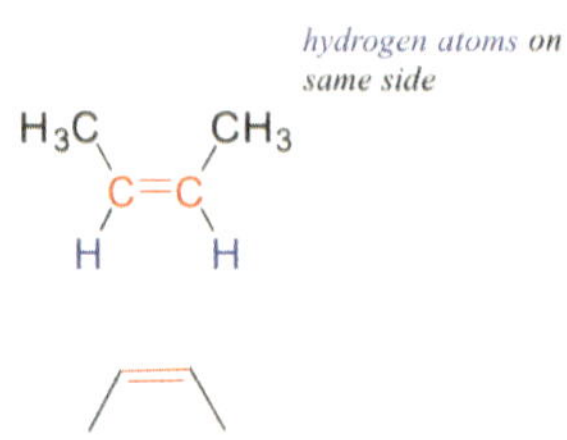

cis-alkene

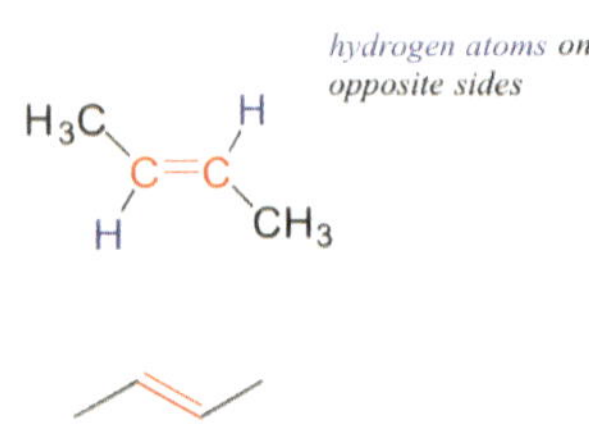

trans-alkene

Alkanes, alkenes, and alkynes are all classified as hydrocarbons, because they are composed solely of carbon and hydrogen atoms. Alkanes are said to be saturated hydrocarbons, because the carbons are bonded to the maximum possible number of hydrogen- in other words, they are saturated with hydrogen atoms. The double and triple-bonded carbons in alkenes and alkynes have fewer hydrogen atoms bonded to them, they are thus referred to as unsaturated hydrocarbons. Hydrogen can be added to double and triple bonds, in a type of reaction called 'hydrogenation'.

The aromatic group is exemplified by benzene (which used to be a commonly used solvent on the organic lab, but which was shown to be carcinogenic), and naphthalene, a compound with a distinctive ‹mothball' smell. Aromatic groups are planar (flat) ring structures, and are widespread in nature.

Aromatics

When the carbon of an alkane is bonded to one or more halogens, the group is referred to as a alkyl halide or haloalkane. Chloroform is a useful solvent in the laboratory, and was one of the earlier anesthetic drugs used in surgery. Chlorodifluoromethane was used as a refrigerant and in aerosol sprays until the late twentieth century, but its use was discontinued after it was found to have harmful effects on the ozone layer. Bromoethane is a simple alkyl halide often used in organic synthesis. Alkyl halides groups are quite rare in biomolecules.

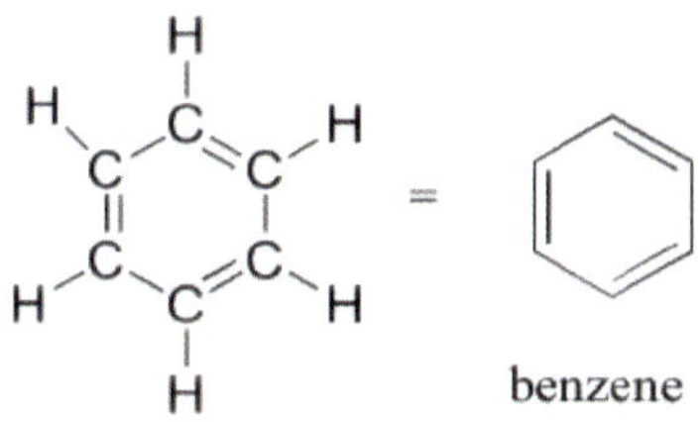

benzene

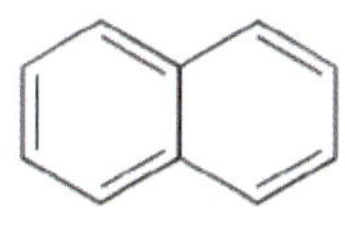

napthalene

Haloalkanes

In the alcohol functional group, a carbon is single-bonded to an OH group (the OH group, by itself, is referred to as a hydroxyl). Except for methanol, all alcohols can be classified as primary, secondary, or tertiary. In a primary alcohol, the carbon bonded to the OH group is also bonded to only one other carbon. In a secondary alcohol and tertiary alcohol, the carbon is bonded to two or three other carbons, respectively. When the hydroxyl group is directly attached to an aromatic ring, the resulting group is called a phenol. The sulfur analog of an alcohol is called a thiol.

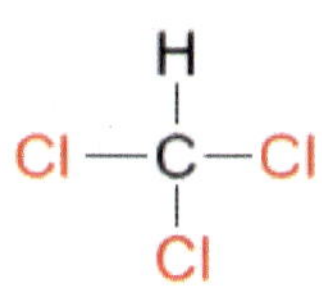

trichloromethane (chloroform)

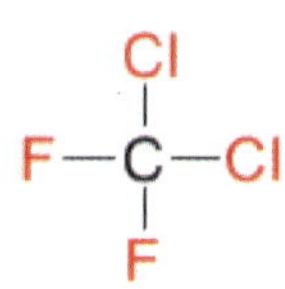

dichlorodifluoromethane (Freon-12)

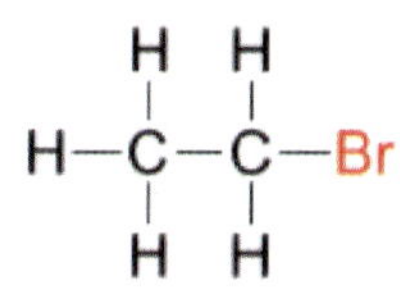

bromoethane

Alcohols, Phenols, and Thiols

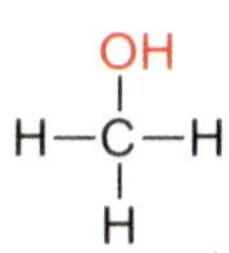

methanol

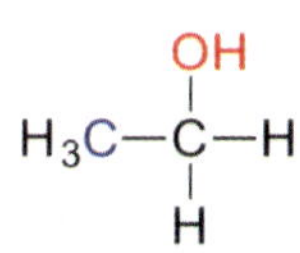

primary alcohol

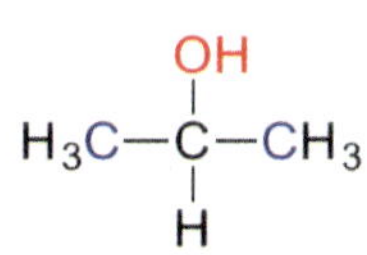

secondary alcohol

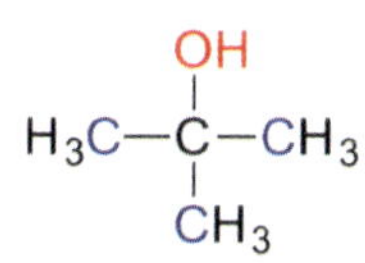

tertiary alcohol

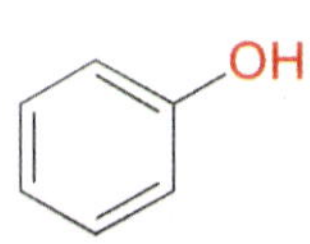

phenol

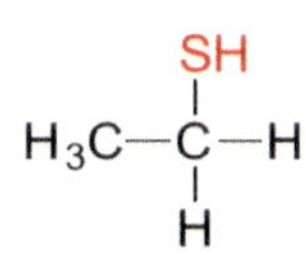

thiol

Note that the definition of a phenol states that the hydroxyl oxygen must be *directly* attached to one of the carbons of the aromatic ring. The compound below, therefore, is *not* a phenol - it is a primary alcohol.

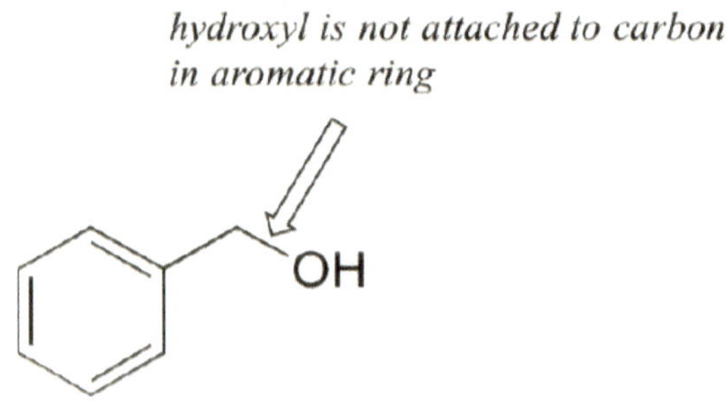

primary alcohol
(*not* a phenol)

The distinction is important, because as we will see later, there is a significant difference in the reactivity of alcohols and phenols.

The deprotonated forms of alcohols, phenols, and thiols are called alkoxides, phenolates, and thiolates, respectively. A protonated alcohol is an oxonium ion.

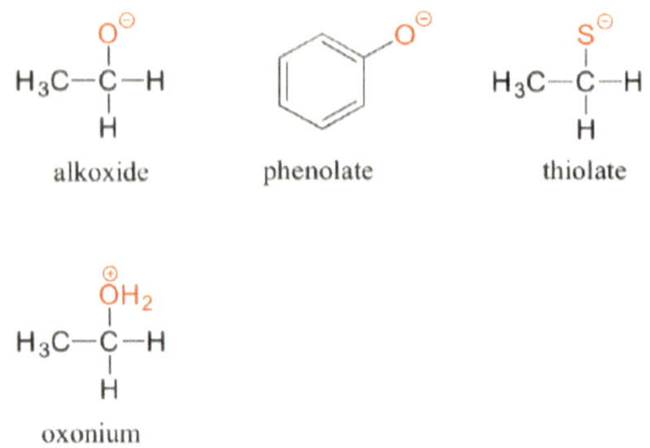

In an ether functional group, central oxygen is bonded to two carbons. Below is the structure of diethyl ether, a common laboratory solvent and also one of the first compounds to be used as an anesthetic during operations. The sulfur analog of an ether is called a thioether or sulfide.

Ethers and Sulfides

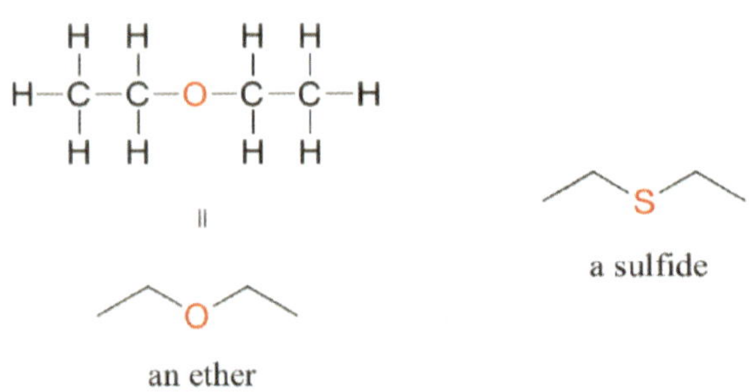

Amines are characterized by nitrogen atoms with single bonds to hydrogen and carbon. Just as there are primary, secondary, and tertiary alcohols, there are primary, secondary, and tertiary amines. Ammonia is a special case with no carbon atoms.

One of the most important properties of amines is that they are basic, and are readily protonated to form ammonium cations. In the case where a nitrogen has four bonds to carbon (which is somewhat unusual in biomolecules), it is called a quaternary ammonium ion.

Amines

ammonia — a primary amine — a secondary amine — a tertiary amine

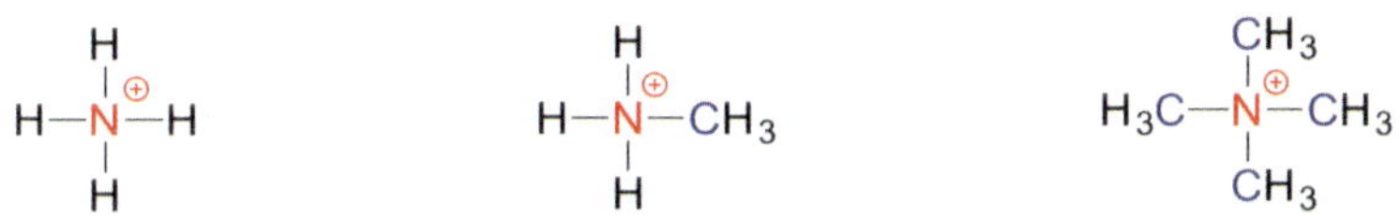

ammonium ion — a primary ammonium ion — a quaternary ammonium ion

There is a difference in the usage of the terms 'primary', 'secondary', and 'tertiary' when they are applied to alcohols and amines for the definitions are different. In alcohols, what matters is how many other carbons, the alcohol *carbon* is bonded to, while in amines, what matters is how many carbons the *nitrogen* is bonded to.

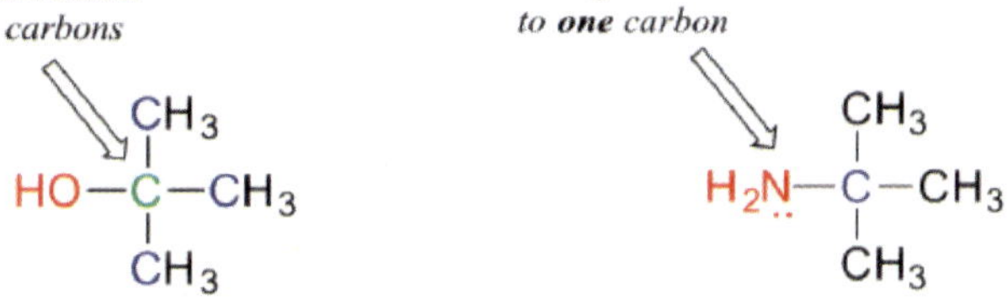

a *tertiary* alcohol — a *primary* amine

Phosphate and its derivative functional groups are ubiquitous in biomolecules. Phosphate linked to a single organic group is called a phosphate ester; when it has two links to organic groups it is called a phosphate diester. A linkage between two phosphates creates a phosphate anhydride.

Organic phosphates

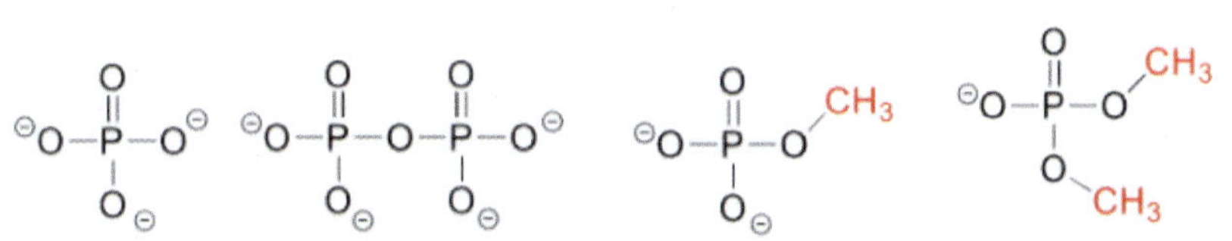

phosphate — a phosphate anhydride — a phosphate ester — a phosphate diester

There are a number of functional groups that contain a carbon-oxygen double bond, which is commonly referred to as a carbonyl. Ketones and aldehydes are two closely related carbonyl-based functional groups that react in very similar ways. In a ketone, the carbon atom of a carbonyl is bonded to two other carbons. In an aldehyde, the carbonyl carbon is bonded on one side to a hydrogen, and on the other side to a carbon. The exception to this definition is formaldehyde, in which the carbonyl carbon has bonds to two hydrogens.

A group with a carbon-nitrogen double bond is called an imine, or sometimes a Schiff base (in this book we will use the term 'imine').

Aldehydes, Ketones, and Imines

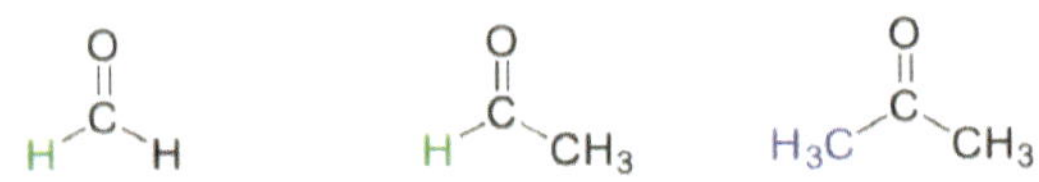

formaldehyde an aldehyde a ketone

imines

When a carbonyl carbon is bonded on one side to a carbon (or hydrogen) and on the other side to an oxygen, nitrogen, or sulfur, the functional group is considered to be one of the 'carboxylic acid derivatives', a designation that describes a set of related functional groups. The eponymous member of this family is the carboxylic acid functional group, in which the carbonyl is bonded to a hydroxyl group. The conjugate base of a carboxylic acid is a carboxylate. Other derivatives are carboxylic esters (usually just called 'esters'), thioesters, amides, acyl phosphates, acid chlorides, and acid anhydrides. With the exception of acid chlorides and acid anhydrides, the carboxylic acid derivatives are very common in biological molecules and metabolic pathways.

Carboxylic Acid Derivatives

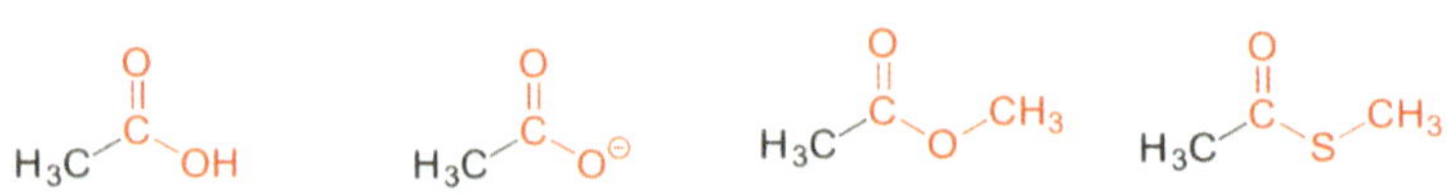

a carboxylic acid a carboxylate ion a carboxylic ester a thioester

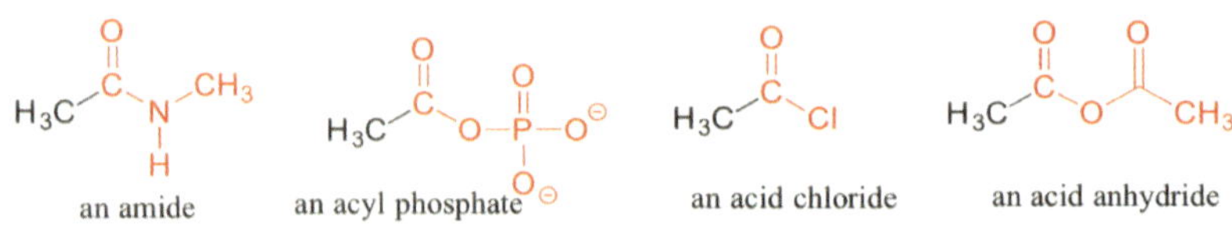

an amide an acyl phosphate an acid chloride an acid anhydride

Finally, a nitrile group is characterized by a carbon triple-bonded to nitrogen.

Nitriles

$$:N\equiv C-CH_3$$

acetonitrile

A single compound often contains several functional groups, particularly in biological organic chemistry. The six-carbon sugar molecules glucose and fructose, for example, contain aldehyde and ketone groups, respectively, and both contain five alcohol groups (a compound with several alcohol groups is often referred to as a 'polyol').

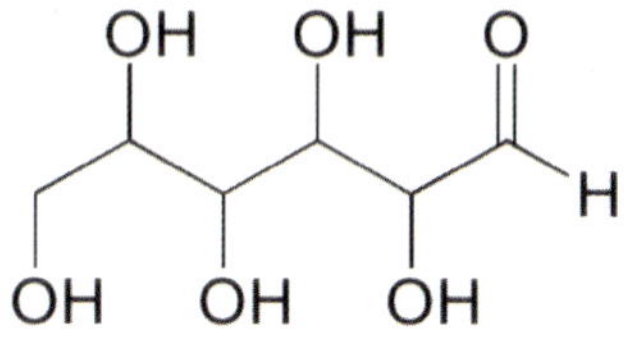

glucose

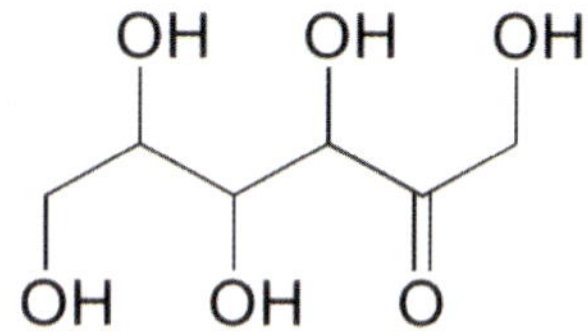

fructose

The hormone testosterone, the amino acid phenylalanine, and the glycolysis metabolite dihydroxy-acetone phosphate all contain multiple functional groups, as labeled below:

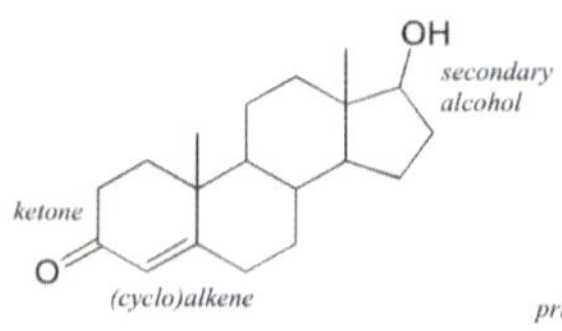

testosterone

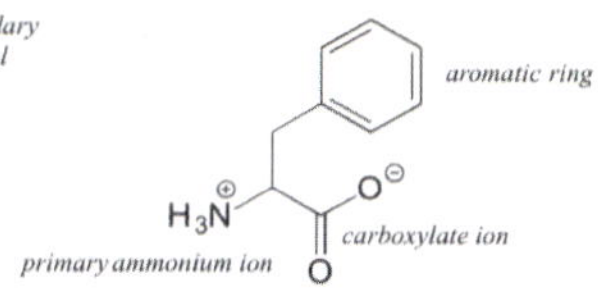

phenylalanine

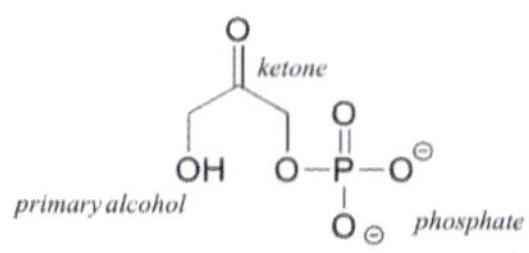

dihydroxyacetone phosphate (DHAP)

While not in any way a complete list, the subject matter here covers most of the important functional groups that we encounter in biological organic chemistry.

Hydrocarbons

Hydrocarbons are obtained by refining petroleum at refineries such as this one.

A hydrocarbon is any chemical compound that is constituted of just the elements carbon (C) and hydrogen (H). Each hydrocarbon molecule consists of a carbon backbone, or "carbon skeleton," with hydrogen atoms attached to that backbone.

Hydrocarbons are among the Earth's most important natural resources. They are currently the main source of the world's electric energy and heat energy (such as for heating buildings) because they produce large amounts of heat when burned. The gasoline that serves as fuel for automobiles consists primarily of hydrocarbons. In addition, many hydrocarbons serve as base materials for the synthesis of organic chemicals used in the production of consumer products and industrial materials.

Natural Occurrence and Extraction

Hydrocarbons are the main constituents of petroleum (literally, "rock oil"), also called "oil," and natural gas. They are commonly found in and extracted from the Earth's subsurface. Petroleum is a mixture of liquid hydrocarbons, while natural gas is mainly constituted of methane gas.

The extraction of liquid hydrocarbon fuel from a number of sedimentary basins has been integral to modern energy development. Hydrocarbons are mined from tar sands and oil shale. These reserves require distillation and upgrading to produce synthetic crude and petroleum. A future source of methane may be methane hydrates found on ocean floors.

Types of Hydrocarbons

	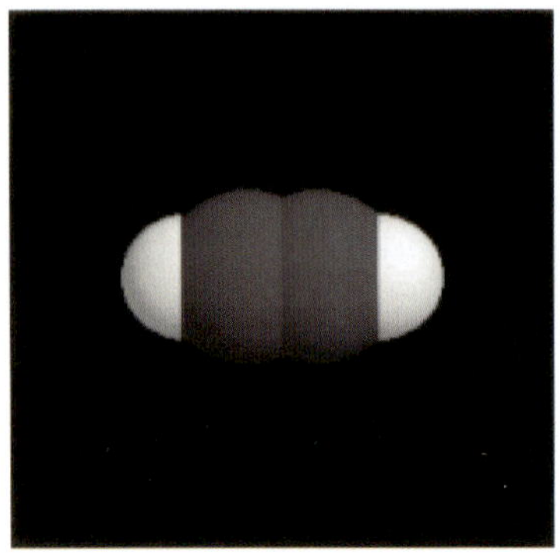	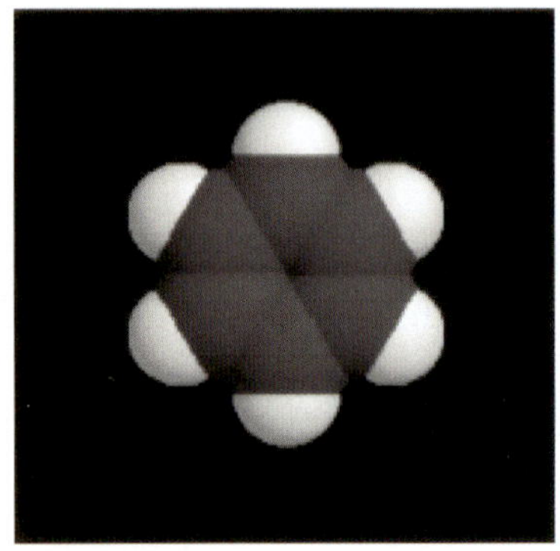
A model of the alkane known as *hexane.*	A model of the alkyne called *ethyne.*	A model of an arene called *benzene.*

There are essentially three types of hydrocarbons: Saturated hydrocarbons, also known as alkanes: In each molecule of an alkane, the chemical bonds that join the carbon atoms are single covalent bonds. If the alkane molecule includes a ring of carbon atoms (all connected by single covalent bonds), it is called a cycloalkane. Unsaturated hydrocarbons, which are subdivided into two groups:

- Alkenes: Each molecule of an alkene contains at least one double covalent bond between carbon atoms.
- Alkynes: Each molecule of an alkyne contains at least one triple covalent bond between carbon atoms.
- Aromatic hydrocarbons, or arenes: Each molecule of an aromatic hydrocarbon contains at least one aromatic ring, in which the bonds between carbon atoms are aromatic bonds.

When organic compounds are considered in general, saturated and unsaturated hydrocarbons are placed in the category known as aliphatic compounds, while aromatic hydrocarbons are categorized as aromatic compounds.

Some Simple Hydrocarbons

The simplest hydrocarbon is methane, the main constituent of natural gas. Its chemical formula, CH_4, indicates that each molecule of methane consists of four hydrogen atoms attached to one carbon atom. Ethane, with the formula C_2H_6, is a hydrocarbon (more specifically, an alkane) in which each molecule has two carbon atoms held together with a single covalent bond, and three hydrogen atoms are bound to each carbon atom. Each molecule of propane (C_3H_8) has three carbon atoms, and each molecule of butane (C_4H_{10}) has four carbons.

General Chemical Formulas

If a hydrocarbon molecule contains "n" carbon atoms, and the type of hydrocarbon is known, its general chemical formula can be written as follows:

- Alkane: C_nH_{2n+2}.
- Cycloalkane: C_nH_{2n} (assuming a single ring, with all the carbon atoms in the ring structure).
- Alkene: C_nH_{2n} (assuming only one double bond in each molecule).
- Alkyne: C_nH_{2n-2} (assuming only one triple bond in each molecule).
- Aromatic ring: C_nH_n (assuming a single ring, with all the carbon atoms in the ring structure).

The above chemical formulas are based on the assumption that each carbon atom forms four covalent bonds, including bonds with hydrogen atoms and other carbon atoms. A double covalent bond counts as two bonds; a triple covalent bond counts as three bonds. A carbon atom in an aromatic ring is a special case and can have only one hydrogen atom attached to it.

Molecular Graph

A "molecular graph" of a hydrocarbon is a simple representation of the carbon skeleton of the molecule under consideration. Each line in the molecular graph represents a covalent bond that connects two carbon atoms. Thus, although the graph shows no symbols for carbon atoms, carbon atoms are (by convention) located at each end of each line. The symbols for hydrogen atoms are not shown either, but each carbon atom holds as many hydrogen atoms as it needs to form four covalent bonds.

Uses of Hydrocarbons

Most hydrocarbons are combustible. When burned, they produce carbon dioxide, water, and large quantities of heat. This energy is used to heat homes and other buildings and to generate electricity. When heating a home, for example, oil or natural gas is burned and the energy released is used to heat water or air. The hot water or air is then circulated around the building.

A similar principle is used to create electric energy in power plants. A hydrocarbon fuel (such as natural gas) is burned, the energy released is used to convert water into steam, and the steam is used to drive turbines that generate electricity.

Hydrocarbons are also raw materials that serve as feedstock for the production of a wide range of organic chemicals, which in turn are used for such products as plastics, pigments, solvents, pharmaceuticals, and explosives.

Effects on Health and the Environment

Hydrocarbon vapors can be harmful if inhaled. Moreover, hydrocarbons contribute to the formation of ozone in the troposphere. For these reasons, hydrocarbons in the atmosphere are considered to be pollutants.

Ideally, the combustion of hydrocarbons should produce only carbon dioxide, water, and heat. Yet, incomplete combustion leads to the production of carbon monoxide, a toxic gas. Carbon monoxide binds to hemoglobin (in the blood) more readily than oxygen does; so, when carbon monoxide is inhaled, it blocks oxygen from being absorbed and leads to suffocation. Moreover, the hydrocarbon fuel being burned may contain other substances that are harmful when released.

Moreover, carbon dioxide, a product of the combustion of hydrocarbons, is a greenhouse gas—a gas that helps trap heat in the Earth's atmosphere. This has led to concerns that the excessive burning of hydrocarbon fuels contributes to global warming.

IUPAC Nomenclature of Organic Chemistry

The base part of the name reflects the number of carbons in what you have assigned to be the parent chain. The suffix of the name reflects the type(s) of functional group(s) present on (or within) the parent chain. Other groups which are attached to the parent chain are called substituents.

Alkanes - Saturated Hydrocarbons

The names of the straight chain saturated hydrocarbons for up to a 12 carbon chain are shown below. The names of the substituents formed by the removal of one hydrogen from the end of the chain is obtained by changing the suffix -ane to -yl.

Number of Carbons	Name
1	methane
2	ethane
3	propane
4	butane
5	pentane
6	hexane
7	heptane
8	octane

9	nonane
10	decane
11	undecane
12	dodecane

There are a few common branched substituents which you should memorize. These are shown below:

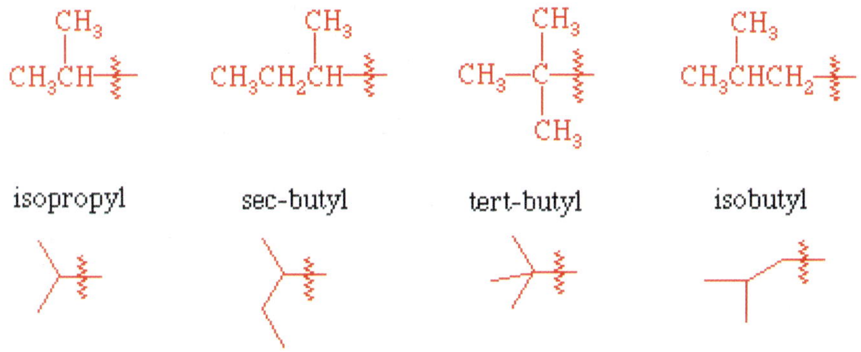

Here is a simple list of rules to follow, some examples are given at the end of the list:

- Identify the longest carbon chain. This chain is called the parent chain.
- Identify all of the substituents (groups appending from the parent chain).
- Number the carbons of the parent chain from the end that gives the substituents the lowest numbers. When compairing a series of numbers, the series that is the "lowest" is the one which contains the lowest number at the occasion of the first difference. If two or more side chains are in equivalent positions, assign the lowest number to the one which will come first in the name.
- If the same substituent occurs more than once, the location of each point on which the substituent occurs is given. In addition, the number of times the substituent group occurs is indicated by a prefix (di, tri, tetra, etc.).
- If there are two or more different substituents they are listed in alphabetical order using the base name (ignore the prefixes). The only prefix which is used when putting the substituents in alphabetical order is iso as in isopropyl or isobutyl. The prefixes sec- and tert- are not used in determining alphabetical order except when compared with each other.
- If chains of equal length are competing for selection as the parent chain, then the choice goes in series to:
    - The chain which has the greatest number of side chains;
    - The chain whose substituents have the lowest- numbers;
    - The chain having the greatest number of carbon atoms in the smaller side chain;
    - The chain having the least branched side chains.

- A cyclic (ring) hydrocarbon is designated by the prefix cyclo-which appears directly in front of the base name.

In summary, the name of the compound is written out with the substituents in alphabetical order followed by the base name (derived from the number of carbons in the parent chain). Commas are used between numbers and dashes are used between letters and numbers. There are no spaces in the name.

Here are some examples:

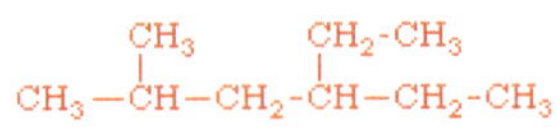

4-ethyl-2-methylhexane

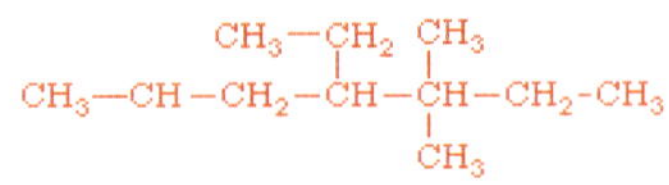

4-ethyl-3,3-dimethylheptane

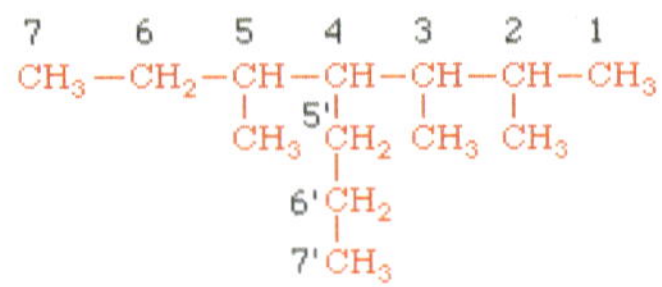

2,3,5-trimethyl-4-propylheptane
(NOT: 2,3-dimethyl-4-sec-butylheptane)

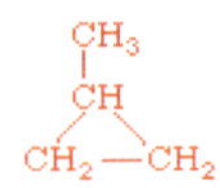

methylcyclopropane

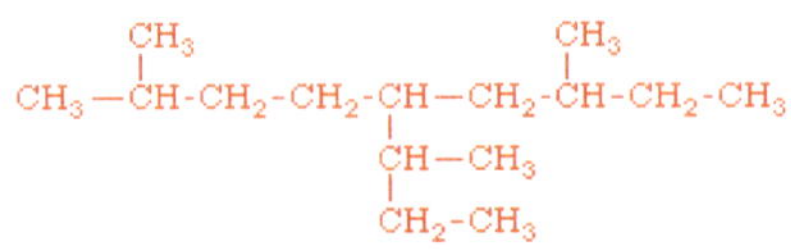

5-sec-butyl-2,7-dimethylnonane

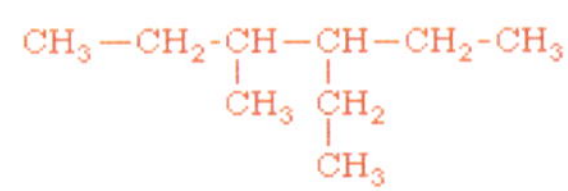

3-ethyl-4-methylhexane

Alkyl Halides

The halogen is treated as a substituent on an alkane chain. The halo- substituent is considered of equal rank with an alkyl substituent in the numbering of the parent chain. The halogens are represented as follows:

F	fluoro-
Cl	chloro-
Br	bromo-
I	iodo-

Here are some examples:

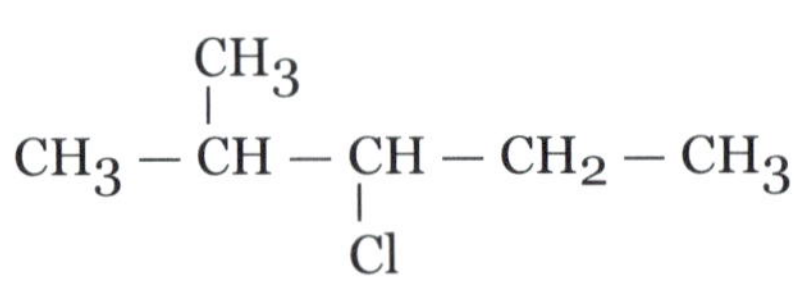

3-chloro-2-methylpentane

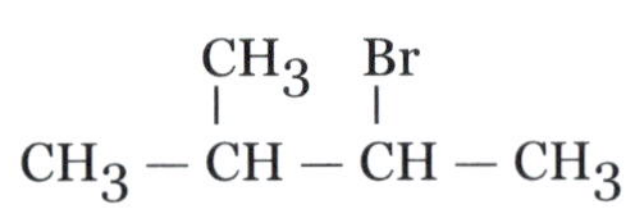

2-bromo-3-methylbutane

Alkenes and Alkynes - Unsaturated Hydrocarbons

Double bonds in hydrocarbons are indicated by replacing the suffix -anewith -ene. If there is more than one double bond, the suffix is expanded to include a prefix that indicates the number of double bonds present (-adiene, -atriene, etc.). Triple bonds are named in a similar way using the suffix -yne. The position of the multiple bond(s) within the parent chain is indicated by placing the number(s) of the first carbon of the multiple bond(s) directly in front of the base name.

Here is an important list of rules to follow:

- The parent chain is numbered so that the multiple bonds have the lowest numbers (double and triple bonds have priority over alkyl and halo substituents).
- When both double and triple bonds are present, numbers as low as possible are given to double and triple bonds even though this may at times give "-yne" a lower number than "-ene". When there is a choice in numbering, the double bonds are given the lowest numbers.
- When both double and triple bonds are present, the -en suffix follows the parent chain directly and the -yne suffix follows the -en suffix (notice that the e is left off, -en instead of -ene). The location of the double bond(s) is indicated before the parent name as before, and the location of the triple bond(s) is indicated between the -en and -yne suffixes.
- For a branched unsaturated acyclic hydrocarbon, the parent chain is the longest carbon chain that contains the maximum number of double and triple bonds. If there are two or more chains competing for selection as the parent chain (chain with the most multiple bonds), the choice goes to:
    - The chain with the greatest number of carbon atoms,
    - The carbon atoms being equal, the chain containing the maximum number of double bonds.
- If there is a choice in numbering not previously covered, the parent chain is numbered to give the substituents the lowest number at the first point of difference.

Here are some examples:

$CH_3{-}CH{=}CH{-}CH_2{-}CH{=}CH_2$

1,4-hexadiene

$CH{\equiv}C{-}CH{=}CH{-}CH{=}CH_2$

1,3-hexadien-5-yne

$CH_3{-}CH{=}CH{-}C{\equiv}CH$

3-penten-1-yne

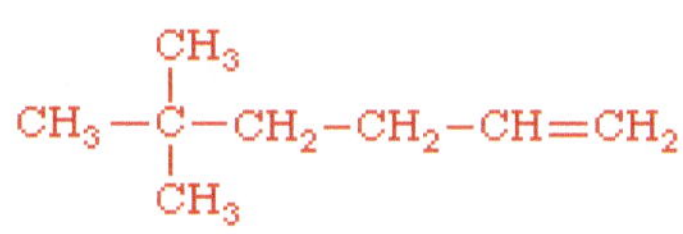

5,5-dimethyl-1-hexene

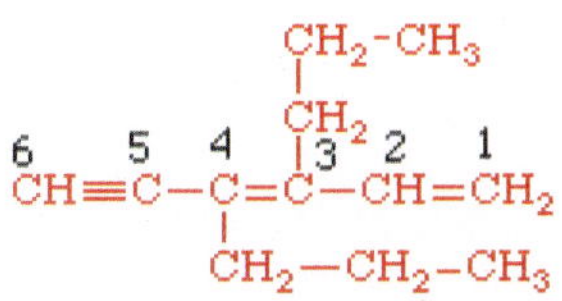

3,4-dipropyl-1,3-hexadien-5-yne

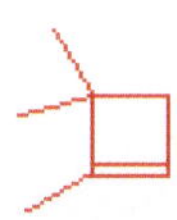

1,4,4-trimethylcyclobutene
(NOT: 2,3,3-trimethylcyclobutene)

Alcohols

Alcohols are named by replacing the suffix -ane with -anol. If there is more than one hydroxyl group (-OH), the suffix is expanded to include a prefix that indicates the number of hydroxyl groups present (-anediol, -anetriol, etc.). The position of the hydroxyl group(s) on the parent chain is indicated by placing the number(s) corresponding to the location(s) on the parent chain directly in front of the base name (same as alkenes).

Here is an important list of rules to follow:

- The hydroxyl group takes precedence over alkyl groups and halogen substituents, as well as double bonds, in the numbering of the parent chain.
- When both double bonds and hydroxyl groups are present, the -en suffix follows the parent chain directly and the -ol suffix follows the -en suffix (notice that the e is left off, -en instead of -ene). The location of the double bond(s) is indicated before the parent name as before, and the location of the hydroxyl group(s) is indicated between the -en and -ol suffixes. See below for examples. Again, the hydroxyl gets priority in the numbering of the parent chain.
- If there is a choice in numbering not previously covered, the parent chain is numbered to give the substituents the lowest number at the first point of difference.

Here are some examples:

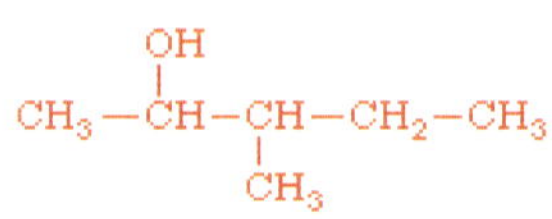
3-methyl-2-pentanol

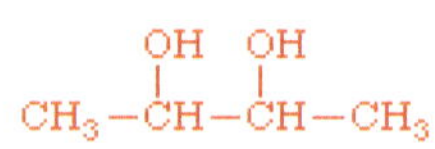
2,3-butanediol

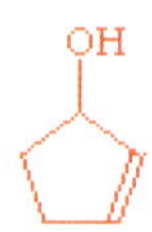
2-cyclopenten-1-ol

Ethers

You are only expected to know how to name ethers by their commmon names. The two alkyl groups attached to the oxygen are put in alphabetical order with spaces between the names and they are followed by the word ether. The prefix di- is used if both alkyl groups are the same.

Here are some examples:

$$CH_3 - CH_2 - O - CH_2 - CH_3$$
diethyl ether

$$CH_3 - O - CH_2 - CH_3$$
ethyl methyl ether

Aldehydes

Aldehydes are named by replacing the suffix -ane with -anal. If there is more than one -CHO group, the suffix is expanded to include a prefix that indicates the number of -CHO groups present

(-anedial - there should not be more than 2 of these groups on the parent chain as they must occur at the ends). It is not necessary to indicate the position of the -CHO group because this group will be at the end of the parent chain and its carbon is automatically assigned as C-1.

Here is an important list of rules to follow:

- The carbonyl group takes precedence over alkyl groups and halogen substituents, as well as double bonds, in the numbering of the parent chain.
- When both double bonds and carbonyl groups are present, the -en suffix follows the parent chain directly and the -al suffix follows the -en suffix (notice that the e is left off, -en instead of -ene). The location of the double bond(s) is(are) indicated before the parent name as before, and the -al suffix follows the -en suffix directly. Remember it is not necessary to specify the location of the carbonyl group because it will automatically be carbon 1. Again, the carbonyl gets priority in the numbering of the parent chain.
- There are a couple of common names which are acceptable as IUPAC names. They are shown in the examples at the end of this list but at this point these names will not be accepted by the computer. Eventually they will be accepted.
- If there is a choice in numbering not previously covered, the parent chain is numbered to give the substituents the lowest number at the first point of difference.

Here are some examples:

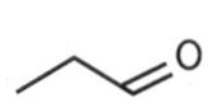

propanal

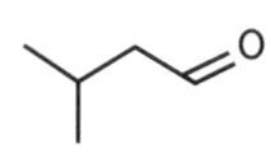

3-methylbutanal

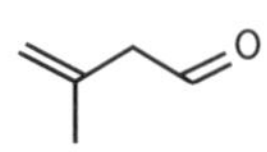

3-methyl-3-butanal

methanal
(common name: formaldehyde)

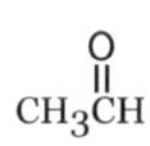

ethanal
(common name: acetaldehyde)

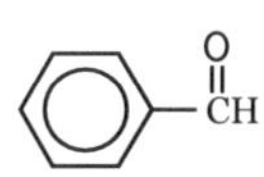

Benzaldehyde

Ketones

Ketones are named by replacing the suffix -ane with -anone. If there is more than one carbonyl group (C=O), the suffix is expanded to include a prefix that indicates the number of carbonyl groups present (-anedione, -anetrione, etc). The position of the carbonyl group(s) on the parent chain is indicated by placing the number(s) corresponding to the location(s) on the parent chain directly in front of the base name (same as alkenes).

Here is an important list of rules to follow:

- The carbonyl group takes precedence over alkyl groups and halogen substituents, as well as double bonds, in the numbering of the parent chain.

- When both double bonds and carbonyl groups are present, the -en suffix follows the parent chain directly and the -one suffix follows the -en suffix (notice that the e is left off, -en instead of -ene). The location of the double bond(s) is indicated before the parent name as before, and the location of the carbonyl group(s) is(are) indicated between the -en and -one suffixes.
- If there is a choice in numbering not previously covered, the parent chain is numbered to give the substituents the lowest number at the first point of difference.

Here are some examples,

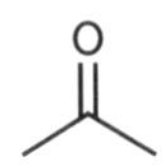

propanone
(common name : acetone

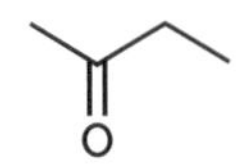

2-butanone

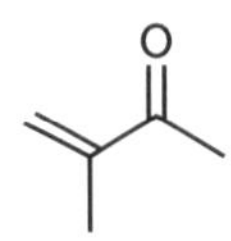

3-methyl-3-buten-2-one

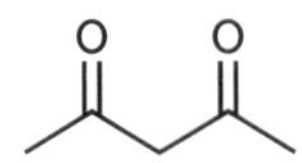

2, 4-pentanedione

Carboxylic Acids

Carboxylic acids are named by counting the number of carbons in the longest continuous chain including the carboxyl group and by replacing the suffix -ane of the corresponding alkane with -anoic acid. If there are two -COOH groups, the suffix is expanded to include a prefix that indicates the number of -COOH groups present (-anedioic acid - there should not be more than 2 of these groups on the parent chain as they must occur at the ends). It is not necessary to indicate the position of the -COOH group because this group will be at the end of the parent chain and its carbon is automatically assigned as C-1.

Here is an important list of rules to follow:

- The carboxyl group takes precedence over alkyl groups and halogen substituents, as well as double bonds, in the numbering of the parent chain.
- If the carboxyl group is attached to a ring the parent ring is named and the suffix -carboxylic acid is added.
- When both double bonds and carboxyl groups are present, the -en suffix follows the parent chain directly and the -oic acid suffix follows the -en suffix (notice that the e is left off, -en instead of -ene). The location of the double bond(s) is indicated before the parent name as before, and the -oic acid suffix follows the -en suffix directly. Remember it is not necessary to specify the location of the carboxyl group because it will automatically be carbon.
- There are several common names which are acceptable as IUPAC names. They are shown in the examples at the end of this list butat this point these names will not be accepted by the computer. Eventually they will be accepted.

- If there is a choice in numbering not previously covered, the parent chain is numbered to give the substituents the lowest number at the first point of difference.

Here are some examples:

methanoic acid
(common name : formic acid)

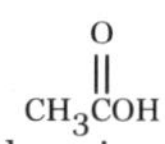
ethanoic acid
(common name : acetic acid)

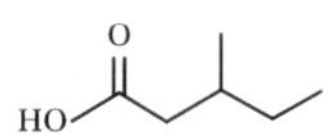
3-methylpentanoic acid

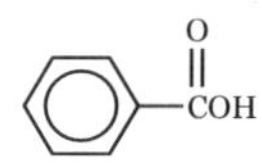
benzoic acid

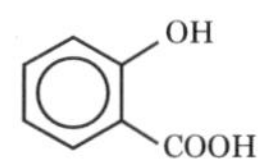
salicylic acid
(common name)

ethanedioic acid
(common name : oxalic acid)

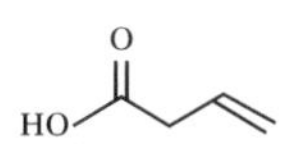
3-butenoic acid

Esters

Systematic names of esters are based on the name of the corresponding carboxylic acid. Remember esters look like this:

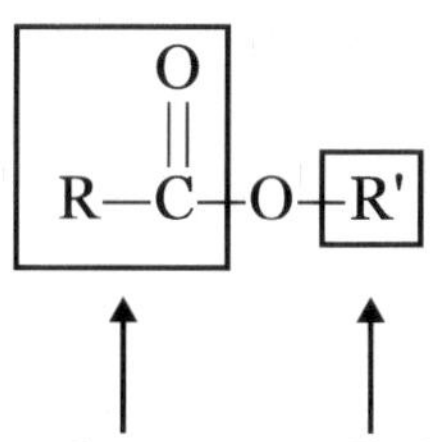
acyl group alkyl group

The alkyl group is named like a substituent using the -yl ending. This is followed by a space. The acyl portion of the name (what is left over) is named by replacing the -ic acid suffix of the corresponding carboxylic acid with -ate.

Here are some examples:

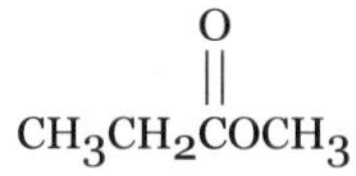
methyl propanoate

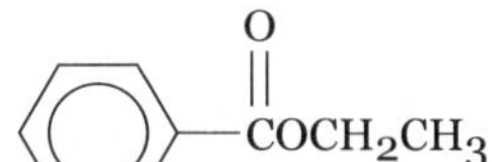
ethyl benzoate

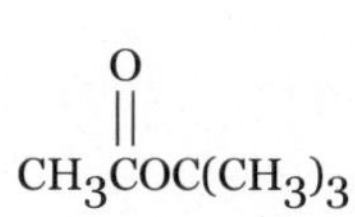
tert-butyl acetate

Amines

You are only expected to know how to name amines by their common names. They are named like ethers, the alkyl (R) groups attached to the nitrogen are put in alphabetical order with no spaces between the names and these are followed by the word amine. The prefixes di- and tri- are used if two or three of the alkyl groups are the same.

Note: Some books put spaces between the parts of the name, but we will not.

Here are some examples:

$NH(CH_3)_2$

dimethylamine

CH_3CH_2NH
$\qquad\qquad |$
$\qquad\quad CH_3$

ethylmethylamine

Heterocyclic Compound

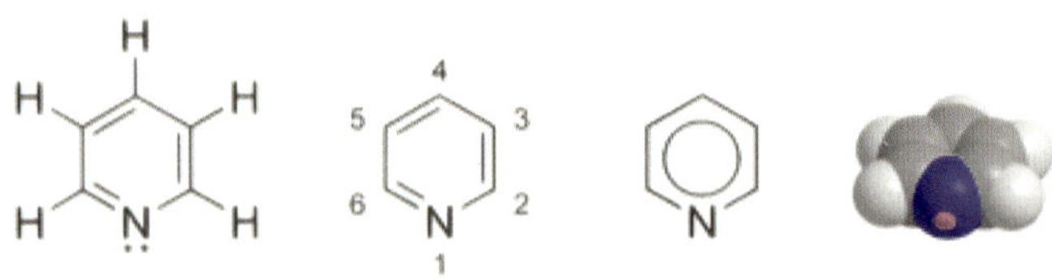

Various representations of a molecule of pyridine. The hexagonal ring is composed of five carbon.

Heterocyclic compounds are organic compounds that contain a ring structure containing atoms in addition to carbon, such as sulfur, oxygen or nitrogen, as part of the ring. They may be either simple aromatic rings or non-aromatic rings. Some examples are pyridine (C_5H_5N), pyrimidine ($C_4H_4N_2$) and dioxane ($C_4H_8O_2$).

Note that compounds such as cyclopropane, an anaesthetic with explosive properties, and cyclohexane, a solvent, are not heterocyclic, they are merely cycloalkanes. The suffix '-cyclic' implies a ring structure, while 'hetero' refers to an atom other than carbon, as above. Many heterocyclic compounds, including some amines, are carcinogenic.

Heterocyclic chemistry is the chemistry branch dealing exclusively with synthesis, properties and applications of heterocycles especially vital to drug design.

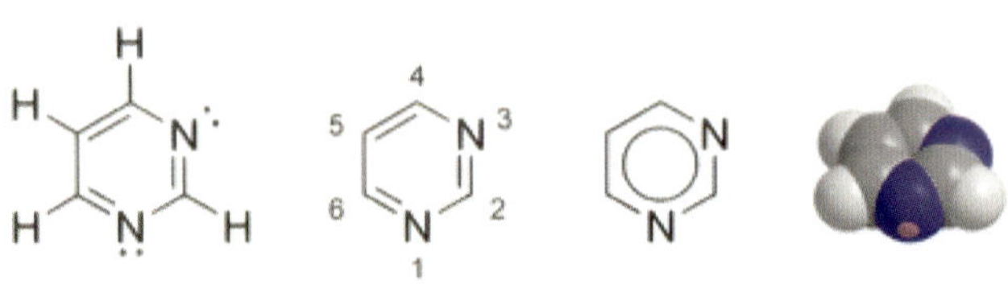

Various representations of a molecule of pyrimidine. The hexagonal ring is composed of four carbon atoms (at the four unmarked corners) and two nitrogen (N) atoms.

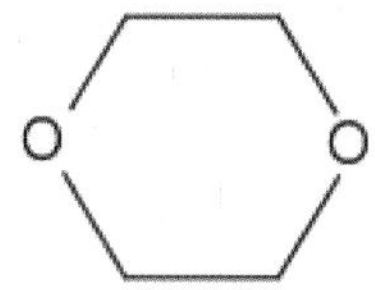

A molecule of 1,4-dioxane has two oxygen (O) atoms in its ring structure.

3-membered Rings

Heterocycles with three atoms in the ring are more reactive because of ring strain. Those containing one heteroatom are generally stable. Those with two heteroatoms are more likely to occur as reactive intermediates. Common 3-membered heterocycles are:

heteroatom	saturated	unsaturated
Nitrogen	aziridine	
Oxygen	ethylene oxide (epoxides, oxiranes)	oxirene
Sulphur	thiirane (episulfides)	

4-membered Rings

heteroatom	saturated	unsaturated
Nitrogen	azetidine	
Oxygen	oxetane	

5-membered Rings

With heterocycles containing five atoms, the unsaturated compounds are frequently more stable because of aromaticity.

heteroatom	saturated	unsaturated
Nitrogen	dihydropyrrole (pyrroline) & tetrahydropyrrole (pyrrolidine)	pyrrole
Oxygen	dihydrofuran & tetrahydrofuran	furan
Sulphur	dihydrothiophene & tetrahydrothiophene	thiophene (thiole)
Arsenic		arsole

With two heteroatoms:

- The azoles:
    - Two N: pyrazole & imidazole;
    - Three N: triazole;

    - Four N: tetrazole;
    - One N and one O: oxazole & isoxazole;
    - One N and one S: thiazole & isothiazole;
- Two S: Dithiolane.

6-membered Rings

heteroatom	saturated	unsaturated
Nitrogen	piperidine	pyridine
Oxygen	tetrahydropyran	pyran
Sulphur	Thiane	Thiine aka thiapyrane

With two heteroatoms:

- Two N: Pyridazine, Pyrimidine, and Pyrazine are the 1,2-, 1,3-, and 1,4-isomers, respectively;
- Two N: Piperazine;
- One N and one O: Oxazines;
- One N and one S: Thiazine;
- Two S: Dithiane;
- Two O: Dioxane.

Heterocyclic Amines and Cancer

Some heterocyclic amines (HCAs) found in cooked meat are known carcinogens. Research has shown that cooking certain meats at high temperatures creates chemicals that are not present in uncooked meats. For example, heterocyclic amines are the carcinogenic chemicals formed from the cooking of muscle meats such as beef, pork, fowl, and fish. HCAs form when amino acids and creatine (a chemical found in muscles) react at high cooking temperatures. Researchers have identified 17 different HCAs resulting from the cooking of muscle meats that may pose human cancer risk. NCI's Division of Cancer Epidemiology and Genetics found a link between individuals with stomach cancer and the consumption of cooked meat, and other studies for colorectal, pancreatic, and breast cancer is associated with high intakes of well-done, fried, or barbecued meats. Other sources of protein (milk, eggs, tofu, and organ meats such as liver) have very little or no HCA content naturally or when cooked.

References

- Caminiti, R.; Pieretti, A.; Bencivenni, L.; Ramondo, F.; Sanna, N. (1996). "Amidine N–C(N)–N Skeleton: Its Structure in Isolated and Hydrogen-Bonded Guanidines from ab Initio Calculations". J. Phys. Chem. 100: 10928–10935. doi:10.1021/jp960311p
- Properties-of-aromatic-compounds: lumenlearning.com, Retrieved 19 May 2018

- Campbell, Neil A.; Williamson, Brad; Heyden, Robin J. (2006). Biology: Exploring Life. Boston, MA: Pearson Prentice Hall. ISBN 0-13-250882-6. Retrieved 2012-02-05
- Covalent-bond, science: britannica.com, Retrieved 11 July 2018
- Langmuir, Irving (1919-06-01). "The Arrangement of Electrons in Atoms and Molecules". Journal of the American Chemical Society. 41 (6): 868–934. doi:10.1021/ja02227a002
- Formation-of-sigma-a-and-pi-n-bonds: chemistry-assignment.com, Retrieved 22 June 2018
- March, Jerry (1992). Advanced Organic Chemistry: Reactions, Mechanisms, and Structure. John Wiley & Sons. ISBN 0-471-60180-2
- Hydrocarbon: newworldencyclopedia.org, Retrieved 29 March 2018
- Wiberg, Kenneth B. (1990). "Resonance interactions in acyclic systems. 2. Y-Conjugated anions and cations". J. Am. Chem. Soc. 112: 4177–4182. doi:10.1021/ja00167a011

7

Branches of Organic Chemistry

Organic chemistry is a vast subject that branches out into a number of significant sub-disciplines such as organometallic chemistry, biomolecular chemistry, heterocyclic chemistry, organosulfur chemistry and organophosphorus chemistry, which have been thoroughly discussed in this chapter.

Organometallic Chemistry

Organometallic chemistry is the study of chemical compounds containing bonds between carbon and metal atoms. In more general terms, it is the study of compounds containing metal-element bonds that are largely covalent in character. Organometallic chemistry combines aspects of inorganic chemistry and organic chemistry. Living systems contain a variety of organometallic compounds, prominent examples being hemoglobin and chlorophyll. The specialized field focused on the study of such compounds is known as bioinorganic chemistry.

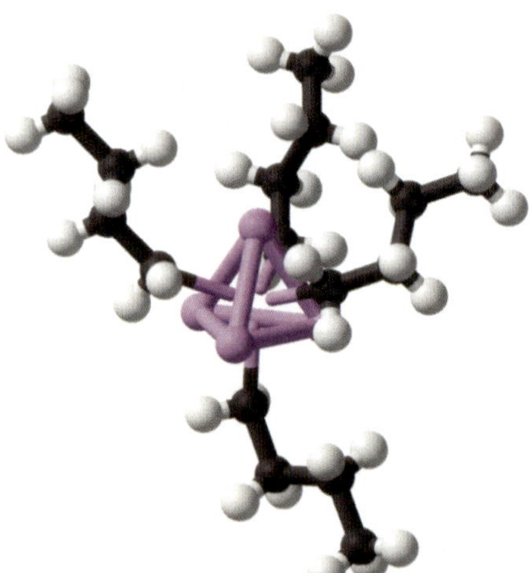

A model of a molecule of *n*-butyllithium, an organometallic compound.

Organometallic compounds have a variety of practical uses. For example, ferrocene and MMT (methylcyclopentadienyl manganese tricarbonyl) are added to gasoline as antiknock agents. The industrial manufacture of acetic acid from methanol and carbon monoxide (by the Monsanto process) involves the use of a rhodium-carbonyl complex. A titanium-based organometallic compound, called the Ziegler-Natta catalyst, is used in the production of polyethylene and other polymers. Also, the ruthenium-BINAP complex is used in the production of fine chemicals and pharmaceuticals.

A number of researchers have been awarded the Nobel Prize in Chemistry for their work in the area of organometallic chemistry. For example, the 1973 Nobel Prize was awarded to Ernst Fischer and Geoffrey Wilkinson for their work on metallocenes. In 2005, Yves Chauvin, Robert H. Grubbs, and Richard R. Schrock shared the Nobel Prize for their work on metal-catalyzed olefin metathesis.

Organometallic Compounds

Organometallic compounds are often distinguished by the prefix "organo-," as in organopalladium compounds. They are also known as organo-inorganics, metallo-organics, and metal organics. Examples of such organometallic compounds include all Gilman reagents that contain lithium and copper, and Grignard reagents that contain magnesium. Tetracarbonyl nickel and ferrocene are examples of organometallic compounds containing transition metals.

In addition to the traditional metals and semimetals, elements such as boron, silicon, arsenic, and selenium are included to form organometallic compounds. Examples include organomagnesium compounds such as iodo(methyl)magnesium MeMgI, diethylmagnesium (Et_2Mg); organolithium compounds such as butyllithium (BuLi), organozinc compounds such as chloro(ethoxycarbonylmethyl)zinc ($ClZnCH_2C(=O)OEt$); organocopper compounds such as lithium dimethylcuprate ($Li^+[CuMe_2]^-$); and organoborane compounds such as triethylborane (Et_3B).

Biological systems contain a variety of organometallic compounds. Examples include hemoglobinand myoglobin, each of which contains an iron center bonded to a porphyrin ring; and chlorophyll, with magnesium at the center of a chlorin ring.

Structure and Properties

In the absence of direct structural evidence for a carbon–metal bond, some compounds are not considered to be organometallic. Thus, the status of compounds in which the canonical anion has a delocalized structure in which the negative charge is shared with an atom more electronegative than carbon (as in enolates) may vary with the nature of the anionic moiety, the metal ion, and possibly the medium.

Depending mostly on the nature of the metallic ion and somewhat on the nature of the organic compound, the character of the bond may either be ionic or covalent. Organic compounds bonded to sodium or potassium are primarily ionic. Those bonded to lead, tin, or mercury, are considered as having covalent bonds, and those bonded to magnesium or lithium have bonds with intermediate properties.

Organometallic compounds with bonds that have characters in between ionic and covalent are very important in industry, as they are both relatively stable in solution and relatively ionic to undergo reactions. Two important classes are organolithium compounds and Grignard reagents. In certain organometallic compounds, such as ferrocene or dibenzenechromium, the pi orbitals of the organic moiety are believed to be involved in binding to the metal.

Electron Counting

Electron counting is a key to understanding organometallic chemistry. The 18 electron rule is helpful in predicting the stabilities of organometallic compounds. Organometallic compounds that have 18 electrons (filled s, p, and penultimate d orbitals) are relatively stable. This suggests the compound is isolobal, but it can result in the compound being inert.

To understand chemical bonding and reactivity in organometallic compounds the isolobal principle should be used. NMR and infrared spectroscopy are common techniques used to determine structure and bonding in this field.

Reactions

Organometallic compounds undergo several important reactions. They include the following:

- Oxidative addition: A metal within an organometallic complex and at a relatively low oxidation state is oxidized by inserting the metal into a covalent bond (X-Y).
- Reductive elimination: The reverse of oxidative addition.
- Transmetalation (or transmetallation): The exchange of ligands between two metal centers.
- Carbometalation: The nucleophilic addition to alkenes and alkynes of a range of organometallic compounds.
- Hydrometalation (or hydrometallation): A chemical compound with a hydrogen-to-metal bond (M-H, metal hydride) adds to a compound with an unsaturated bond such as an alkene (RC=CR).
- Electron transfer: The transfer of an electron from one atom or molecule to another.
- Beta-hydride elimination: An alkyl group bonded to a metal center is converted into the corresponding metal-bonded hydride and an alkene.
- Carbon-hydrogen bond activation (or CH activation): A carbon-hydrogen cleavage reaction with an organometallic "MX" species.
- Cyclometalation: Formation of a metallocycle, that is, a cyclic compound with at least one carbon atom replaced by a metal atom.

Applications

Organometallic compounds find practical use in stoichiometric and catalytically active compounds. Tetraethyl lead previously was combined with gasoline as an antiknock agent. Due to the toxicity of lead, it is no longer used, and it has been replaced by other organometallic compounds such as ferrocene and methylcyclopentadienyl manganese tricarbonyl (MMT).

The Monsanto process utilizes a rhodium-carbonyl complex to manufacture acetic acid from methanol and carbon monoxide industrially. The Ziegler-Natta catalyst is a titanium-based organometallic compound used in the production of polyethylene and other polymers.

In the production of fine chemicals and pharmaceuticals, Ryoji Noyori's chiral ruthenium-BINAP complex catalytically reduces beta-ketoesters to secondary alcohols.

Elements that form Organometallics

A wide variety of elements of the periodic table can form organometallic compounds. Examples are given below:

- Period 2 elements: Organolithium chemistry, organoberyllium chemistry, organoborane chemistry.
- Period 3 elements: Organomagnesium chemistry, organoaluminum chemistry, organosilicon chemistry.
- Period 4 elements: Organotitanium chemistry, organochromium chemistry, organomanganese chemistry, organoiron chemistry, organocobalt chemistry organonickel chemistry, organocopper chemistry, organozinc chemistry, organogallium chemistry, organogermanium chemistry.
- Period 5 elements: Organopalladium chemistry, organosilver chemistry, organocadmium chemistry, organoindium chemistry, organotin chemistry.
- Period 6 elements: Organoplatinum chemistry, organogold chemistry, organomercury chemistry, organothallium chemistry, organolead chemistry.

Organogold Chemistry

Organogold chemistry is enjoying a resurgence of interest as useful applications, largely physical or medicinal, are being found. Organic derivatives of gold are known for both of the common oxidation states, +1 and +3, and there is an increasing number of gold(II) derivatives. The normal and predominant coordination numbers of gold are two for gold(I) and four for gold(III); as is common in this part of the Periodic table, this gives stable configurations with 14 and 16 electrons respectively.

Homoleptic compounds AuR or AuR_3 are therefore electron deficient and are unstable and highly reactive. This is true even though they achieve the 'normal' coordination number through polymerization via bridging alkyl or aryl groups and three-center two electron bonds. The principal examples of stable compounds of empirical formula AuR are those in which the R-groups carry a functionality which allows polymerization by coordination to neighboring gold atoms (e.g. $[AuC \equiv CR]_n$)'. Compounds formulated originally as 'AuR_3' or 'AuR_3(O E_{t2}), are probably Li[AuR$_3$BrJ] (e.g. R = CH_3, C_6F_5).

Gold(I)		Gold(III)			
AuR(L)	AuR_2^-	$AuRX_2L$	$AuR_2(X)L$	AuR_3L	AuR_4^-
$AuR(X)^-$		$AuRX_3^-$	$AuR_2X_2^-$	AuR_3X^-	

Figure: The possible stoichiometries (L = neutralligand; X =anionic ligand; R = organic group)

Provided additional ligands are present to make up the coordination number, compounds of all possible stoichiometry's are known, with one or two R-groups bonded to gold(I) and from one to four to gold(III). The structures of these compounds are as expected, linear and square planar; where appropriate, geometrical isomers can usually be obtained. In effect, the R-groups are carnations acting as conventional, nonnegative, two-electron ligands. Being amongst the softest ligands known", they transfer much electron density to the gold atom; they exert a high trans-influence.

Compounds in which two R-groups are mutually trans are therefore generally of lower stability, and other soft ligands usually adopt positions cis to the R-group. For the same reason, $[AuR_2]^-$ and $[AuR_4]^-$ are highly reactive. In all cases, derivatives of the harder perfluoro-organic groups -CF_3 and -C_6F_5 show considerably greater stability.

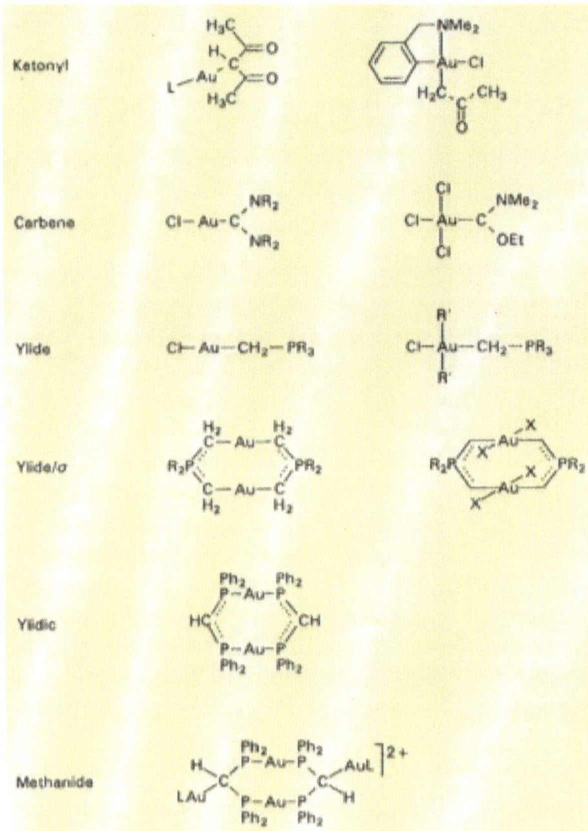

Gold compounds with less conventional ligand systems

σ-Bonded organic ligands can be of all common types: alkyl, aryl, vinylic, acetylenic. The aryl derivatives in particular lend themselves to the formation of chelates when suitable substituents are present; naturally, the stereochemistry required means that these are limited to gold(III); examples are shown as I-IV.

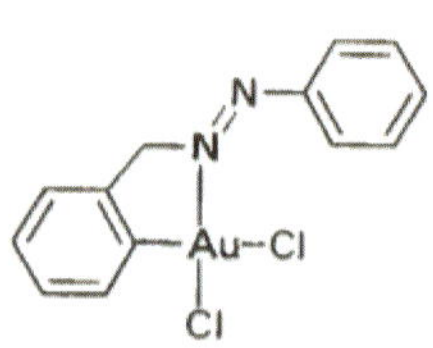

I

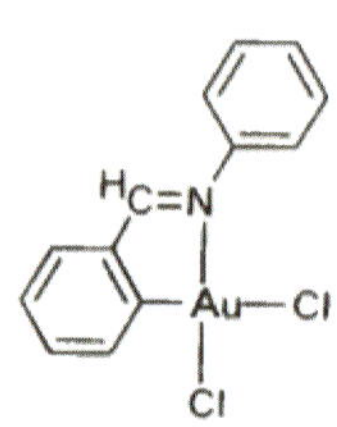

II

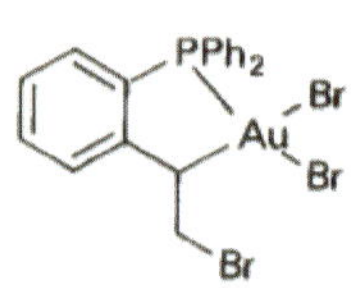

III

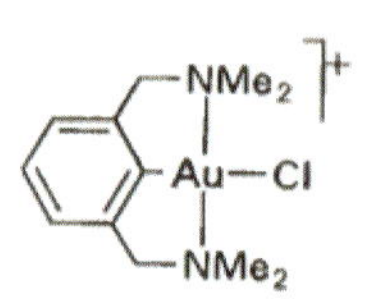

IV

Considerable attention has also been given to the synthesis of derivatives containing less conventional ketonyl, diketonyl, carbene, or ylide ligands. The distinction between the different types is not always obvious, and more than one can occur in the same molecule. A carbene may be considered, formally at least, as a neutral divalent carbon ligand: CY_2> where Y is often an alkoxy or amido group. The most common ylide ligands are effectively deprotonated phosphonium cations, e.g. CH_2PPh_3 (derived from the methyltriphenylphosphonium cation, $CH_3Ph_3P^+$), therefore also neutral. The dimethyldiphenylphosphonium cation can undergo two deprotonation stages to give: $CH_2PPh_2CH_2^-$, which can act simultaneously as a conventional cbonded ligand and as an ylide; however, the negative charge is delocalized, and this ligand forms symmetrical bridges: two such ligands bind to two gold atoms forming an eight-membered ring system (V; VI). The dppm ligand, $Ph_2PCH_2PPh_2$, forms a similar cyclic bis-gold(I) complex which can be deprotonated to give VII. Although ylidic, this complex does not contain gold-carbon bonds. It can, however, use the residual electron density on the bridging carbon atom to bind extra gold atoms (VIII); to distinguish complexes of this type from conventional σ-bonded organogold compounds, they are often called methanide complexes. Despite the unconventionality of some of these organic ligands, and the curiosity and complexity of some of the structures, the stereochemistry at gold is highly normal.

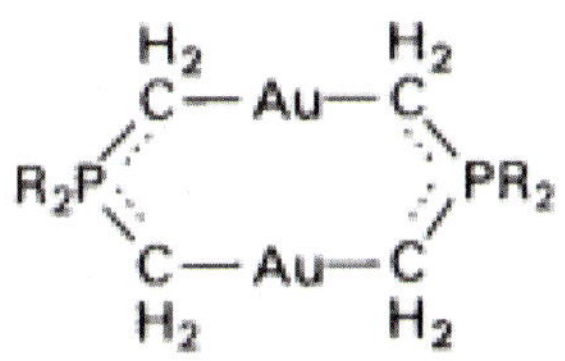

V

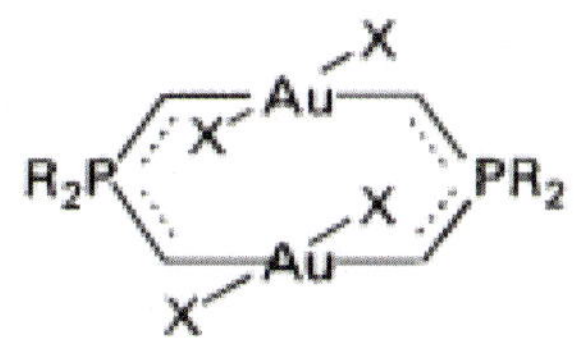

VI

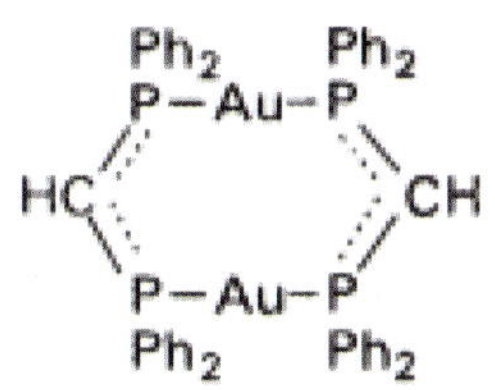

VII

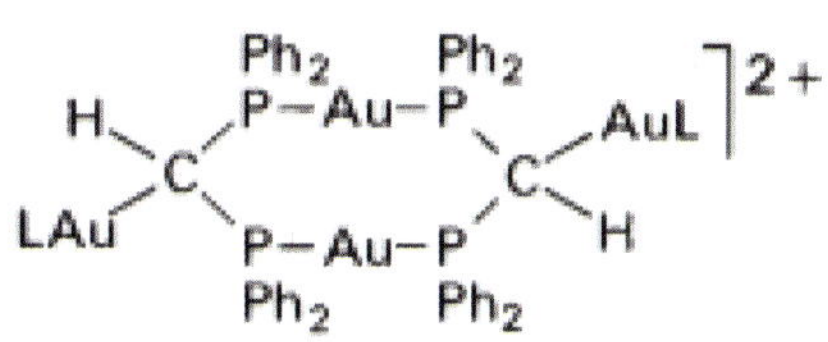

VIII

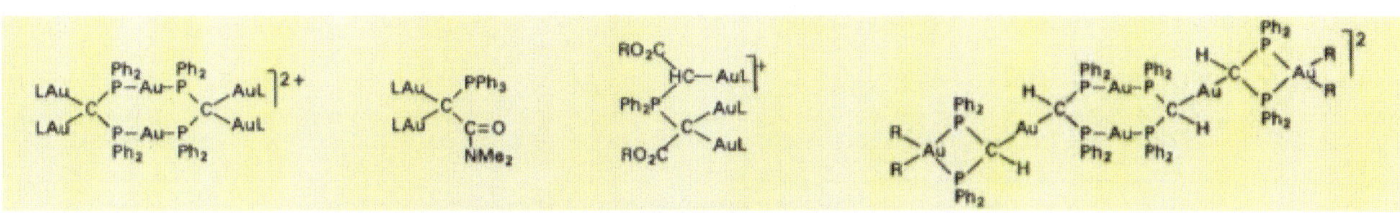

Some 'exotic'ylidic and methanidic structures

The cyclic digold(l) ylide complexes are readily oxidized to give gold(III) complexes with conventional square planar AUC_2X_2 coordination but retaining the ring structure, VI, IX. With a single molar equivalent of halogen, gold(II) derivatives (e.g. X) are obtained. Although the gold atoms now formally have a d^9 configuration, the complexes are all diamagnetic; the coordination is square planar and includes a short bond to the second gold atom. This behavior is, of course, analogous to the long established mercury(I) compounds.

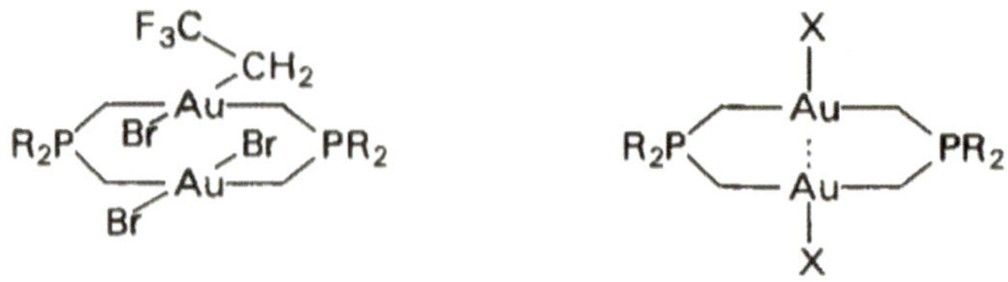

IX X

In recent years a series of compounds has been discovered in which carbon atoms are σ -bonded to one to six Au'L units, where L is a tertiary phosphine. In those involving four coordinate carbon, the structures are basically tetrahedral but the Au-C-Au angles are often smaller than the tetrahedral angle rather than larger as might have been expected on steric grounds. Even more curiously, the coordination number of the carbon atom increases readily above four, to five and even six; $C(AuL)_4$ can be obtained only when L is a bulky ligand, such as $P(C_6H_{11})_3$. It should also be mentioned that analogous 'hypervalent' compounds are known with central atoms other than carbon (e.g. boron, oxygen, nitrogen, phosphorus, arsenic).

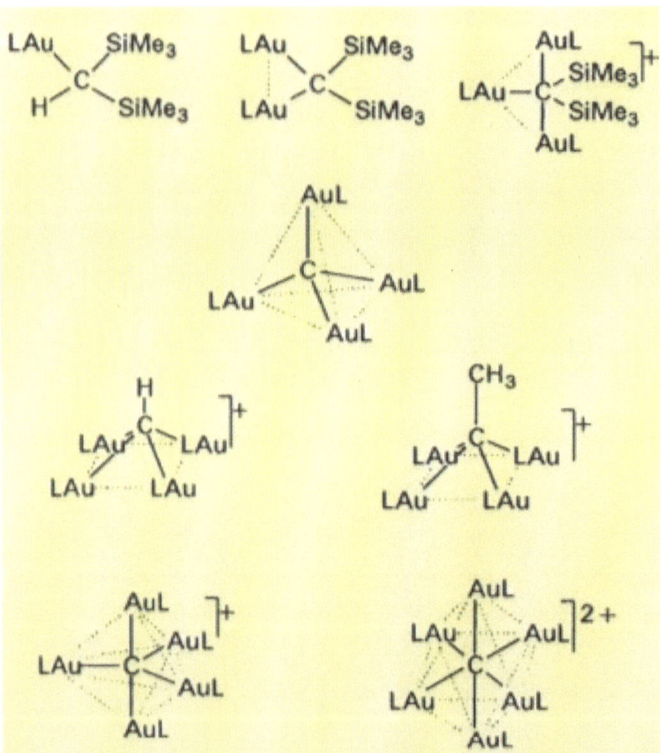

Examples of carbon-centered and hypervalent gold compounds

Satisfactory molecular orbital schemes may be constructed, in which the available valence electrons just fill the bonding molecular orbitals. The driving force for the formation of these remarkable compounds is the tendency for gold atoms to attract each other and for weak Au. Au bonding interactions to stabilize the compounds. Such interactions appear in a wide range of compounds, and are estimated to have a strength comparable to hydrogen-bonds, *ca* 30 kJ mol^{-1}". This effect has been named 'aurophilicity', and is thought to be due to relativistic effects arising from the high nuclear charge. Under its influence, electrons are accelerated to velocities comparable to that of light and they experience an increase of mass. This affects their energies and, for the valence shell, allows much more orbital mixing than would otherwise be expected. These effects operate for all heavy atoms, but just happen to have a maximum at gold.

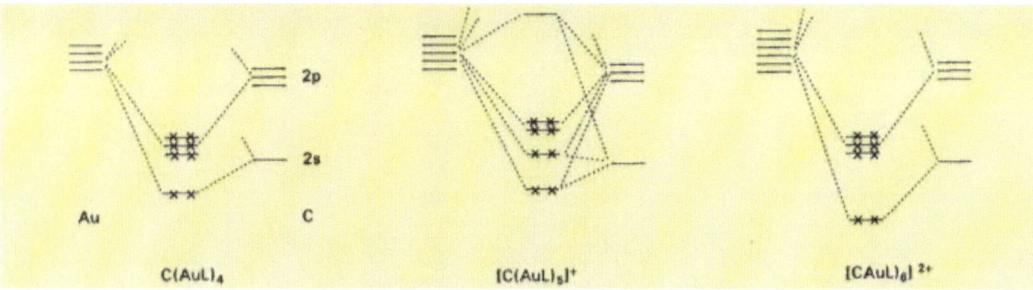

Figure: MO schemes

Other electron-deficient gold(I) compounds are best exemplified by $AU_2Zn_2Ph_6$, for which structure XI has been proposed. This type of bridging has been confirmed in the ferrocene derivative XII, $(C_5H_5)Fe[C_5H_4(AuPPh_3)_2]^+$. The well-established gold(I) alkynyl, $\left[AuC\equiv CBu^t\right]_n$ was for a long time assumed to have a tetrameric square structure (n =4), which would allow the triple bond to occupy the second coordination site at an adjacent gold atom, and thus remove the electron deficiency. This material has recently been crystallized and found to have a much more fascinating, complicated structure involving all possible bonding situations: one gold atom has two o-bonded alkynyl groups, one is 'π -bonded to two triple bonds, and a third has one σ - and one 'π - bond. Two of each of these units are bound together in a ring in which the gold atoms form a flat regular hexagon, and two rings are interlaced. A similar mix of (σ - and 'π -bonded groups may occur in the gold(I) derivative of a property-substituted calixarene, XIII, and is also seen in XIV and XV. Apart from the polymeric alkynyl derivatives, there is little evidence for 'IT-bonded complexes. The highly unstable (explosive) cyclopentadienyl $[Au(C_5H_5)]$, could polymerize either via three-center bonds or by 'π -bonding analogous to that of the alkynyls, but nothing is known of its structure. In the substituted monomeric derivatives, such as $Ph_3PAu(C_5H_5)$, the cyclopentadienyl group is expected to be bound as a (Jbonded monohapto ligand (η^1, two-electron donor). However, these molecules are fluxional on the NMR timescale and all carbon atoms are equivalent; as in other cases, the 1, 2-shift may involve transient 'π -bonded intermediates.

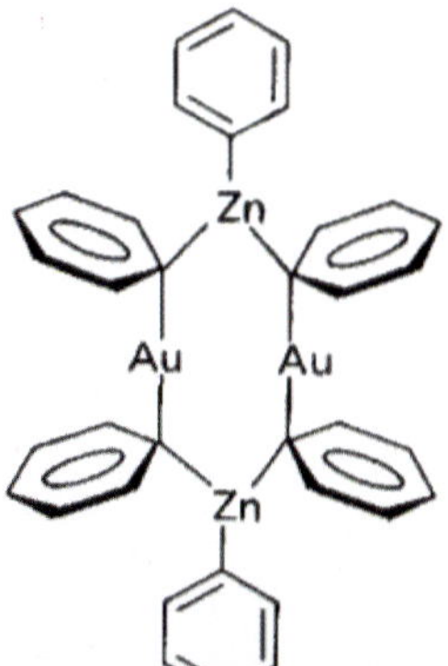
XI

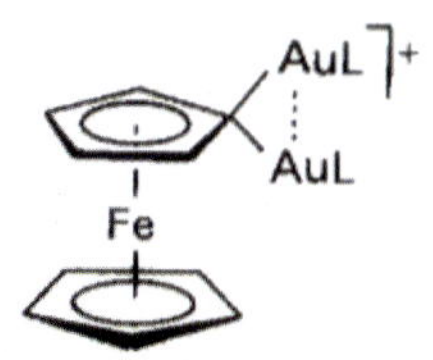
XII

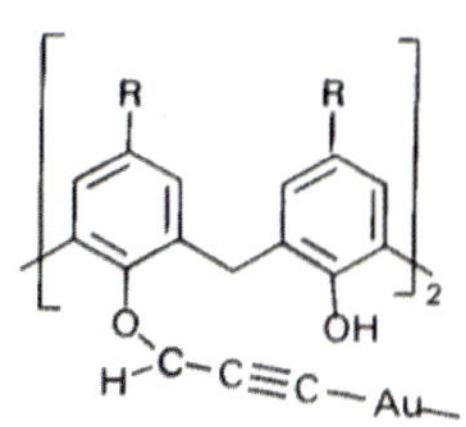
XIII

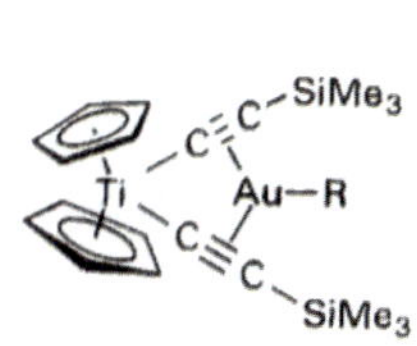
XIV

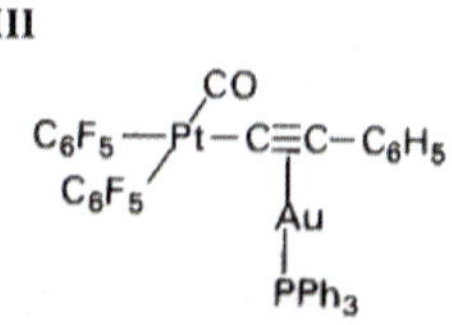
XV

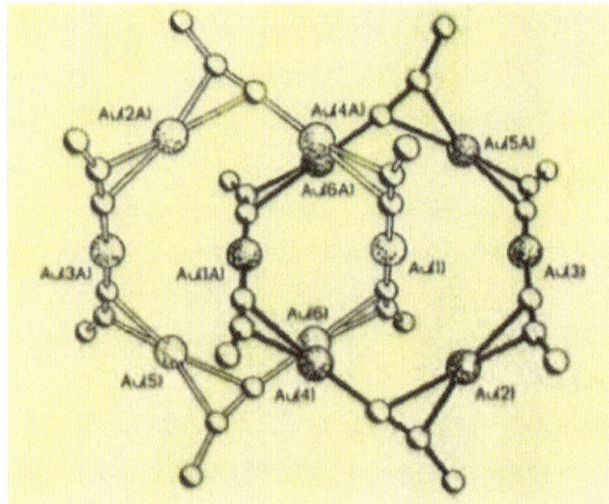

Figure: The molecular structure of $\left[\mathrm{AuC{\equiv}CBu^t}\right]_n$

Synthesis

The most versatile and most often used method for the preparation of organogold compounds is the transmetallation reaction. Organolithium or Grignard reagents readily replace halide bound to gold. For gold(I), it is necessary to start with a complexed halide, AuX(L) (Equation), and to avoid an excess of the organometallic reagent unless it is intended to obtain

$$\text{L-Au-X} \xrightarrow{RMgX} \text{L-Au-R}$$

$$\text{L-Au-R} \xrightarrow{LiR} \text{Li}\left[\text{AuR}_2\right]$$

the di-organoaurate(I) ion (Equation). Satisfactory Grignard reagents can be made from fluoro-aryls, but trilluoromerhyl derivatives are best formed from $Cd(CF_3)_2$. With gold(III) halides, the diorgano derivative AuR_2X is the usual first product when starting from AU_2X_6 or $H[AuX_4;]$. Again further reaction occurs readily with lithium reagents, to give $[AuR_3X]^-$ or $[AuR_4]^-$; this is also a convenient way of making mixed organic-ligand derivatives, which do not appear to undergo disproportionarion. Monoorganogold(III) derivatives are usually obtained from organomercury reagents, and this is most commonly done with potential chelate ligands such as azobenzene or dimethylaminobenzylamine, e.g. to obtain I or IV.

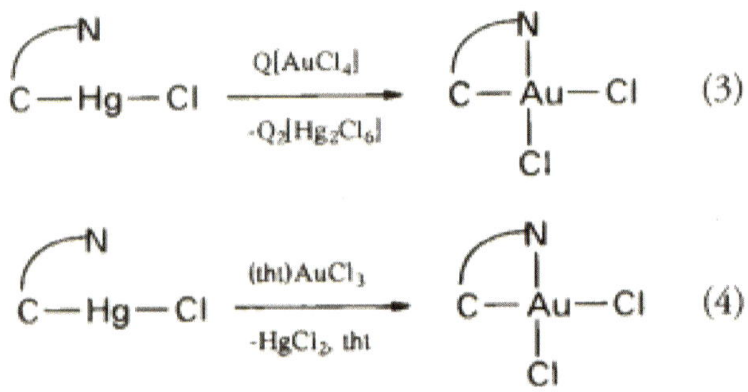

In the normal solvents (acetone or acetonitrile), the reactions represented by Equations (Q = Na or Me_4N; tht = rerrahydrorhiophene) are generally more efficient, but they appear to be finely balanced equilibria; it is sometimes necessary to add Me_4NCl to encourage the precipitation of $(Me_4N)_2[Hg_2CI_6]$. For some ligands (e.g. as in XVI, XVII) even this is not effective, but reaction will proceed if a more polar solvent is chosen, such as DMSO. It has been postulated that these reactions proceed via a bridged intermediate, XVIII, and that the solvent may inhibit the coordination of the N-donor group to the mercury atom and allow it to bind to the gold. This is consistent with the failure of the lithium derivatives of these and similar ligands to react with $[AuCl_4]^-$; presumably the organolithium is chelated, again discouraging the formation of the Au-N bond.

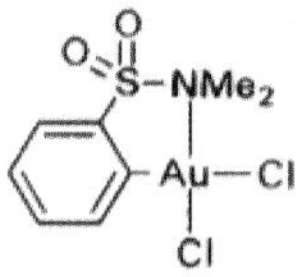

XVI

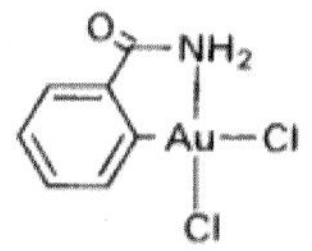

XVII

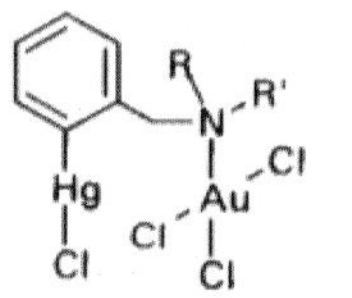

XVIII

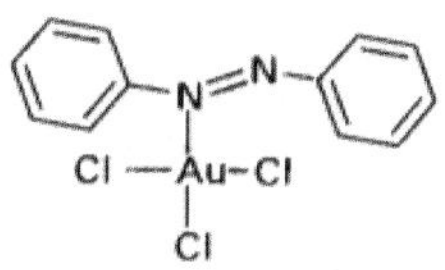

XIX

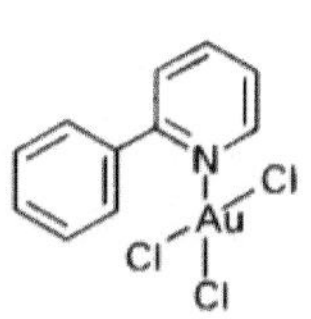

XX

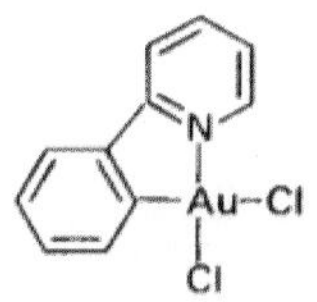

XXI

Solvent effects are also important in direct auration reactions of aromatic systems by gold(III) halides. The first arylgold derivative, $[AuPhCl_2]_2$, reported in 1931, was obtained by direct reaction between benzene and Au_2Cl_6. Alkyl benzenes would also react but not benzenes with potentially coordinating substituents; for instance, azobenzene gave XIX. More recently it was found that some of these adducts will undergo an ortho-metallation reaction, such as the transformation of: XX into XXI. This is effected by refluxing the adduct in aqueous acetonitrile (3:1), and similar reactions have been used recently to prepare a range of C, N-bonded gold (III) chelates containing sixmembered rings such as XXII, including one in which an alkyl carbon atom has been metallated (XXIII). Although it has not been commented on, the role of the solvent must be critical here, in allowing the escape or, more probably, the solvation of the liberated HCl.

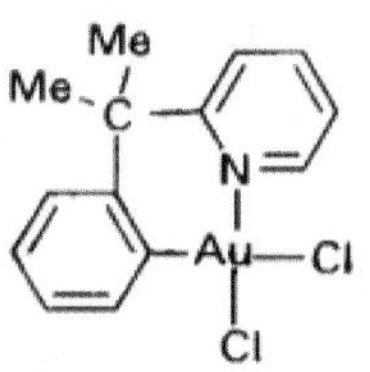

XXII

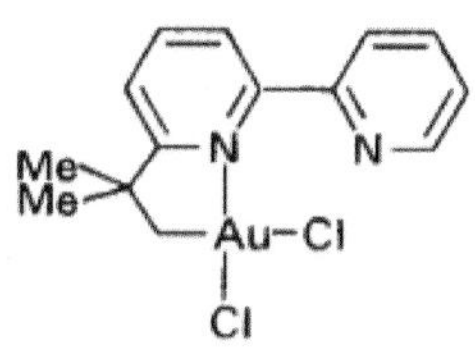

XXIII

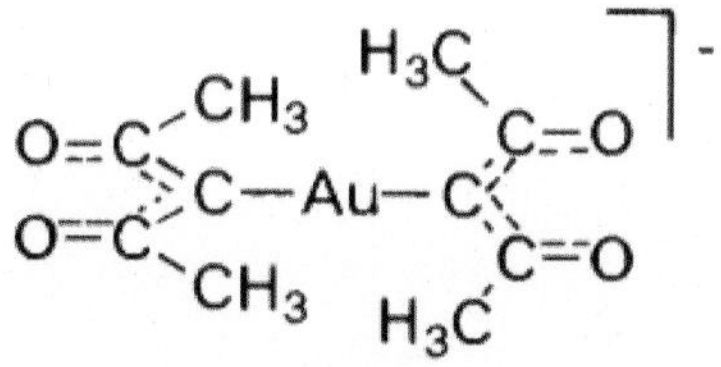

XXIV

A versatile method involving exchange of organic groups has been developed. Reaction of $[AuCl_2]^-$ with thallium(I) acetylacetonate produces $\left[Au(acac)_2\right]$" (XXIV). This complex reacts with organic molecules having a moderately acidic hydrogen (HR) to liberate acetylacetone and form the corresponding $[AuR_2]^-$. For instance, the ethynyl complex $\left[Au(C\equiv CH)_2\right]^-$ is readily made this way; as the $[N(PPh_3)_2]^+$ salt, this anion is much more stable than previously reported. Simple gold(I) alkynyls are still made by the traditional method, which was effectively the reaction of $[AuBr_2]^-$ with $HC\equiv CR$ in the presence of sodium acetate. Similar proton-elimination reactions are used to prepare the substituted derivatives $LAuC\equiv CR$. From LAuCl alternatively, $\left[AuC\equiv CR\right]_n$ is treated with the ligand.

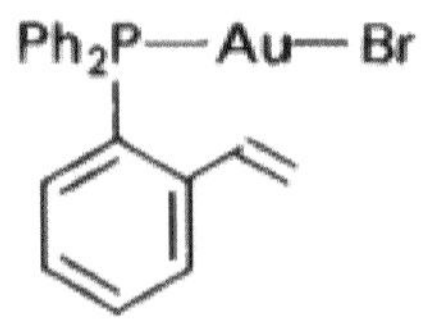

XXV

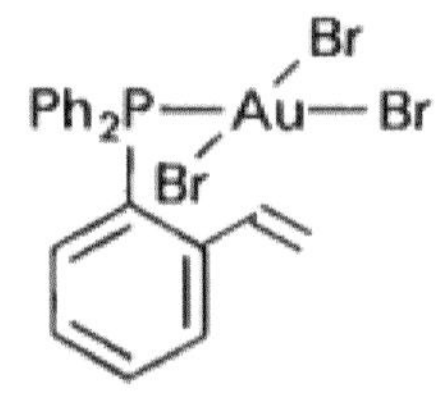

XXVI

The unusual compound III is prepared by bromination of a gold(I) derivative of sryryldiphenylphosphine (XXV). Oxidation occurs rapidly at the gold atom to give XXVI which undergoes slow rearrangement to III. Analogous reactions occur with the corresponding allyl-substituted phosphine.

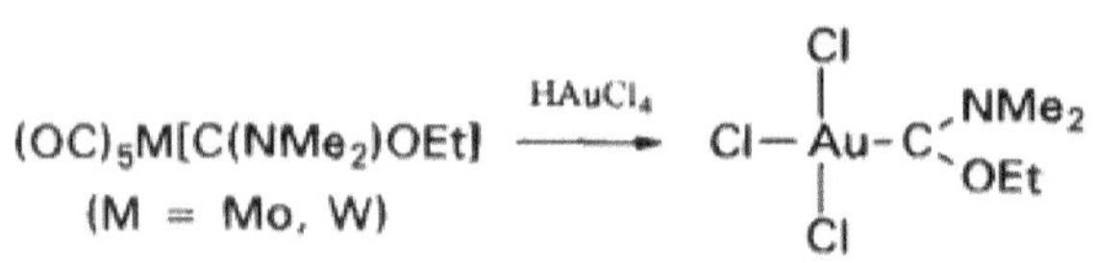

(5)

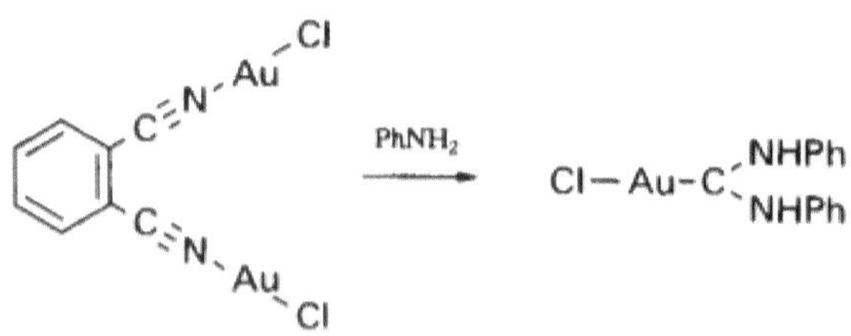

(6)

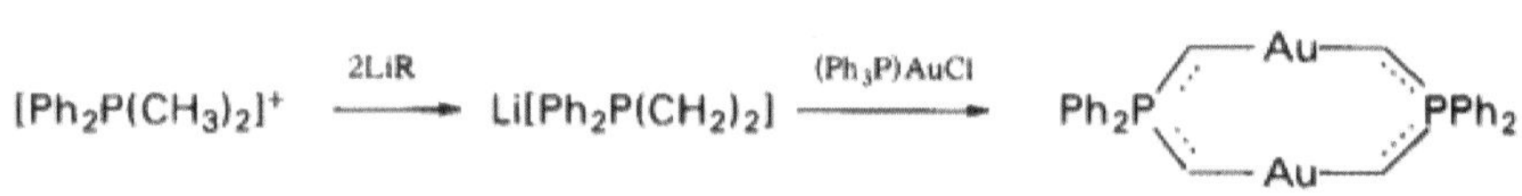

(7)

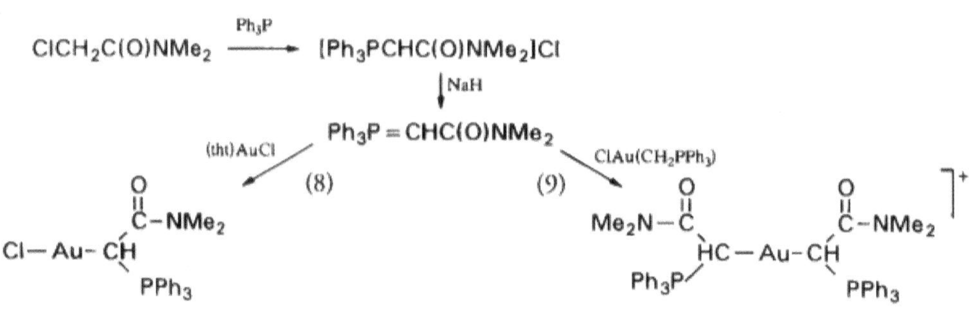

Carbene complexes can be obtained by transmetallation, involving metal carbonyl derivatives (equation) or by nucleophilic attack on an isonitrile. Ylides are usually obtained by deproronarion of a methyl phosphonium salt, using butyl lithium or sodium hydride. They are strong ligands and readily replace halides, or other weakly bound ligands from gold(l) complexes. In some cases a ligand already bound may act as the deprotonating agent. A range of methanide complexes can be obtained by making use of the basicity of an ylide carbono atom in equations.

Finally, many of the hypervalent compounds are made by an ingenious reaction in which a dimethoxyboryl compound undergoes displacement of the $B(OMe)_2$ group by an AuL^+ group and simultaneously becomes transformed into a BF4- anion by the addition of CsF (in THF solution).

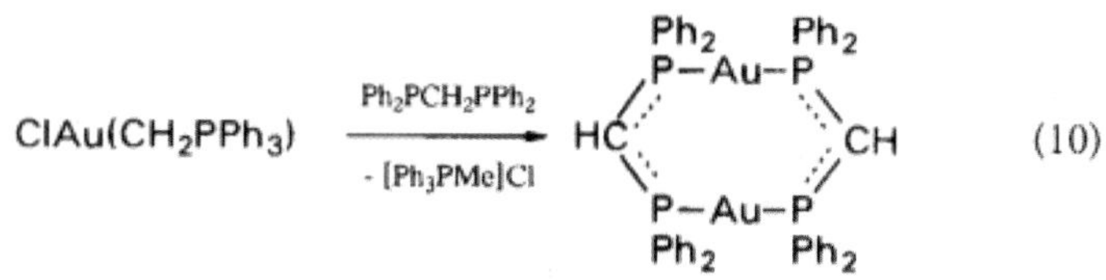

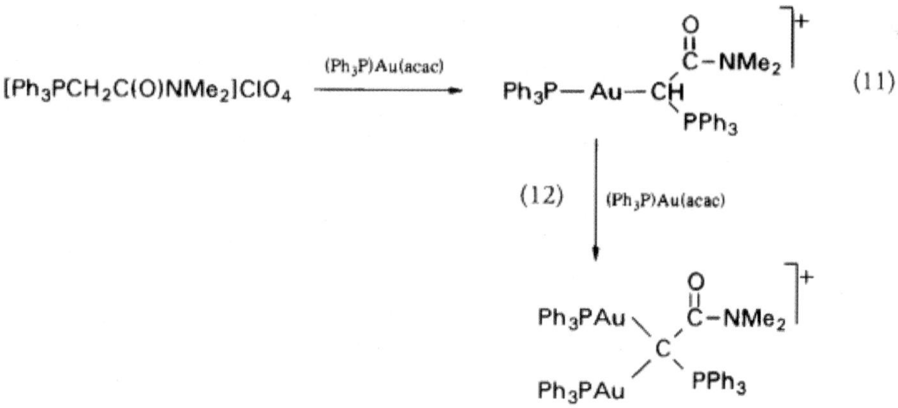

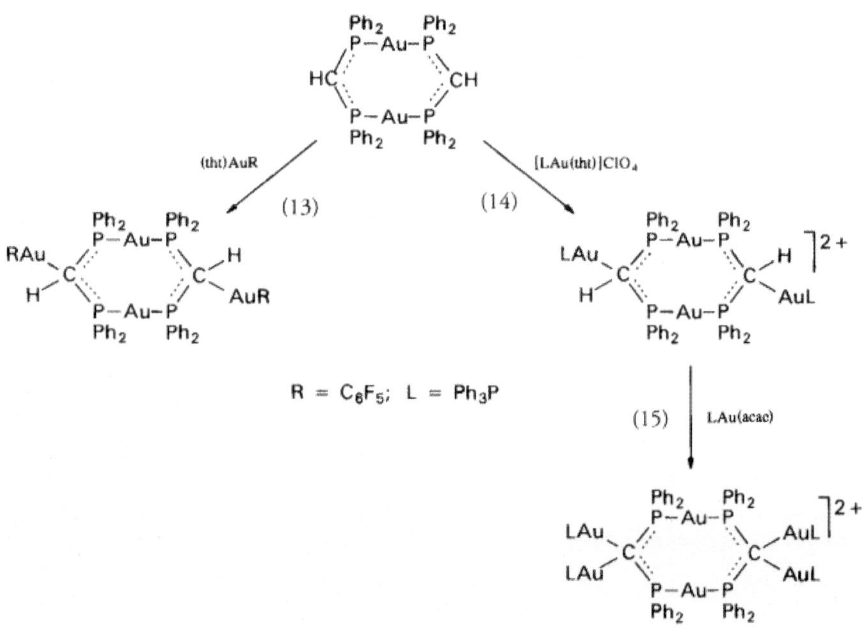

Organogallium Chemistry

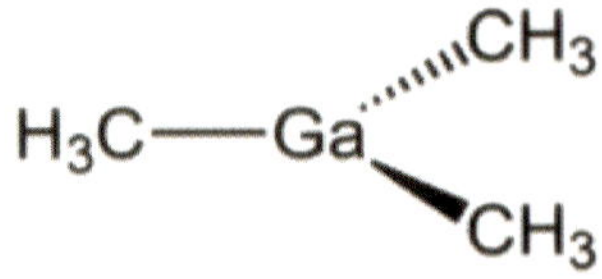

Organogallium chemistry is the chemistry of organometallic compounds containing a carbon to gallium (Ga) chemical bond. Despite their high toxicity, organogallium compounds have some use in organic synthesis. The compound trimethylgallium is of some relevance to MOCVD as a precursor to gallium arsenide via its reaction with arsine at 700° C:

$$Ga(CH_3)_3 + AsH_3 \rightarrow GaAs + 3CH_4$$

Gallium trichloride is an important reagent for the introduction of gallium into organic compounds.

The main gallium oxidation state is Ga(III), as in all lower group 13 elements (such as aluminium).

Compounds of the type R_3Ga are monomeric. Lewis acidity decreases in the order Al > Ga > In and as a result organogallium compounds do not form bridged dimers as organoaluminum compounds do. Organogallium compounds are also less reactive than organoaluminum compounds. They do form stable peroxides.

Organogallium compounds can be synthesized by transmetallation, for example the reaction of gallium metal with dimethylmercury:

$$2Ga + 3Me_2Hg \rightarrow 2Me_3Ga + 3Hg$$

or via organolithium compounds or Grignards:

$$GaCl_3 + 3MeMgBr \rightarrow Me_3Ga + 3MgBrCl$$

The electron-deficient nature of gallium can be removed by complex formation, for example

$$Me_2GaCl + NH_3 \rightarrow [Me_2Ga(NH_3)Cl]^+Cl^-$$

Pi complex formation with alkynes is also known.

Organogallium compounds are reagents or intermediates in several classes of organic reactions:

- Barbier-type reactions with elemental gallium, allylic substrates and carbonyl compounds.
- Carbometallation (carbogallation) reactions.

Higher Group 13 Organometallic Chemistry

The chemistry of organoindium (In) and organothallium (Tl) compounds parallel that of organogallium in many regards. Indium and thallium in oxidation state +1 are more common, for example the metallocenes cyclopentadienylindium(I) and cyclopentadienylthallium. Trimethylindium is important in the semiconductor industry. A special thallium feature is electrophilic thallation of arene compounds, reminiscent of mercuration. A common reagent for this purpose is thallium(III) trifluoroacetate. The intermediate arylthallium bisfluoroacetate can be isolated and converted to an aryl halide, aryl cyanide, aryl thiol or nitroarene. An example is the iodation of para-xylene.

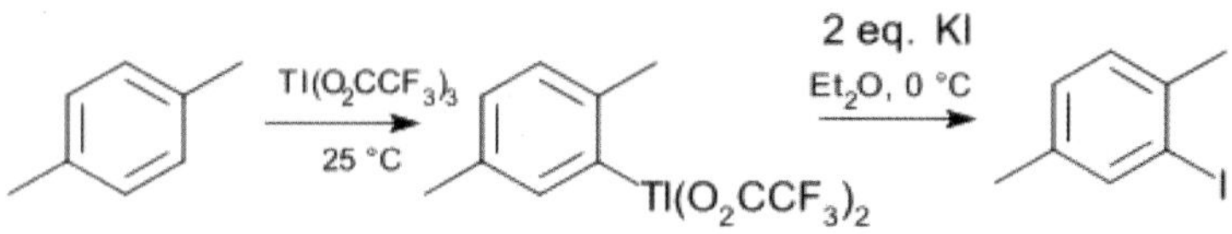

A specific niche indium research topic is indium mediated allylation.

Organorhenium Chemistry

Organorhenium chemistry describes the compounds with Re–C bonds. Because rhenium is a rare element, relatively few applications exist, but the area has been a rich source of concepts and a few useful catalysts.

General Features

Re exists in ten known oxidation states from −3 to +7 except −2, and all but Re(−3) are represented by organorhenium compounds. Most are prepared from salts of perrhenate and related binary oxides. The halides, e.g., $ReCl_5$ are also useful precursors as are certain oxychlorides.

A noteworthy feature of organorhenium chemistry is the coexistence of oxide and organic ligands in the same coordination sphere.

Carbonyl Compounds

Dirhenium decacarbonyl is a common entry point to other rhenium carbonyls. The general patterns are similar to the related manganese carbonyls. It is possible to reduce this dimer with sodium amalgam to $Na[Re(CO)_5]$ with rhenium in the formal oxidation state −1. Bromination of dirhenium decacarbonyl gives bromopentacarbonylrhenium(I), then reduced with zinc and acetic acid to pentacarbonylhydridorhenium:

$$Re_2(CO)_{10} + Br_2 \rightarrow 2Re(CO)_5Br$$

$$Re(CO)_5Br + Zn + HOAc \rightarrow Re(CO)_5H + ZnBr(OAc)$$

Bromopentacarbonylrhenium(I) is readily decarbonylated. In refluxing water, it forms the triaquo cation:

$$Re(CO)_5Br + 3H_2O \rightarrow [Re(CO)_3(H_2O)_3]Br + 2CO$$

With tetraethylammonium bromide $Re(CO)_5Br$ reacts to give the anionic tribromide:

$$Re(CO)_5Br + 2NEt_4Br \rightarrow [NEt_4]_2[Re(CO)_3Br_3] + 2CO.$$

Cyclopentadienyl Complexes

One of the first transition metal hydride complexes to be reported was $(C_5H_5)_2ReH$. A variety of half-sandwich compounds have been prepared from $(C_5H_5)Re(CO)_3$ and $(C_5Me_5)Re(CO)_3$. Notable derivatives include the electron-precise oxide $(C_5Me_5)ReO_3$ and $(C_5H_5)_2Re_2(CO)_4$.

Re-alkyl and aryl Compounds

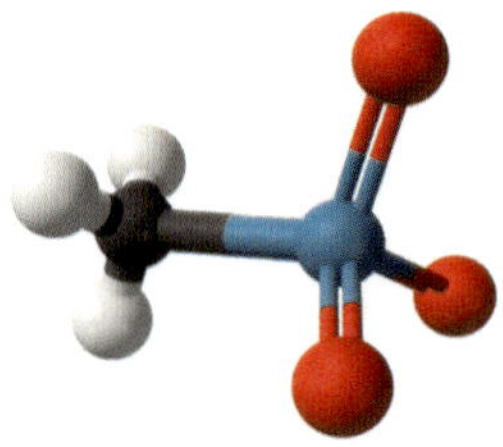

Structure of methylrhenium trioxide.

Rhenium forms a variety of alkyl and aryl derivatives, often with pi-donor coligands such as oxide. Well known is methylrhenium trioxide ("MTO"), CH_3ReO_3 a volatile, colourless solid. This compound has been used as a catalyst in some laboratory experiments. It can be prepared by many routes, a typical method is the reaction of Re_2O_7 and tetramethyltin:

$$Re_2O_7 + (CH_3)_4Sn \rightarrow CH_3ReO_3 + (CH_3)_3SnOReO_3$$

Analogous alkyl and aryl derivatives are known. MTO catalyses for the oxidations with hydrogen peroxide. Terminal alkynes yield the corresponding acid or ester, internal alkynes yield diketones, and alkenes give epoxides. MTO also catalyses the conversion of aldehydes and diazoalkanes into an alkene.

Rhenium is also able to make complexes with fullerene ligands such as $Re_2(PMe_3)_4H_8(\eta^2:\eta^2C_{60})$.

Organoyttrium Chemistry

Organoyttrium chemistry is the study of compounds containing carbon-yttrium bonds. They are studied in academic research, but have not received widespread use otherwise. These compounds use Ycl_3 as a starting material, which is in turn obtained in a reaction of Y_2O_3 with concentrated hydrochloric acid and ammonium chloride.

Organosilver Chemistry

Organosilver chemistry is an area of chemistry that is concerned with the study of organometallic compounds that contain a carbon-silver chemical bond and with the role of silver as a catalyst in organic reactions. In organosilver compounds, the oxidation state for silver is +1 with the exception of Ag(III) in. Owing to poor thermal stability and relative high price of organosilver compounds, organosilver catalysis is less common compared to organocopper chemistry.

Compounds

Phenylsilver can be obtained from the reaction of silver nitrate with an trialkylphenyllead compound:

$$AgNO_3 + R_3PhPb \rightarrow PhAg.$$

Alternatively, diphenylzinc may be used

$$Ph_2Zn + AgNO_3 \rightarrow PhAg + PhZnNO_3.$$

The silver mesitylene adduct is a tetramer. It can be formed from silver chloride and the Grignard reagent.

$$AgCl + MesMgBr \rightarrow Ag_4(Mes)_4 + MgClBr$$

Silver forms stable complexes with ylides such as triphenylphosphonium methylide.

$$AgCl + Ph_3P{=}CH_2 \rightarrow AgCl(Ph_3P{=}CH_2)$$

The C-Ag bond is stabilized by perfluoroalkyl ligands.

$$AgF + CF_2{=}CF(CF_3) \rightarrow AgCF(CF_3)_2$$

Alkenylsilver compounds are also more stable than their alkylsilver counterparts. Vinylsilver can be obtained by reaction of silver nitrate with tetravinyllead as:

$$AgNO_3 + (CH_2{=}CH)_4Pb \rightarrow (CH_2{=}CH)Ag + (CH_2{=}CH)_3PbNO_3$$

Silver-NHC complexes are easily prepared, and are commonly used to prepare other NHC complexes by displacing labile ligands. For example, the reaction of the bis(NHC)silver(I) complex with bis(acetonitrile)palladium dichloride or chlorido(dimethyl sulfide)gold(I) is given as:

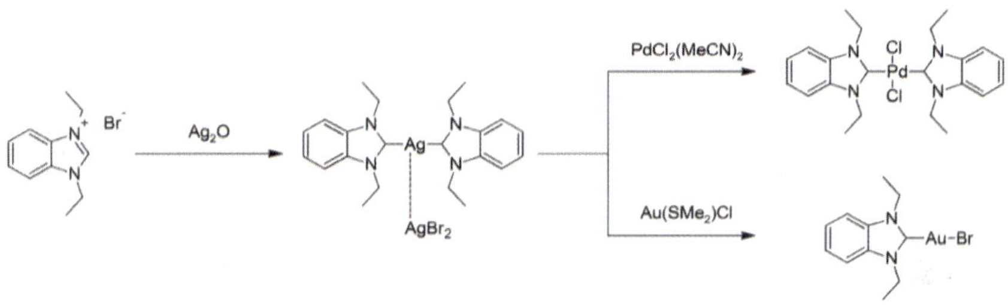

Biomolecular Chemistry

Biomolecular chemistry is the study of chemical processes at the interface of chemistry and biology, is a relatively new and largely unexplored area. Materials of controlled structure are needed for biomaterial, electronic, photonic, and medical applications. Nature makes materials of precisely controlled architectures that perform various functions, including, for example, catalysis, structural support, and information processing and storage. The study of such materials has the potential to provide fundamental knowledge that can be used to advantage in the design and preparation of new systems. A better understanding of chemical reactions underlying biological systems is in turn expected to lead to a better ability to design novel materials with properties that equal or exceed those of the naturally occurring materials upon which they are based. The ability to control surface functional groups, chain length, topology, and structure is critical to the design of novel materials that can be beneficial for naval applications.

Recent advances in molecular genetics, combinatorial chemistry, and the design of semisynthetic enzymes and catalytic antibodies provide unique opportunities for new applications. An attractive possibility is to use DNA templates to prepare polymers having controlled molecular dimensions and judiciously placed functional groups and chain folding patterns. Such materials are difficult if not impossible to prepare by current synthetic polymerization methods. This potential capability to control the folding of macromolecular chains will lead to materials with improved strength and elasticity, as well as other desired properties. Application of the technology to the incorporation of "unnatural" amino acids and carbohydrates will significantly expand the scope of available materials. For example, it will be possible to precisely vary by design the hydrophilic and hydrophobic surface characteristics of materials.

Another area of research where biological chemistry is poised to have a significant impact is in the development of high-strength materials. For example, some forms of silk, such as spider dragline, have been synthesized in the laboratory using DNA templates and have been found to be stronger than steel but much more elastic. Thus, it may be possible to develop materials based on naturally occurring biopolymers to use as lightweight reinforcement for superior composites. However, the molecular-level understanding of the parameters that control the properties of such materials is lacking. Hence, a vigorous research program in the area is highly desirable. An added benefit lies in the design of new synthetic routes to producing protective clothing, composites, and materials that have a better potential for being compatible with substances in living systems. An intriguing prospect in this area is the possibility of combining naturally occurring segments with synthetic polymer segments to give hybrid materials that will synergistically incorporate the useful properties of both systems. Opportunities exist here for property-directed synthesis of novel materials that can be tailored to naval applications.

Research at the interface of biology and chemistry will lead to a better understanding of the factors that control bioadhesion and ultimately biocorrosion, two areas of major concern to the Navy. Adhesion of marine organisms to naval structures leads to fouling, and known methods of combating the problem are increasingly limited by environmental concerns. Biogenetic synthesis can be used to incorporate appropriate functional groups, which may be designed to migrate to the surface of a protective coating, thereby controlling marine biofilm formation. Mechanistic understanding of bioadhesion is critically needed for control of fouling and biocorrosion and represents the critical input for this material design effort. Continued research into the mechanisms of bioadhesion is an area of opportunity.

Another area in which research at the interface of chemistry and biology is expected to have an impact is in the design of physical methods for fast genetic analysis. Combinatorial synthesis is making available arrays of materials for numerous applications, including coatings and films. However, the design of appropriate processes for synthesizing molecular arrays and physical methods for measurement and analysis of surface properties are lagging. Hence, a concerted effort in both synthetic methods research and research designed to exploit the atomiclevel characterization offered by the new forms of microscopy scanning tunneling microscopy, atomic force microscopy, and near-field optical microscopy is timely. These techniques will lead to new, robust, and fast analytical methods.

New biomaterials are needed in applications such as wound healing, bone replacement, and controlled delivery of biologically active species. Biomolecular chemistry is poised to have a significant impact on the availability of biocompatible materials. It is important to know that current biomaterials have not been designed for such uses but have been chosen empirically from materials developed for nonbiological uses. In some cases this empirical approach has led to serious immunogenic complications. Understanding the fundamentals of biological performance at the molecular level will facilitate incorporation of biocompatible segments into synthetic systems for use in medical applications.

Heterocyclic Chemistry

Heterocyclic chemistry is the chemistry which emphasis on the correlations between the methods of preparation and the properties of the various ring systems. It also describes the concepts of the electronic theory of organic chemistry and the relationship of heterocyclic and carbocyclic aromatic compounds.

Heterocyclic Compounds

Compounds classified as heterocyclic probably constitute the largest and most varied family of organic compounds. After all, every carbocyclic compound, regardless of structure and functionality, may in principle be converted into a collection of heterocyclic analogs by replacing one or more of the ring carbon atoms with a different element. Even if we restrict our consideration to oxygen, nitrogen and sulfur (the most common heterocyclic elements), the permutations and combinations of such a replacement are numerous.

Nomenclature

Devising a systematic nomenclature system for heterocyclic compounds presented a formidable challenge, which has not been uniformly concluded. Many heterocycles, especially amines, were identified early on, and received trivial names which are still preferred. Some monocyclic compounds of this kind are shown in the following chart, with the common (trivial) name in bold and a systematic name based on the Hantzsch-Widman system given beneath it in blue. The rules for using this system will be given later. For most students, learning these common names will provide an adequate nomenclature background.

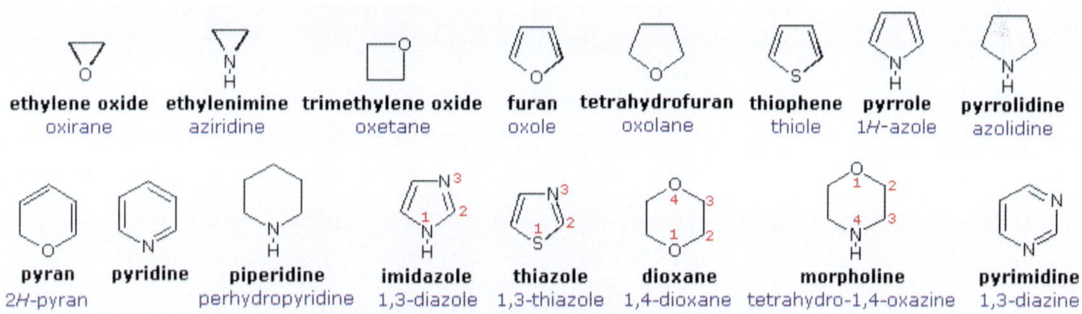

An easy to remember, but limited, nomenclature system makes use of an elemental prefix for the heteroatom followed by the appropriate carbocyclic name. A short list of some common prefixes is given in the following table, priority order increasing from right to left. Examples of this nomenclature are: ethylene oxide = oxacyclopropane, furan = oxacyclopenta-2, 4-diene, pyridine = azabenzene, and morpholine = 1-oxa-4-azacyclohexane.

Element	oxygen	sulfur	selenium	nitrogen	phosphorous	silicon	boron
Valence	II	II	II	III	III	IV	III
Prefix	Oxa	Thia	Selena	Aza	Phospha	Sila	Bora

The Hantzsch-Widman system provides a more systematic method of naming heterocyclic compounds that is not dependent on prior carbocyclic names. It makes use of the same hetero atom prefix defined above (dropping the final "a"), followed by a suffix designating ring size and saturation. As outlined in the following table, each suffix consists of a ring size root (blue) and an ending intended to designate the degree of unsaturation in the ring. In this respect, it is important to recognize that the saturated suffix applies only to completely saturated ring systems, and the unsaturated suffix applies to rings incorporating the maximum number of non-cumulated double bonds. Systems having a lesser degree of unsaturation require an appropriate prefix, such as "dihydro"or "tetrahydro".

Ring Size	3	4	5	6	7	8	9	10
Suffix								
Unsaturated	irene	ete	ole	ine	epine	ocine	onine	ecine
Saturated	irane	etane	olane	inane	epane	ocane	onane	ecane

Despite the general systematic structure of the Hantzsch-Widman system, several exceptions and modifications have been incorporated to accommodate conflicts with prior usage. Some examples are:

- The terminal "e" in the suffix is optional though recommended.
- Saturated 3, 4 & 5-membered nitrogen heterocycles should use respectively the traditional "iridine", "etidine" & "olidine" suffix.
- Unsaturated nitrogen 3-membered heterocycles may use the traditional "irine" suffix.
- Consistent use of "etine" and "oline" as a suffix for 4 & 5-membered unsaturated heterocycles is prevented by their former use for similar sized nitrogen heterocycles.
- Established use of oxine, azine and silane for other compounds or functions prohibits their use for pyran, pyridine and silacyclohexane respectively.

Examples of these nomenclature rules are written in blue, both in the previous diagram and that shown below. Note that when a maximally unsaturated ring includes a saturated atom, its location may be designated by a "*H* " prefix to avoid ambiguity, as in pyran and pyrrole above and several examples below. When numbering a ring with more than one heteroatom, the highest priority atom is 1 and continues in the direction that gives the next priority atom the lowest number.

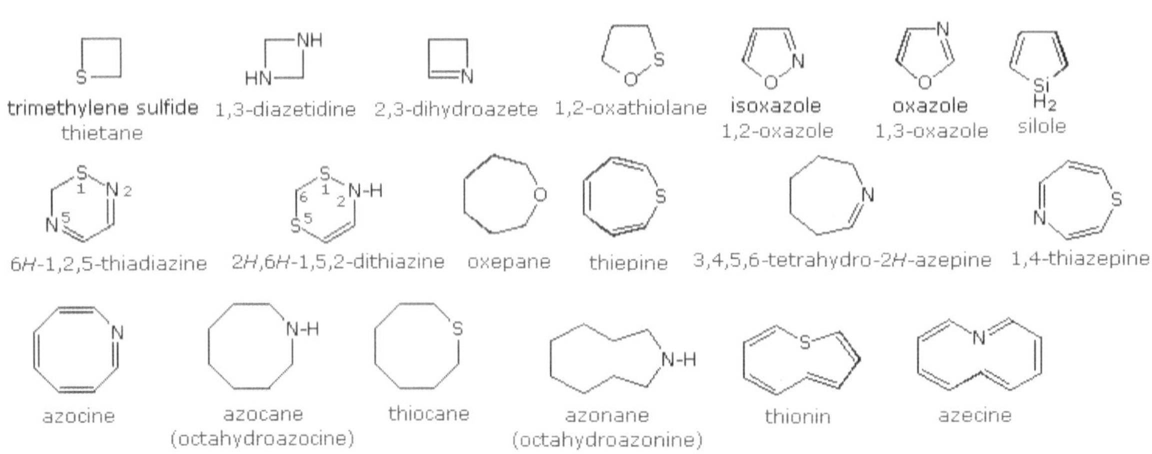

All the previous examples have been monocyclic compounds. Polycyclic compounds incorporating one or more heterocyclic rings are well known. A few of these are shown in the following diagram. As before, common names are in black and systematic names in blue. The two quinolines illustrate another nuance of heterocyclic nomenclature. Thus, the location of a fused ring may be indicated by a lowercase letter which designates the edge of the heterocyclic ring involved in the fusion, as shown by the pyridine ring in the green shaded box.

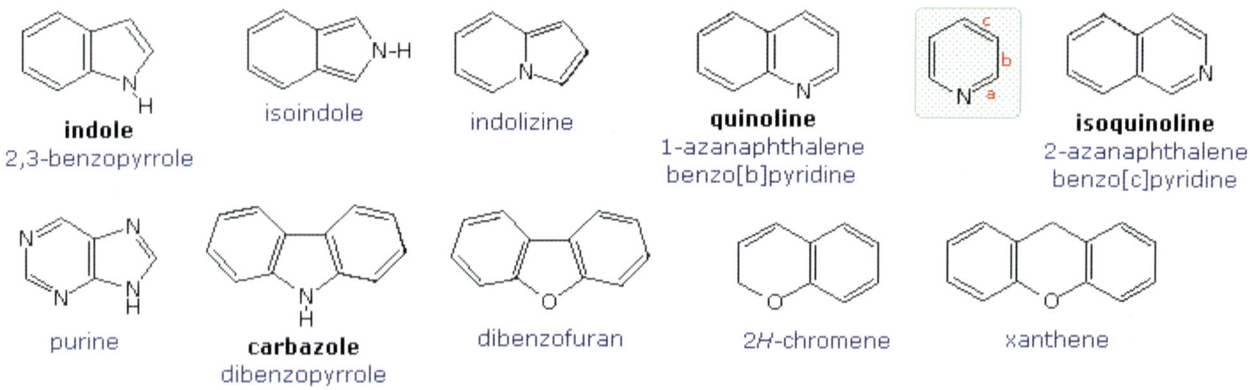

Heterocyclic rings are found in many naturally occurring compounds. Most notably, they compose the core structures of mono and polysaccharides, and the four DNA bases that establish the genetic code.

Preparation and Reactions

3-membered Rings

Oxiranes (epoxides) are the most commonly encountered three-membered heterocycles. Epoxides are easily prepared by reaction of alkenes with peracids, usually with good stereospecificity. Because of the high angle strain of the three-membered ring, epoxides are more reactive that unstrained ethers. Addition reactions proceeding by electrophilic or nucleophilic opening of the ring constitute the most general reaction class. Example in the following diagram shows one such transformation, which is interesting due to subsequent conversion of the addition intermediate into the corresponding thiirane. The initial ring opening is stereoelectronically directed in a trans-diaxial fashion, the intermediate relaxing to the diequatorial conformer before cyclizing to a 1, 3-oxathiolane intermediate. Other examples show similar addition reactions to thiiranes and aziridines. The acid-catalyzed additions in examples, illustrate the influence of substituents on the regioselectivity of addition. Example reflects the S_N2 character of nucleophile (chloride anion) attack on the protonated aziridine (the less substituted carbon is the site of addition). The phenyl substituent in example serves to stabilize the developing carbocation to such a degree that S_N1 selectivity is realized. The reduction of thiiranes to alkenes by reaction with phosphite esters (example) is highly stereospecific, and is believed to take place by an initial bonding of phosphorous to sulfur.

Examples are thermal reactions in which both the heteroatom and the strained ring are important factors. The α-lactone intermediate shown in the solvolysis of optically active 2-bromopropanoic acid accounts both for the 1st-order kinetics of this reaction and the retention of configuration in the product.

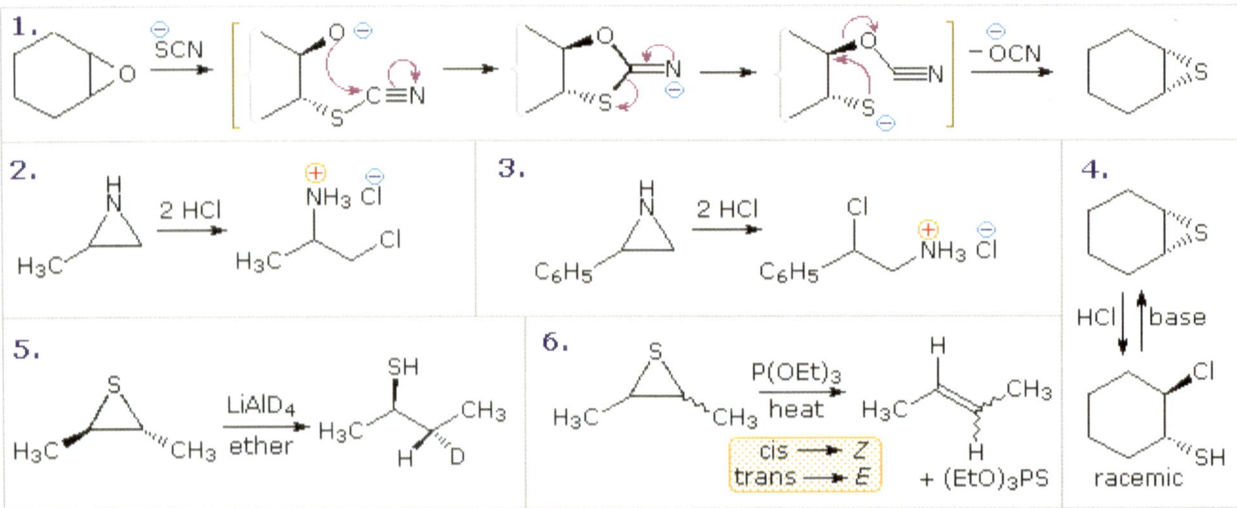

An interesting regioselectivity in the intramolecular ring-opening reactions of disubstituted epoxides having a pendant γ-hydroxy substituent has been noted. As illustrated below, acid and base-catalyzed reactions normally proceed by 5-exo-substitution (reaction 1), yielding a tetrahydrofuran product. However, if the oxirane has an unsaturated substituent (vinyl or phenyl), the acid-catalyzed opening occurs at the allylic (or benzylic) carbon (reaction 2) in a 6-endo fashion. The π-electron system of the substituent assists development of positive charge at the adjacent oxirane carbon, directing nucleophilic attack to that site.

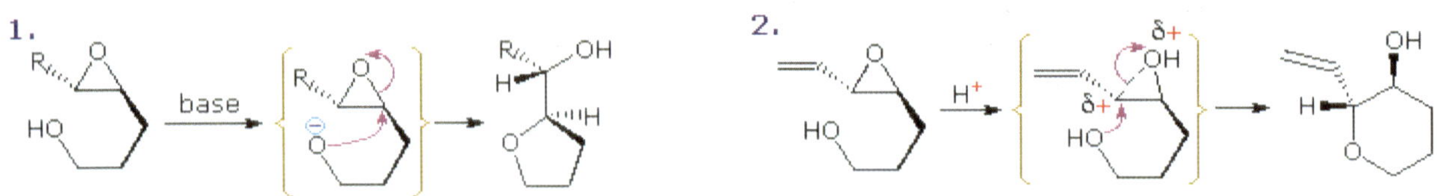

4-Membered Rings

Preparation

Several methods of preparing four-membered heterocyclic compounds are shown in the following diagram. The simple procedure of treating a 3-halo alcohol, thiol or amine with base is generally effective, but the yields are often mediocre. Dimerization and elimination are common side reactions, and other functions may compete in the reaction. In the case of example, cyclization to an oxirane competes with thietane formation, but the greater nucleophilicity of sulfur dominates, especially if a weak base is used. In example both aziridine and azetidine formation are possible, but only the former is observed. This is a good example of the kinetic advantage of three-membered ring formation. Example demonstrates that this approach to azetidine formation works well in the absence of competition. Indeed, the exceptional yield of this product is attributed to the gem-dimethyl substitution, the thorpe-ingold effect, which is believed to favor coiled chain conformations. The relatively rigid configuration of the substrate in example, favors oxetane formation and prevents an oxirane cyclization from occurring. Finally, the Paterno-Buchi photocyclizations in examples are particularly suited to oxetane formation.

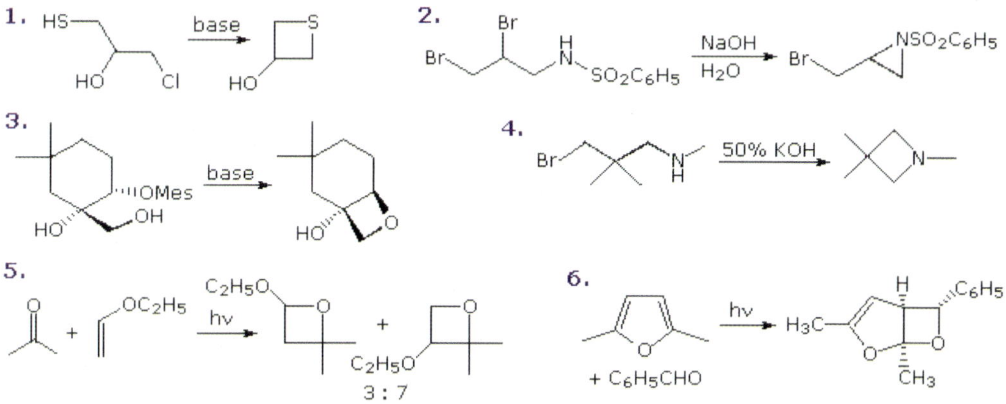

Reactions

Reactions of four-membered heterocycles also show the influence of ring strain. Some examples are given in the following diagram. Acid-catalysis is a common feature of many ring-opening reactions, as shown by examples. In the thietane reaction (2), the sulfur undergoes electrophilic chlorination to form a chlorosulfonium intermediate followed by a ring-opening chloride ion substitution. Strong nucleophiles will also open the strained ether, as shown by reaction 3b. Cleavage reactions of β-lactones may take place either by acid-catalyzed acyl exchange, or by alkyl-O rupture by nucleophiles. Example is an interesting case of intramolecular rearrangement to an ortho-ester. Finally, the β-lactam cleavage of penicillin G (reaction 6) testifies to the enhanced acylating reactivity of this fused ring system. Most amides are extremely unreactive acylation reagents, thanks to stabilization by p-π resonance. Such electron pair delocalization is diminished in the penicillins, leaving the nitrogen with a pyramidal configuration and the carbonyl function more reactive toward nucleophiles.

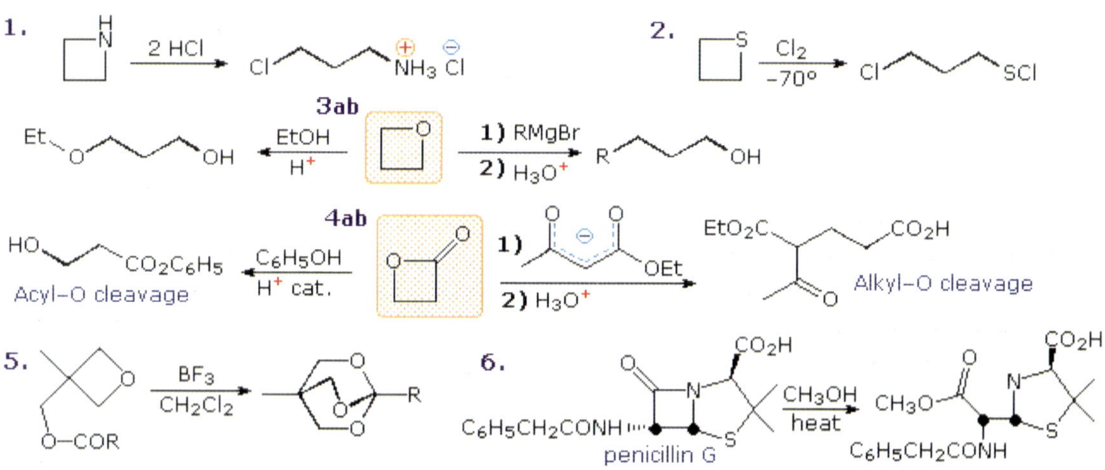

5-Membered Rings

Preparation

Commercial preparation of furan proceeds by way of the aldehyde, furfural, which in turn is generated from pentose containing raw materials like corncobs, as shown in the uppermost equation below. Similar preparations of pyrrole and thiophene are depicted in the second row equations. Equation in the third row illustrates a general preparation of substituted furans, pyrroles and thiophenes from 1, 4-dicarbonyl compounds, known as the Paal-Knorr synthesis. Many other procedures leading to substituted heterocycles of this kind have been devised. Two of these are shown in reactions 2 and 3. Furan is reduced to tetrahydrofuran by palladium-catalyzed hydrogenation. This cyclic ether is not only a valuable solvent, but it is readily converted to 1, 4-dihalobutanes or 4-haloalkylsulfonates, which may be used to prepare pyrrolidine and thiolane.

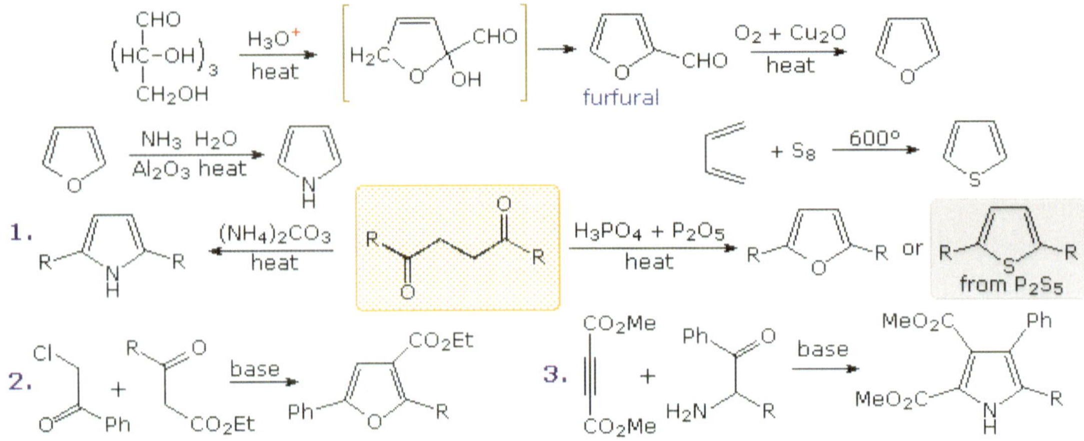

Dipolar cycloaddition reactions often lead to more complex five-membered heterocycles.

Indole is probably the most important fused ring heterocycle in this class. The first proceeds by an electrophilic substitution of a nitrogen-activated benzene ring. The second presumably takes place by formation of a dianionic species in which the $ArCH_2(-)$ unit bonds to the deactivated carbonyl group. Finally, the Fischer indole synthesis is a remarkable sequence of tautomerism, sigmatropic rearrangement, nucleophilic addition, and elimination reactions occurring subsequent to phenylhydrazone formation. This interesting transformation involves the oxidation of two carbon atoms and the reduction of one carbon and both nitrogen atoms.

Reactions

The chemical reactivity of the saturated members of this class of heterocycles: tetrahydrofuran, thiolane and pyrrolidine, resemble that of acyclic ethers, sulfides, and 2°-amines, and will not be described here. 1, 3-Dioxolanes and dithiolanes are cyclic acetals and thioacetals. These units are commonly used as protective groups for aldehydes and ketones, and may be hydrolyzed by the action of aqueous acid.

It is the “aromatic” unsaturated compounds, furan, thiophene and pyrrole that require our attention. In each case the heteroatom has at least one pair of non-bonding electrons that may combine with the four π-electrons of the double bonds to produce an annulene having an aromatic sextet of electrons. This is illustrated by the resonance description at the top of the following diagram. The heteroatom Y becomes sp^2-hybridized and acquires a positive charge as its electron pair is delocalized around the ring. An easily observed consequence of this delocalization is a change in dipole moment compared with the analogous saturated heterocycles, which all have strong dipoles with the heteroatom at the negative end. As expected, the aromatic heterocycles have much smaller dipole moments, or in the case of pyrrole a large dipole in the opposite direction. An important characteristic of aromaticity is enhanced thermodynamic stability, and this is usually demonstrated by relative heats of hydrogenation or heats of combustion measurements. By this standard, the three aromatic heterocycles under examination are stabilized, but to a lesser degree than benzene.

Additional evidence for the aromatic character of pyrrole is found in its exceptionally weak basicity (pK_a *ca.* 0) and strong acidity (pK_a = 15) for a 2°-amine. The corresponding values for the saturated amine pyrrolidine are: basicity 11.2 and acidity 32.

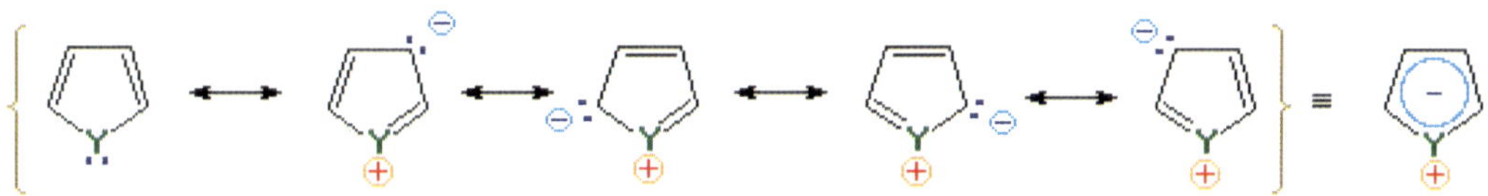

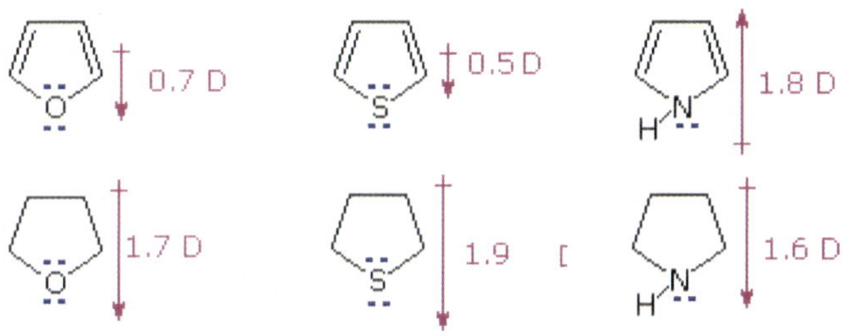

Stabilization Energies

Benzene	Furan	Thiophene	Pyrrole
38	16	11	16

kcal/mole from heats of combustion

Another characteristic of aromatic systems, of particular importance to chemists, is their pattern of reactivity with electrophilic reagents. Whereas simple cycloalkenes generally give addition reactions, aromatic compounds tend to react by substitution. As noted for benzene and its derivatives, these substitutions take place by an initial electrophile addition, followed by a proton loss from the "onium" intermediate to regenerate the aromatic ring. The aromatic five-membered heterocycles all undergo electrophilic substitution, with a general reactivity order: pyrrole >> furan > thiophene > benzene. Some examples are given in the following diagram. The reaction conditions show clearly the greater reactivity of furan compared with thiophene. All these aromatic heterocycles react vigorously with chlorine and bromine, often forming polyhalogenated products together with polymers. The exceptional reactivity of pyrrole is evidenced by its reaction with iodine (bottom left equation), and formation of 2-acetylpyrrole by simply warming it with acetic anhydride (no catalyst).

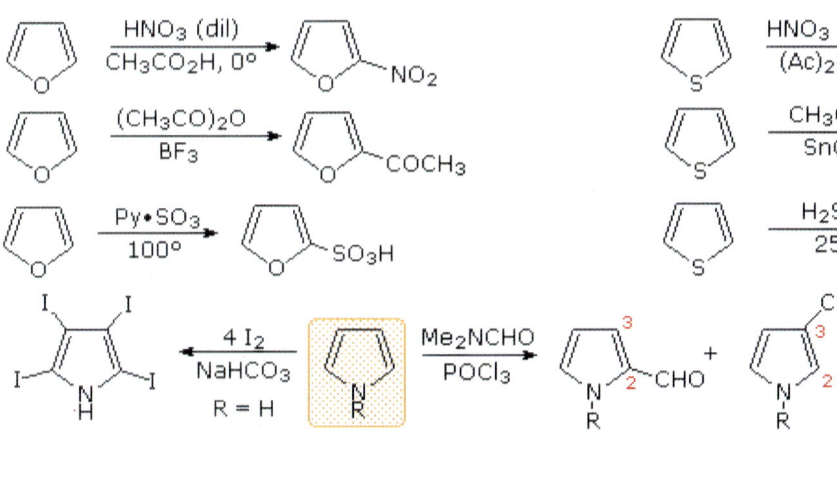

R	Ratio 2:3
CH_3	100 : <1
C_2H_5	12 : 1
$(CH_3)_2CH$	1.9 : 1
$(CH_3)_3C$	1 : 14

There is a clear preference for substitution at the 2-position (α) of the ring, especially for furan and thiophene. Reactions of pyrrole require careful evaluation, since N-protonation destroys its aromatic character. Indeed, N-substitution of this 2°-amine is often carried out prior to subsequent reactions. For example, pyrrole reacts with acetic anhydride or acetyl chloride and triethyl amine to give N-acetylpyrrole. Consequently, the regioselectivity of pyrrole substitution is variable, as noted by the bottom right equation.

An explanation for the general α-selectivity of these substitution reactions is apparent from the mechanism given below. The intermediate formed by electrophile attack at C-2 is stabilized by charge delocalization to a greater degree than the intermediate from C-3 attack. From the Hammond postulate we may then infer that the activation energy for substitution at the former position is less than the latter substitution.

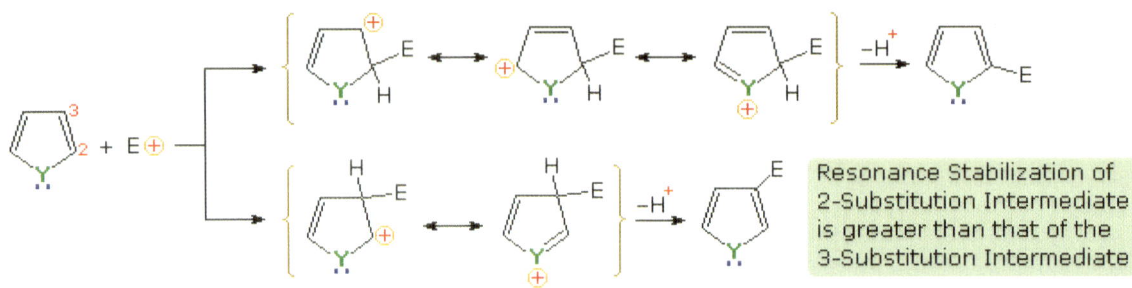

Functional substituents influence the substitution reactions of these heterocycles in much the same fashion as they do for benzene. Indeed, once one understands the ortho-para and meta-directing character of these substituents, their directing influence on heterocyclic ring substitution is not difficult to predict. The following diagram shows seven such reactions. Reactions 1 and 2 are 3-substituted thiophenes, the first by an electron donating substituent and the second by an electron withdrawing group. The third reaction has two substituents of different types in the 2 and 5-positions. Finally, examples illustrate reactions of 1, 2- and 1, 3-oxazole, thiazole and diazole. Note that the basicity of the sp²-hybridized nitrogen in the diazoles is over a million times greater than that of the apparent sp³-hybridized nitrogen, the electron pair of which is part of the aromatic electron sextet.

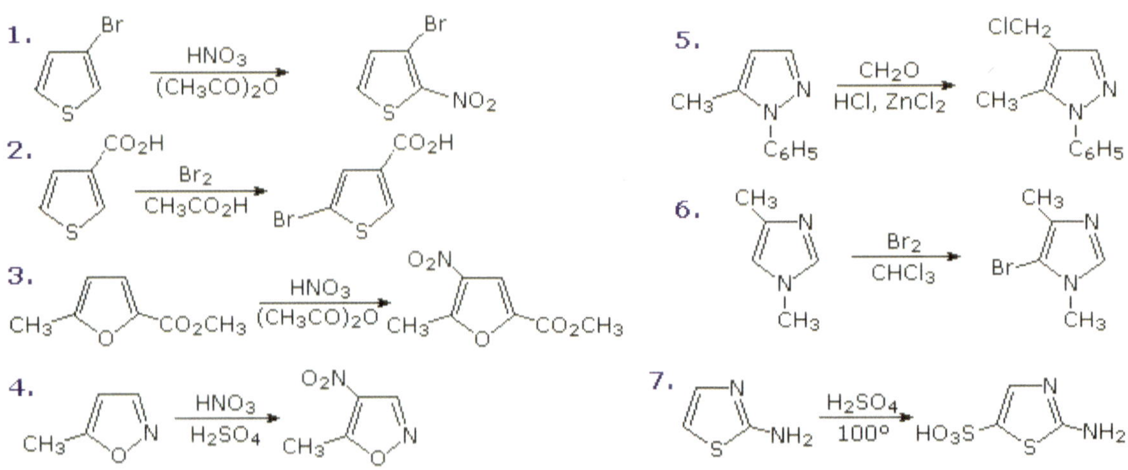

Other possible reactions are suggested by the structural features of these heterocycles. For example, furan could be considered an enol ether and pyrrole an enamine. Such functions are known to undergo acid-catalyzed hydrolysis to carbonyl compounds and alcohols or amines. Since these compounds are also heteroatom substituted dienes, we might anticipate Diels-Alder cycloaddition reactions with appropriate dienophiles. As noted in the upper example, furans may indeed be hydrolyzed to 1, 4-dicarbonyl compounds, but pyrroles and thiophenes behave differently. The second two examples, shown in the middle, demonstrate typical reactions of furan and pyrrole with the strong dienophile maleic anhydride. The former participates in a cycloaddition reaction; however, the pyrrole simply undergoes electrophilic substitution at C-2. Thiophene does not easily react with this dienophile.

The bottom line of the new diagram illustrates the remarkable influence that additional nitrogen units have on the hydrolysis of a series of N-acetylazoles in water at 25 °C and pH=7. The pyrrole compound on the left is essentially unreactive, as expected for an amide, but additional nitrogens markedly increase the rate of hydrolysis. This effect has been put to practical use in applications of the acylation reagent 1,1'-carbonyldiimidazole (Staab's reagent).

Another facet of heterocyclic chemistry was disclosed in the course of investigations concerning the action of thiamine. As its pyrophosphate derivative, thiamine is a coenzyme for several biochemical reactions, notably decarboxylations of pyruvic acid to acetaldehyde and acetoin. Early workers speculated that an "active aldehyde" or acyl carbanion species was an intermediate in these reactions. Many proposals were made, some involving the aminopyrimidine moiety, and others, ring-opened hydrolysis derivatives of the thiazole ring, but none were satisfactory. This puzzle was solved when R. Breslow (Columbia) found that the C-2 hydrogen of thiazolium salts was unexpectedly acidic (pK_a *ca.* 13), forming a relatively stable ylide conjugate base. As shown, this rationalizes the facile decarboxylation of thiazolium-2-carboxylic acids and deuterium exchange at C-2 in neutral heavy water.

Appropriate thiazolium salts catalyze the conversion of aldehydes to acyloins in much the same way that cyanide ion catalyzes the formation of benzoin from benzaldehyde, the benzoin condensation. Note that in both cases an acyl anion equivalent is formed and then adds to a carbonyl function in the expected manner. The benzoin condensation is limited to aromatic aldehydes, but the use of thiazolium catalysts has proven broadly effective for aliphatic and aromatic aldehydes. This approach to acyloins employs milder conditions than the reduction of esters to enediol intermediates by the action of metallic sodium.

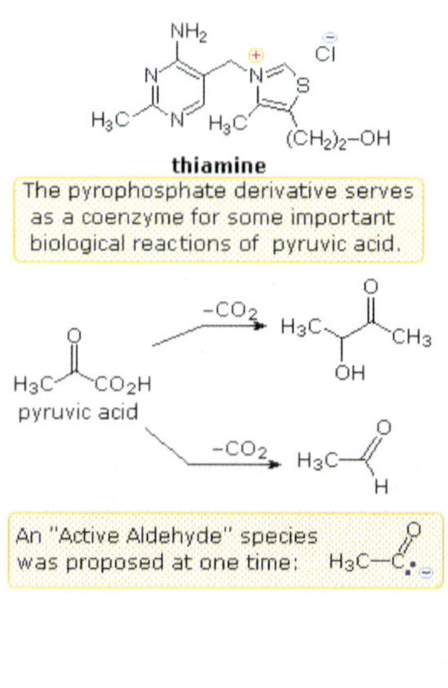

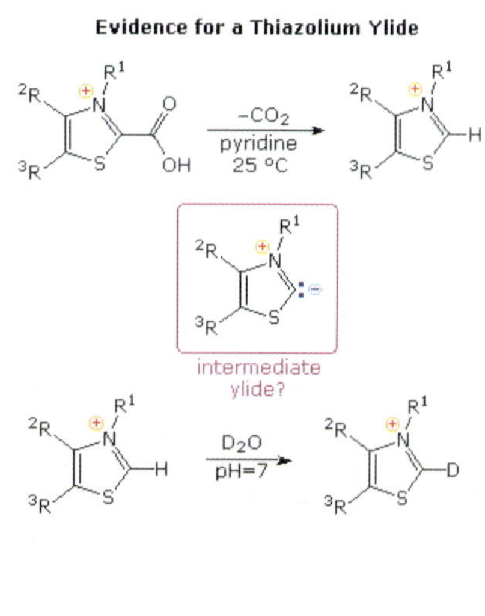

The most important condensed ring system related to these heterocycles is indole. Some electrophilic substitution reactions of indole are shown in the following diagram. Whether the indole nitrogen is substituted or not, the favored site of attack is C-3 of the heterocyclic ring. Bonding of the electrophile at that position permits stabilization of the onium-intermediate by the nitrogen without disruption of the benzene aromaticity.

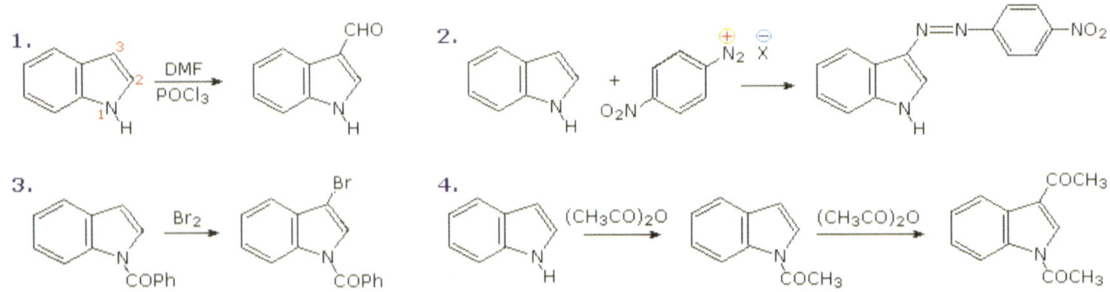

6-Membered Rings

Properties

The chemical reactivity of the saturated members of this class of heterocycles: tetrahydropyran, thiane and piperidine, resemble that of acyclic ethers, sulfides, and 2º-amines, and will not be described here. 1,3-Dioxanes and dithianes are cyclic acetals and thioacetals. These units are commonly used as protective groups for aldehydes and ketones, as well as synthetic intermediates, and may be hydrolyzed by the action of aqueous acid. The reactivity of partially unsaturated compounds depends on the relationship of the double bond and the heteroatom (*e.g.* 3,4-dihydro-2H-pyran is an enol ether).

Fully unsaturated six-membered nitrogen heterocycles, such as pyridine, pyrazine, pyrimidine and pyridazine, have stable aromatic rings. Oxygen and sulfur analogs are necessarily positively charged, as in the case of 2, 4, 6-triphenylpyrylium tetrafluoroborate.

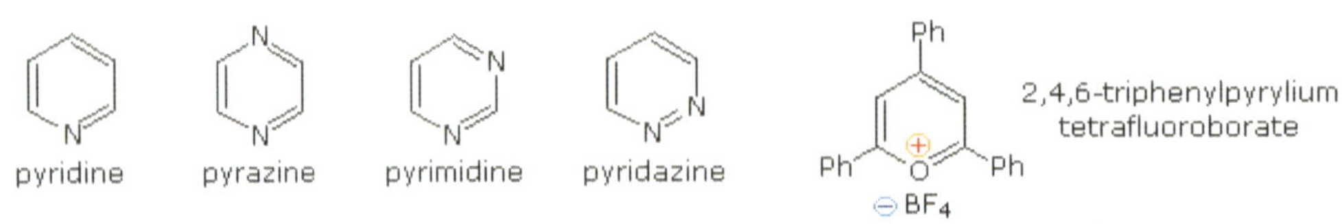

From heat of combustion measurements, the aromatic stabilization energy of pyridine is 21 kcal/mole. The resonance description drawn at the top of the following diagram includes charge separated structures not normally considered for benzene. The greater electronegativity of nitrogen (relative to carbon) suggests that such canonical forms may contribute to a significant degree. Indeed, the larger dipole moment of pyridine compared with piperidine supports this view. Pyridine and its derivatives are weak bases, reflecting the sp^2 hybridization of the nitrogen. From the polar canonical forms shown here, it should be apparent that electron donating substituents will increase the basicity of a pyridine, and that substituents on the 2 and 4-positions will influence this basicity more than an equivalent 3-substituent. The pK_a values given in the table illustrate a few of these substituent effects. Methyl substituted derivatives have the common names picoline (methyl pyridines), lutidine (dimethyl pyridines) and collidine (trimethyl pyridines). The influence of 2-substituents is complex, consisting of steric hindrance and electrostatic components. 4-Dimethylaminopyridine is a useful catalyst for acylation reactions carried out in pyridine as a solvent. At first glance, the sp^3 hybridized nitrogen might appear to be the stronger base, but it should be remembered that N,N-dimethylaniline has a pK_a slightly lower than that of pyridine itself. Consequently, the sp^2 ring nitrogen is the site at which protonation occurs.

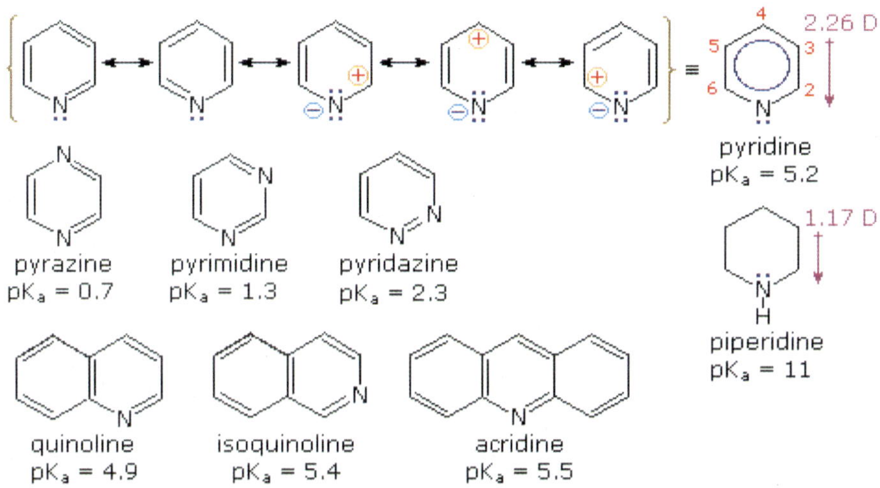

Pyridine Basicities

Substituent(s)	pK_a
4-methyl	6.0
3-methyl	5.7
2,6-dimethyl	6.7
2,4,6-trimethyl	7.5
2,6-di-*t*-butyl	3.6
3-methoxy	4.9
4-methoxy	6.5
4-dimethylamino	9.7
2-chloro	0.7
4-chloro	3.8
2,6-dichloro	-2.9

The diazines pyrazine, pyrimidine and pyridazine are all weaker bases than pyridine due to the inductive effect of the second nitrogen. However, the order of base strength is unexpected. A consideration of the polar contributors helps to explain the difference between pyrazine and pyrimidine, but the basicity of pyridazine seems anomalous. It has been suggested that electron pair repulsion involving the vicinal nitrogens destabilizes the neutral base relative to its conjugate acid.

Electrophilic Substitution of Pyridine

Pyridine is a modest base ($pK_a = 5.2$). Since the basic unshared electron pair is not part of the aromatic sextet, as in pyrrole, pyridinium species produced by N-substitution retain the aromaticity of pyridine. As shown below, N-alkylation and N-acylation products may be prepared as stable crystalline solids in the absence of water or other reactive nucleophiles. The N-acyl salts may serve as acyl transfer agents for the preparation of esters and amides. Because of the stability of the pyridinium cation, it has been used as a moderating component in complexes with a number of reactive inorganic compounds. Several examples of these stable and easily handled reagents are shown at the bottom of the diagram. The poly(hydrogen fluoride) salt is a convenient source of HF for addition to alkenes and conversion of alcohols to alkyl fluorides, pyridinium chlorochromate (PCC) and its related dichromate analog are versatile oxidation agents and the tribromide salt is a convenient source of bromine. Similarly, the reactive compounds sulfur trioxide and diborane are conveniently and safely handled as pyridine complexes.

Amine oxide derivatives of 3°-amines and pyridine are readily prepared by oxidation with peracids or peroxides, as shown by the upper right equation. Reduction back to the amine can usually be achieved by treatment with zinc (or other reactive metals) in dilute acid.

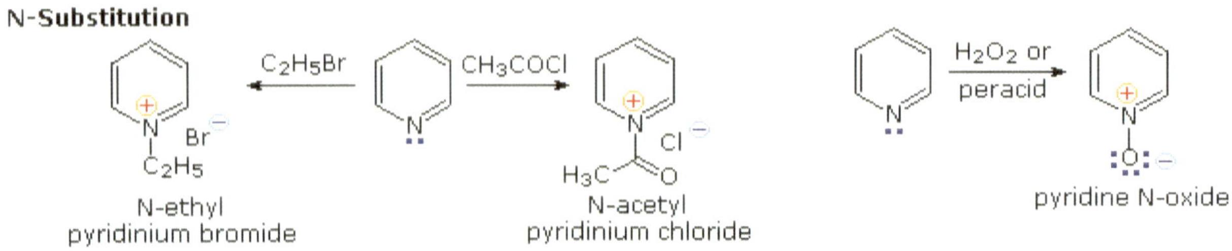

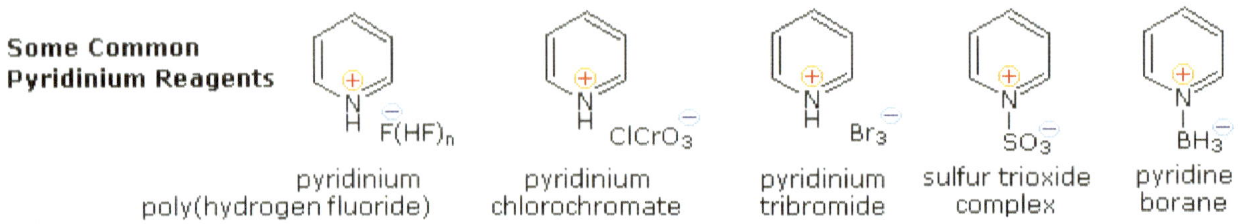

From the previous resonance description of pyridine, we expect this aromatic amine to undergo electrophilic substitution reactions far less easily than does benzene. The electrophilic reagents and catalysts employed in these reactions coordinate with the nitrogen electron pair, exacerbating the positive charge at positions 2, 4 and 6 of the pyridine ring. Three examples of the extreme conditions required for electrophilic substitution are shown on the left. Substituents that block electrophile coordination with nitrogen or reduce the basicity of the nitrogen facilitate substitution, as demonstrated by the examples in the blue-shaded box at the lower right, but substitution at C-3 remains dominant. Activating substituents at other locations also influence the ease and regioselectivity of substitution. By clicking on the diagram a second time, three examples will show on the left. The amine substituent in the upper case directs the substitution to C-2, but the weaker electron donating methyl substituent in the middle example cannot overcome the tendency for 3-substitution. Hydroxyl substituents at C-2 and C-4 tautomerize to pyridones, as shown for the 2-isomer at the bottom left.

Pyridine N-oxide undergoes some electrophilic substitutions at C-4 and others at C-3. The coordinate covalent N–O bond may exert a push-pull influence, as illustrated by the two examples on the right. Although the positively charged nitrogen alone would have a strong deactivating influence, the negatively charged oxygen can introduce electron density at C-2, C-4 and C-6 by π-bonding to the ring nitrogen. This is a controlling factor in the relatively facile nitration at C-4. However, if the oxygen is bonded to an electrophile such as SO_3, the resulting pyridinium ion will react sluggishly and preferentially at C-3.

The fused ring heterocycles quinoline and isoquinoline provide additional evidence for the stability of the pyridine ring. Vigorous permanganate oxidation of quinoline results in predominant attack on the benzene ring; isoquinoline yields products from cleavage of both rings. Note that naphthalene is oxidized to phthalic acid in a similar manner. By contrast, the heterocyclic ring in both compounds undergoes preferential catalytic hydrogenation to yield tetrahydroproducts. Electrophilic nitration, halogenation and sulfonation generally take place at C-5 and C-8 of the benzene ring, in agreement with the preceding description of similar pyridine reactions and the kinetically favored substitution of naphthalene at C-1 (α) rather than C-2 (β).

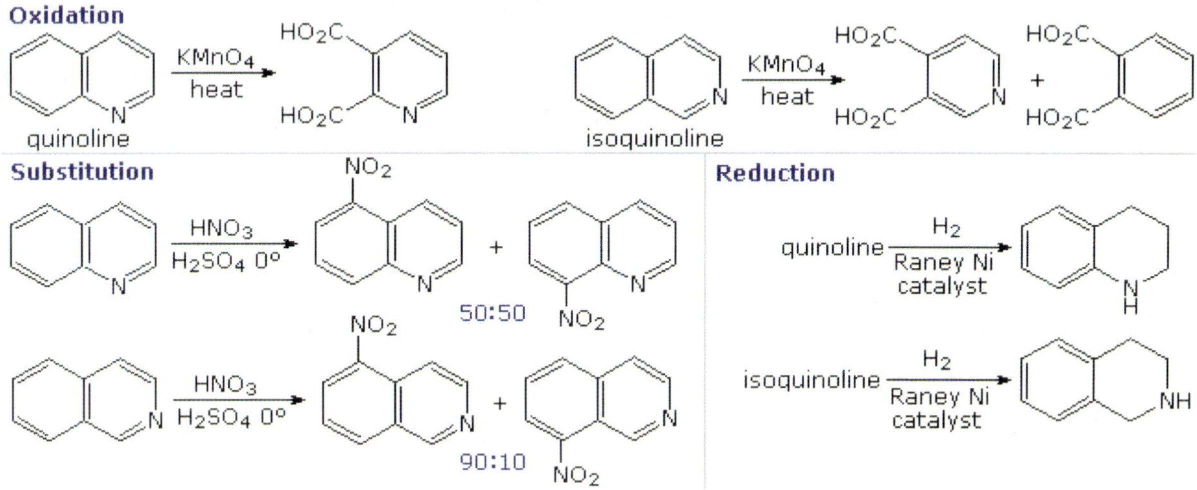

Other Reactions of Pyridine

Thanks to the nitrogen in the ring, pyridine compounds undergo nucleophilic substitution reactions more easily than equivalent benzene derivatives. In the following diagram, reaction 1 illustrates displacement of a -2-chloro substituent by ethoxide anion. The addition-elimination mechanism shown for this reaction is helped by nitrogen's ability to support a negative charge. A similar intermediate may be written for substitution of a -4-halopyridine, but substitution at the 3-position is prohibited by the the failure to create an intermediate of this kind. The two Chichibabin aminations in reactions 2 and 3 are remarkable in that the leaving anion is hydride (or an equivalent). Hydrogen is often evolved in the course of these reactions. In accord with this mechanism, quinoline is aminated at both C-2 and C-4.

Addition of strong nucleophiles to N-oxide derivatives of pyridine proceed more rapidly than to pyridine itself, as demonstrated by reactions 4 and 5. The dihydro-pyridine intermediate easily loses water or its equivalent by elimination of the –Om substituent on nitrogen.

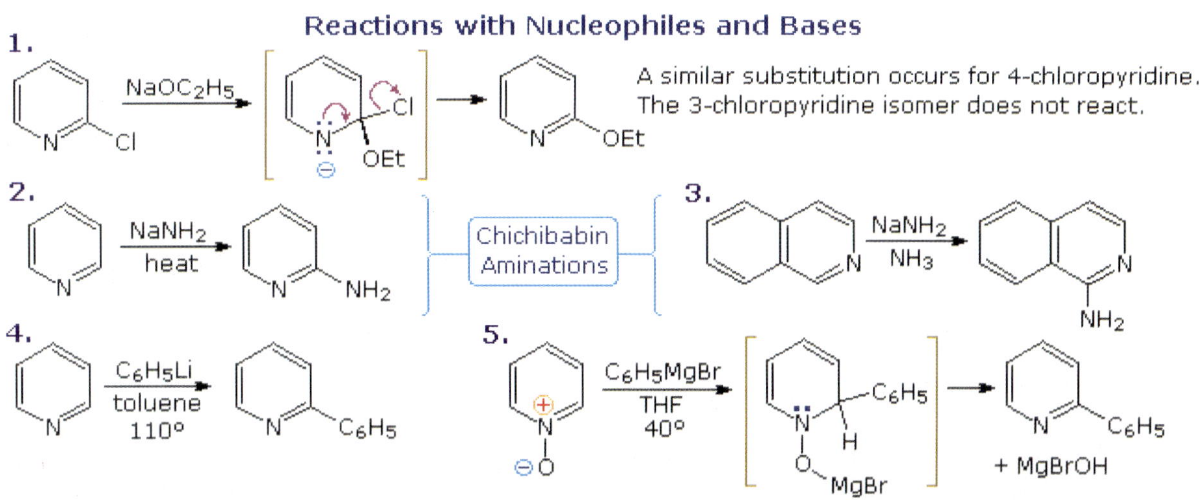

Five additional examples of base or nucleophile reactions with substituted pyridine will be displayed. Because the pyridine ring (and to a greater degree the N-oxide ring) can support a negative charge, alkyl substituents in the 2- and 4-locations are activated in the same fashion as by a carbonyl group. Reactions 6 and 7 show alkylation and condensation reactions resulting from this activation. Reaction 8 is an example of N-alkylpyridone formation by hydroxide addition to an N-alkyl pyridinium cation, followed by mild oxidation. Birch reduction converts pyridines to dihydropyridines that are bis-enamines and may be hydrolyzed to 1,5-dicarbonyl compounds. Pyridinium salts undergo a one electron transfer to generate remarkably stable free radicals. The example shown in reaction 9 is a stable (in the absence of oxygen), distillable green liquid. Although 3-halopyridines do not undergo addition-elimination substitution reactions as do their 2- and 4-isomers, the strong base sodium amide effects amination by way of a pyridyne intermediate. This is illustrated by reaction 10. It is interesting that 3-pyridyne is formed in preference to 2-pyridyne. The latter is formed if C-4 is occupied by an alkyl substituent. The pyridyne intermediate is similar to benzyne.

Some Polycyclic Heterocycles

Heterocyclic structures are found in many natural products. Examples of some nitrogen compounds, known as alkaloids because of their basic properties, were given in the amine. Some other examples are displayed in the following diagram. Camptothecin is a quinoline alkaloid which inhibits the DNA enzyme topoisomerase I. Reserpine is an indole alkaloid, which has been used for the control of high blood pressure and the treatment of psychotic behavior. Ajmaline and strychnine are also indole alkaloids, the former being an antiarrhythmic agent and latter an extremely toxic pesticide. The neurotoxins saxitoxin and tetrodotoxin both have marine origins and are characterized by guanidiniun moieties. Aflatoxin B_1 is a non-nitrogenous carcinogenic compound produced by the Aspergillus fungus.

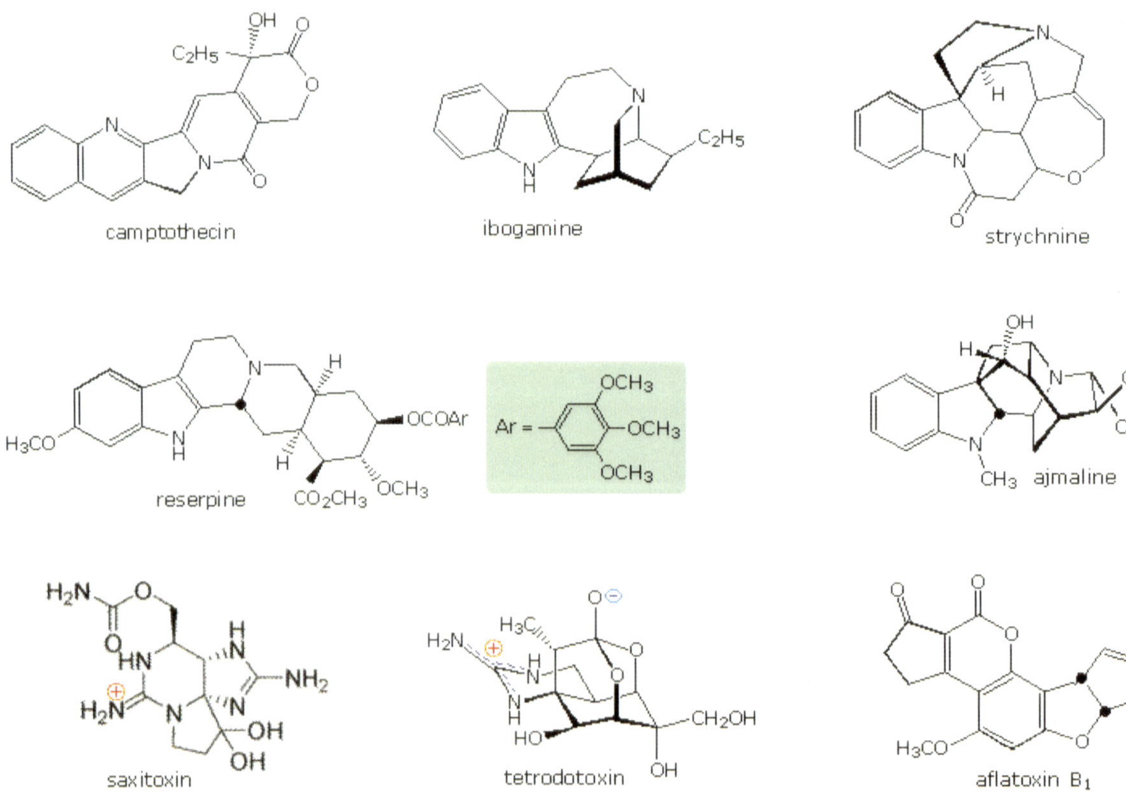

Porphyrin is an important cyclic tertrapyrrole that is the core structure of heme and chlorophyll.

Derivatives of the simple fused ring heterocycle purine constitute an especially important and abundant family of natural products. The amino compounds adenine and guanine are two of the complementary bases that are essential components of DNA. Structures for these compounds are shown in the following diagram. Xanthine and uric acid are products of the metabolic oxidation of purines. Uric acid is normally excreted in the urine; an excess serum accumulation of uric acid may lead to an arthritic condition known as gout.

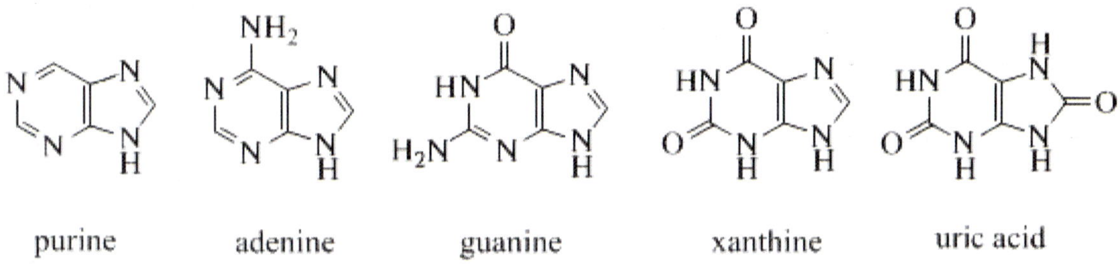

Caffeine, the best known of these, is a bitter, crystalline alkaloid. It is found in varying quantities, along with additional alkaloids such as the cardiac stimulants theophylline and theobromine in the beans, leaves, and fruit of certain plants. Drinks containing caffeine, such as coffee, tea and some soft drinks are arguably the world's most widely consumed beverages. Caffeine is a central nervous system stimulant, serving to ward off drowsiness and restore alertness. Paraxantheine is the chief metabolite of caffeine in the body.

Sulfur heterocycles are found in nature, but to a lesser degree than their nitrogen and oxygen analogs. Two members of the B-vitamin complex, biotin and thiamine, incorporate such heterocyclic moieties. These are shown together with other heterocyclic B-vitamins in the following diagram:

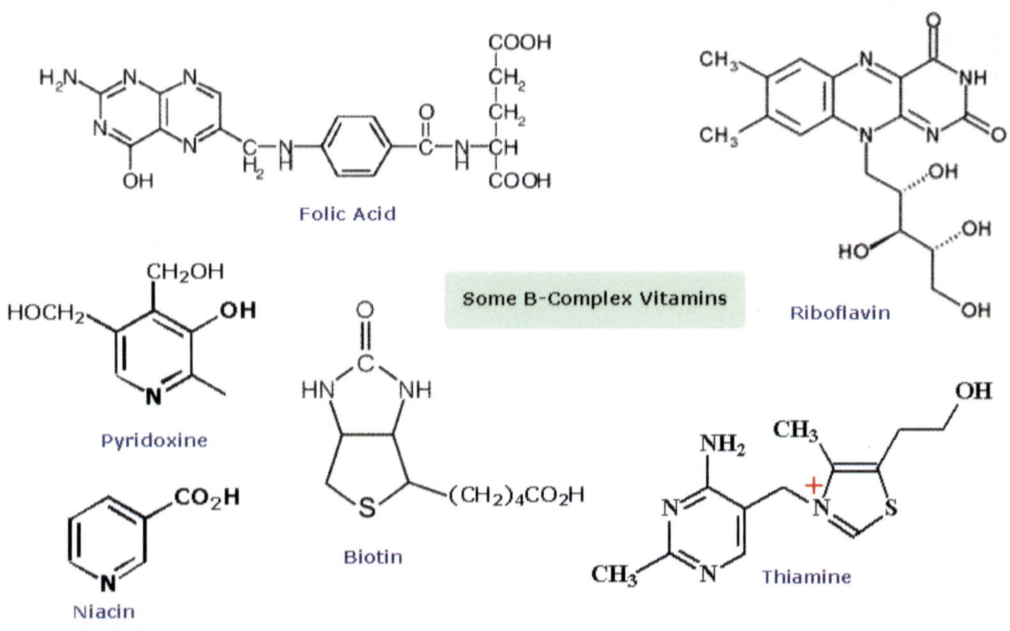

Terthienyl is an interesting thiophene trimer found in the roots of marigolds, where it provides nemicidal activity. Studies have shown that UV irradiation of terthienyl produces a general phototoxicity for many organisms. Polymers incorporating thiophene units and fused systems such as dithienothiophene have interesting electromagnetic properties, and show promise as organic metal-like conductors and photovoltaic materials. The charge transfer complex formed by tetrathiofulvalene and tetracyanoquinodimethane has one of the highest electrical conductivities reported for an organic solid.

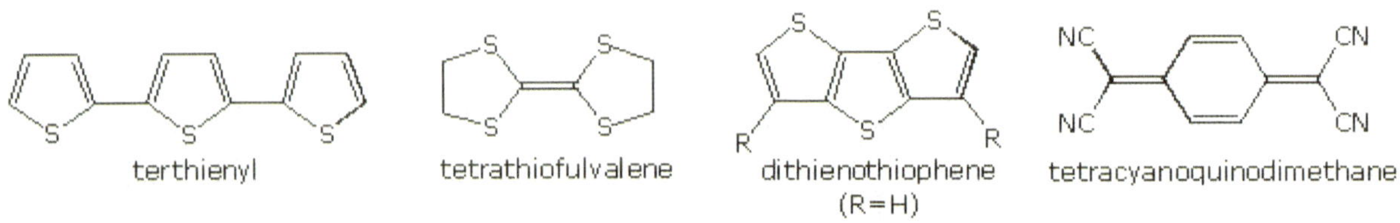

Organosulfur Chemistry

Organosulfur chemistry is the study of the physical and chemical properties of sulfur and the nature of the sulfur-carbon-oxygen bonds. The use of organosulfur compounds has become widespread in medicine and agriculture.

Organosulfur compound is also spelled as organosulphur compound or organic sulfur compound is a subclass of organic substances that contain sulfur and that are known for their varied occurrence and unusual properties. They are found in diverse locations, including in interstellar space, inside hot acidic volcanoes, and deep within the oceans. Organosulfur compounds occur in the bodies of all living creatures in the form of certain essential amino acids (such as cysteine, cystine, and methionine, which are components of proteins), of the tripeptide glutathione, and of enzymes, coenzymes, vitamins, and hormones.

Typical organisms contain 2 percent sulfur dry weight. Coenzyme A (CoA), biotin, thiamin chloride (vitamin B_1), α-lipoic acid, insulin, oxytocin, sulfated polysaccharides, and the nitrogen-fixing nitrogenase enzymes are but a few examples of important natural sulfur-containing compounds. Certain simple organosulfur compounds, such as thiols, are repugnant to humans and most higher animals even at extraordinarily low concentrations; they are used as defensive secretions by a variety of animal species and figure in unpleasant odours associated with polluted air and water, particularly that resulting from the use of sulfur-rich fossil fuels. However, related types of organosulfur compounds found in such foods as garlic, onion, chive, leek, broccoli, cabbage, radish, asparagus, mushroom, mustard, truffle, coffee, and pineapple are sources of olfactory and gustatory delight.

Mustard gas, or bis(β-chloroethyl) sulfide, $(ClCH_2CH_2)_2S$, is a potent chemical warfare agent, whereas other sulfur compounds such as sulfanilamide (a sulfa drug), penicillin, and cephalosporin are valued antibiotics. Synthetic organosulfur compounds include polysulfones, inert polymers used in astronauts' transparent face shields; polythiophenes, materials possessing the metal-like ability to conduct electricity; agricultural chemicals, insecticides, and organic solvents, such as dimethyl sulfoxide, $CH_3S(=O)CH_3$, and carbon disulfide, CS_2; dyes; lubricating oil constituents; food additives; and substances used to make rayon. In chemical research, organosulfur compounds are valued reagents widely used for synthesizing new compounds. A global sulfur cycle exists that interconverts natural organosulfur compounds with either inorganic sulfide or sulfate ions. Sulfide or sulfate ions can also be formed in nature from elemental sulfur.

Analysis of Organosulfur Compounds

In addition to routine methods of analysis that can be used with all classes of organic compounds, certain procedures reflect specific characteristics of sulfur. In a mass spectrometer, organosulfur compounds frequently produce strong molecular ions in which the charge is located predominantly on sulfur. The presence of sulfur is indicated by the occurrence of sulfur-34 (^{34}S) isotope peaks, 4.4 percent of the abundance of ^{32}S. Organically bound sulfur in the form of the natural isotope ^{33}S can be directly examined by nuclear magnetic resonance (NMR) spectroscopy, although the low natural abundance (0.76 percent) and small magnetic and nuclear quadrupole moments make analysis more difficult than for protons (^{1}H) or carbon-13 (^{13}C). Levels of organosulfur compounds in crude petroleum as low as 10 parts per billion or less may have a detrimental effect on metallic catalysis or may cause unpleasant odors. These very low sulfur levels are detected using gas chromatographs with sulfur chemiluminescence or atomic-emission detectors with high sensitivity to detect sulfur compounds in the presence of other compounds.

Classes

Organosulfur compounds can be classified according to the sulfur-containing functional groups, which are listed (approximately) in decreasing order of their occurrence.

Organosulfur Compounds

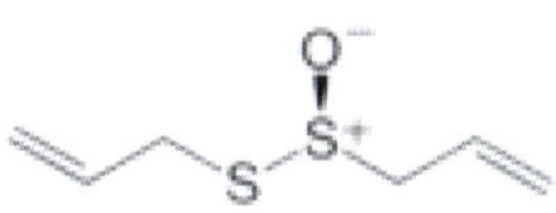

Allicin, the active flavor compound in crushed garlic

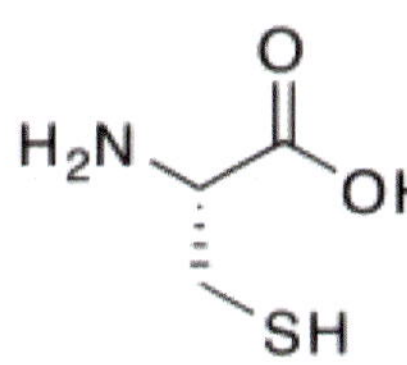

(*R*)-Cysteine, an amino acid containing a thiol group

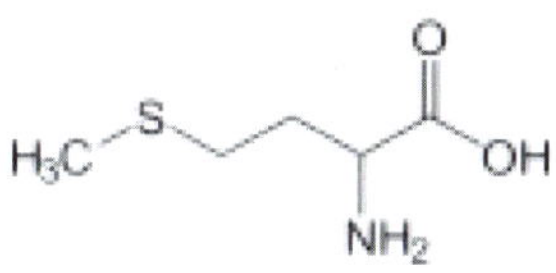

Methionine, an amino acid containing a sulfide

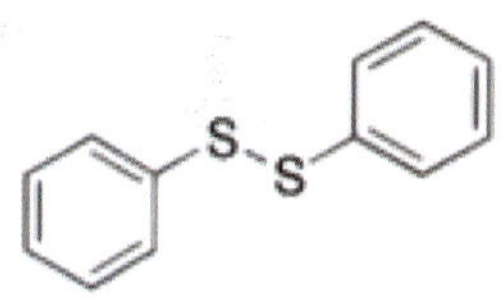

Diphenyl disulfide, a representative disulfide

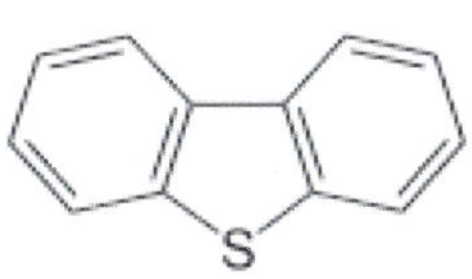

Dibenzothiophene, a component of crude oil

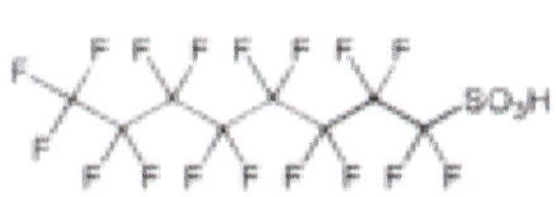

Perfluorooctanesulfonic acid, a controversial surfactant

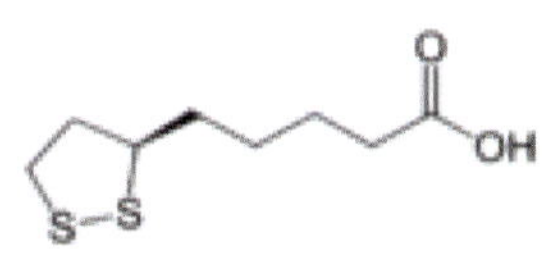

Lipoic acid, an essential cofactor of four mitochondrial enzyme complexes.

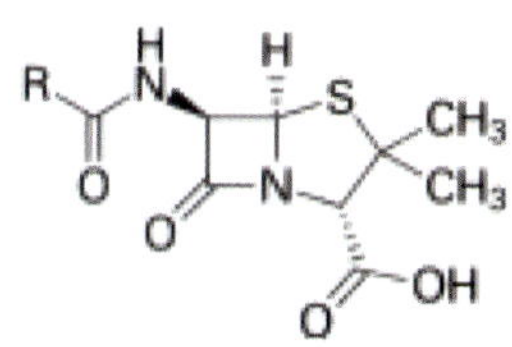

Penicillin core structure, where "R" is the variable group.

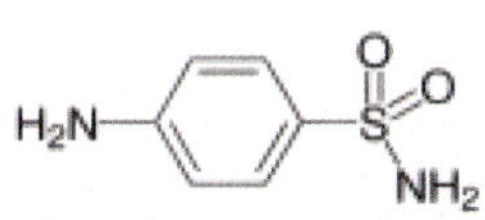

Sulfanilamide, a sulfonamide antibacterial, called a sulfa drug.

Sulfur mustard, a type of sulfide used as a chemical warfare agent.

Sulfides

Sulfides, formerly known as thioethers, are characterized by C–S–C bonds. Relative to C–C bonds, C–S bond are both longer, because S is larger than carbon, and about 10% weaker. Representative bond lengths in sulfur compounds are 183 pm for the S–C single bond in methanethiol and 173 pm in thiophene. The C–S bond dissociation energy for thiomethane is 89 kcal/mol (370 kJ/mol) compared to methane's 100 kcal/mol (420 kJ/mol) and when hydrogen is replaced by a methyl group the energy decreases to 73 kcal/mol (305 kJ/mol). The single carbon to oxygen bond is shorter than that of the C–C bond. The bond dissociation energies for dimethyl sulfide and dimethyl ether are respectively 73 and 77 kcal/mol (305 and 322 kJ/mol).

Sulfides are typically prepared by alkylation of thiols. They can also be prepared via the Pummerer rearrangement. In one named reaction called the Ferrario reaction phenyl ether is converted to phenoxathiin by action of elemental sulfur and aluminium chloride.

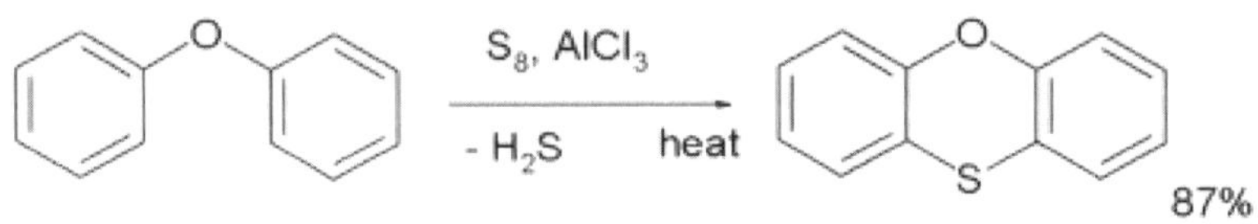

Thioacetals and thioketals feature C–S–C–S–C bond sequence. They represent a subclass of sulfides. The thioacetals are useful in "umpolung" of carbonyl groups. Thioacetals and thioketals can also be used to protect a carbonyl group in organic syntheses.

The above classes of sulfur compounds also exist in saturated and unsaturated heterocyclic structures, often in combination with other heteroatoms, as illustrated by thiiranes, thiirenes, thietanes, thietes, dithietanes, thiolanes, thianes, dithianes, thiepanes, thiepines, thiazoles, isothiazoles, and thiophenes, among others. The latter three compounds represent a special class of sulfur-containing heterocycles that are aromatic. The resonance stabilization of thiophene is 29 kcal/mol (121 kJ/mol) compared to 20 kcal/mol (84 kJ/mol) for the oxygen analogue furan. The reason for this difference is the higher electronegativity for oxygen drawing away electrons to itself at the expense of the aromatic ring current. Yet as an aromatic substituent the thio group is less electron-releasing than the alkoxy group. Dibenzothiophenes, tricyclic heterocycles consisting of two benzene rings fused to a central thiophene ring, occurs widely in heavier fractions of petroleum.

Thiols, Disulfides and Polysulfides

Thiol group contain the functionality R–SH. Thiols are structurally similar to the alcohol group, but these functionalities are very different in their chemical properties. Thiols are more nucleophilic, more acidic, and more readily oxidized. This acidity can differ by 5 pK_a units.

The difference in electronegativity between sulfur (2.58) and hydrogen (2.20) is small and therefore hydrogen bonding in thiols is not prominent. Aliphatic thiols form monolayers on gold, which are topical in nanotechnology.

Certain aromatic thiols can be accessed through a Herz reaction.

Disulfides R–S–S–R with a covalent sulfur to sulfur bond are important for crosslinking: in biochemistry for the folding and stability of some proteins and in polymer chemistry for the cross-linking of rubber.

Longer sulfur chains are also known, such as in the natural product varacin which contains an unusual pentathiepin ring (5-sulfur chain cyclized onto a benzene ring).

Thioesters

Thioesters have general structure R–C(O)–S–R. They are related to regular esters (R–C(O)–O–R) but are more susceptible to hydrolysis and related reactions. Thioesters are prominent in biochemistry, especially in fatty acid synthesis.

Sulfoxides, Sulfones and Thiosulfinates

A sulfoxide, R–S(O)–R, is the *S*-oxide of a sulfide ("sulfide oxide"), a sulfone, R–$S(O)_2$–R, is the *S*, *S*-dioxide of a sulfide, a thiosulfinate, R–S(O)–S–R, is the *S*-oxide of a disulfide, and a thiosulfonate, R–$S(O)_2$–S–R, is the *S*, *S*-dioxide of a disulfide. All of these compounds are well known with extensive chemistry, e.g., dimethyl sulfoxide, dimethyl sulfone, and allicin.

Sulfimides, Sulfoximides and Sulfonediimines

Sulfimides (also called a sulfilimines) are sulfur–nitrogen compounds of structure $R_2S{=}NR'$, the nitrogen analog of sulfoxides. They are of interest in part due to their pharmacological properties. When two different R groups are attached to sulfur, sulfimides are chiral. Sulfimides form stable α-carbanions.

Sulfoximides (also called sulfoximines) are tetracoordinate sulfur–nitrogen compounds, isoelectronic with sulfones, in which one oxygen atom of the sulfone is replaced by a substituted nitrogen atom, e.g., $R_2S(O){=}NR'$. When two different R groups are attached to sulfur, sulfoximides are chiral. Much of the interest in this class of compounds is derived from the discovery that methionine sulfoximide (methionine sulfoximine) is an inhibitor of glutamine synthetase.

Sulfonediimines (also called sulfodiimines, sulfodiimides or sulfonediimides) are tetracoordinate sulfur–nitrogen compounds, isoelectronic with sulfones, in which both oxygen atoms of the sulfone are replaced by a substituted nitrogen atom, e.g., $R_2S({=}NR')_2$. They are of interest because of their biological activity and as building blocks for heterocycle synthesis.

S-Nitrosothiols

S-Nitrosothiols, also known as thionitrites, are compounds containing a nitroso group attached to the sulfur atom of a thiol, e.g. R–S–N=O. They have received considerable attention in biochemistry because they serve as donors of the nitrosonium ion, NO^+, and nitric oxide, NO, which may serve as signaling molecules in living systems, especially related to vasodilation.

Sulfur Halides

A wide range of organosulfur compounds are known which contain one or more halogen atom ("X" in the chemical formulas that follow) bonded to a single sulfur atom, e.g.: sulfenyl halides, RSX; sulfinyl halides, RS(O)X; sulfonyl halides, RSO_2X; alkyl and arylsulfur trichlorides, $RSCl_3$

and trifluorides, RSF_3; and alkyl and arylsulfur pentafluorides, RSF_5. Less well known are dialkylsulfur tetrahalides, mainly represented by the tetrafluorides, e.g., R_2SF_4.

Thioketones, Thioaldehydes and Related Compounds

Compounds with double bonds between carbon and sulfur are relatively uncommon, but include the important compounds carbon disulfide, carbonyl sulfide, and thiophosgene. Thioketones (RC(=S)R′) are uncommon with alkyl substituents, but one example is thiobenzophenone. Thioaldehydes are rarer still, reflecting their lack of steric protection ("thioformaldehyde" exists as a cyclic trimer). Thioamides, with the formula $R_1C(=S)N(R_2)R_3$ are more common. They are typically prepared by the reaction of amides with Lawesson's reagent. Isothiocyanates, with formula R–N=C=S, are found naturally. Vegetable foods with characteristic flavors due to isothiocyanates include wasabi, horseradish, mustard, radish, Brussels sprouts, watercress, nasturtiums, and capers.

S-Oxides and S, S-dioxides of Thiocarbonyl Compounds

The *S*-oxides of thiocarbonyl compounds are known as thiocarbonyl *S*-oxides or sulfines, $R_2C=S=O$, and thiocarbonyl *S, S*-dioxides or sulfenes, $R_2C=SO_2$. These compounds are well known with extensive chemistry.

Triple Bonds between Carbon and Sulfur

Triple bonds between sulfur and carbon in sulfaalkynes are rare and can be found in carbon monosulfide (CS) and have been suggested for the compounds F_3CCSF_3 and F_5SCSF_3. The compound HCSOH is also represented as having a formal triple bond.

Thiocarboxylic Acids and Thioamides

Thiocarboxylic acids (RC(O)SH) and dithiocarboxylic acids (RC(S)SH) are well known. They are structurally similar to carboxylic acids but more acidic. Thioamides are analogous to amides.

Sulfonic, Sulfinic and Sulfenic Acids, Esters, Amides, and Related Compounds

Sulfonic acids have functionality $R–S(=O)_2–OH$. They are strong acids that are typically soluble in organic solvents. Sulfonic acids like trifluoromethanesulfonic acid is a frequently used reagent in organic chemistry. Sulfinic acids have functionality R–S(O)–H while sulfenic acids have functionality R–S–OH. In the series sulfonic—sulfinic—sulfenic acids, both the acid strength and stability diminish in that order. Sulfonamides, sulfinamides and sulfenamides, with formulas $R–SO_2NR'_2$, $R–S(O)NR'_2$, and $R–SNR'_2$, respectively, each have a rich chemistry. For example, sulfa drugs are sulfonamides derived from aromatic sulfonation. Chiral sulfinamides are used in asymmetric synthesis, while sulfenamides are used extensively in the vulcanization process to assist cross-linking. Thiocyanates, R–S–CN, are related to sulfenyl halides and esters in terms of reactivity.

Sulfonium, Oxosulfonium and Related Salts

A sulfonium ion is a positively charged ion featuring three organic substituents attached to sulfur, with the formula $[R_3S]^+$. Together with their negatively charged counterpart, the anion, the compounds are called sulfonium salts. An oxosulfonium ion is a positively charged ion featuring three organic substituents and an oxygen attached to sulfur, with the formula $[R_3S{=}O]^+$. Together with their negatively charged counterpart, the anion, the compounds are called oxosulfonium salts. Related species include alkoxysulfonium and chlorosulfonium ions, $[R_2SOR]^+$ and $[R_2SCl]^+$, respectively.

Sulfonium, Oxosulfonium and Thiocarbonyl Ylides

Deprotonation of sulfonium and oxosulfonium salts affords ylides, of structure $R_2S^+{-}C^-{-}R'_2$ and $R_2S(O)^+{-}C^-{-}R'_2$. While sulfonium ylides, for instance in the Johnson–Corey–Chaykovsky reaction used to synthesize oxiranes, are sometimes drawn with a C=S double bond, e.g., $R_2S{=}CR'_2$, the ylidic carbon–sulfur bond is highly polarized and is better described as being ionic. Sulfonium ylides are key intermediates in the synthetically useful Stevens rearrangement. Thiocarbonyl ylides ($RR'C{=}S^+{-}C^-{-}RR'$) can form by ring-opening of thiiranes, photocyclization of aryl vinyl sulfides, as well as by other processes.

Sulfuranes and Persulfuranes

Sulfuranes are relatively specialized functional group that are tetravalent, hypervalent sulfur compounds, with the formula SR_4 and likewise persulfuranes are hexavalent SR_6. All-carbon hexavalent complexes have been known for the heavier representatives of the chalcogen group, for instance the compound hexamethylpertellurane ($Te(Me)_6$) was discovered in 1990 by reaction of tetramethyltellurium with xenon difluoride to $TeF_2(CH_3)_4$ followed by reaction with dimethylzinc. The sulfur analogue hexamethylpersulfurane ($S(CH_3)_6$) has been predicted to be stable but has not been synthesized yet.

The first ever all-carbon persulfurane actually synthesized in a laboratory has two methyl and two biphenyl ligands:

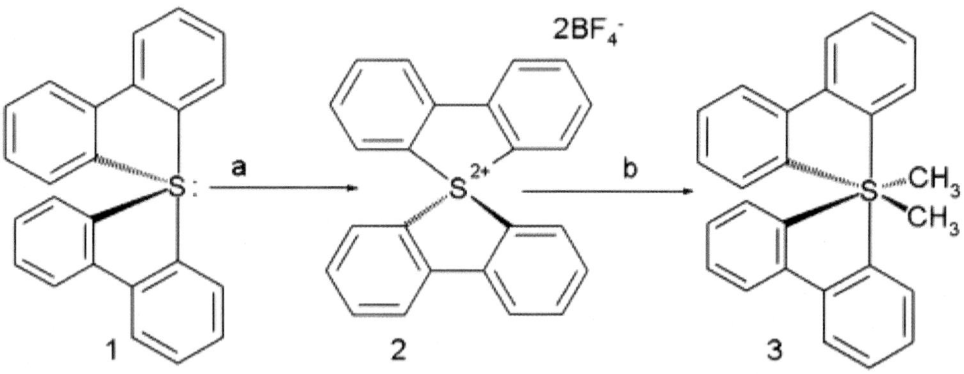

It is prepared from the corresponding sulfurane 1 with xenon difluoride/boron trifluoride in acetonitrile to the sulfuranyl dication 2 followed by reaction with methyllithium in tetrahydrofuran to (a stable) persulfurane 3 as the cis isomer. X-ray diffraction shows C–S bond lengths ranging between 189 and 193 pm (longer than the standard bond length) with the central sulfur atom in a distorted octahedral molecular geometry.

Computer simulation suggests that these bonds are very polar with the negative charges residing on carbon.

Naturally Occurring Organosulfur Compounds

Not all organosulfur compounds are foul-smelling pollutants. Penicillin and cephalosporin are life-saving antibiotics, derived from fungi. Gliotoxin is a sulfur-containing mycotoxin produced by several species of fungi under investigation as an antiviral agent. Compounds like allicin and ajoene are responsible for the odor of garlic, and lenthionine contributes to the flavor of shiitake mushrooms. Volatile organosulfur compounds also contribute subtle flavor characteristics to wine, nuts, cheddar cheese, chocolate, coffee, and tropical fruit flavors. Many of these natural products also have important medicinal properties such as preventing platelet aggregation or fighting cancer.

In pollution

Most organic sulfur compounds in the environment are naturally occurring, as a consequence of the fact that sulfur is essential for life and two amino acids (cysteine and methionine) contain this element.

Some organosulfur compounds in the environment, are generated as minor by-products of industrial processes such as the manufacture of plastics and tires.

Selected smell-producing processes are organosulfur compounds produced by the coking of coal designed to drive out sulfurous compounds and other volatile impurities in order to produce 'clean carbon' (coke), which is primarily used for steel production.

In Fossil Fuels

Odors occur as well in chemical processing of coal or crude oil into precursor chemicals (feed stocks) for downstream industrial uses (e.g. plastics or pharmaceutical production) and the ubiquitous needs of petroleum distillation for gasolines, diesel, and other grades of fuel oils production.

Organosulfur compounds might be understood as aromatic contaminants that need to be removed from natural gas before commercial uses, from exhaust stacks and exhaust vents before discharge. In this latter context, organosulfur compounds may be said to account for the pollutants in sulfurous acid rain, or equivalently, said to be pollutants within most common fossil fuels, especially coal.

The most common organosulfur compound present in all petroleum fractions is thiophene (C_4H_4S), a cyclic and aromatic liquid. In addition, the heavy fractions of oil contain benzothiophene (C_8H_6S, thianaphtene) and dibenzothiophene. Most of the last compounds are solids and smell like naphthalene. Many methylated, dimethyl, diethyl benzothiophene derivatives are present in diesel and fuel oils which make fuel oils very difficult to clean.

All these heterocyclic sulfides account for 200–500 ppm of natural fuel, the heavily substituted dibenzothiophenes remain after HDS and account for 10–20 ppm. These molecules are also found in coals and they must be eliminated before consumption.

Reduced molybdenum together with nickel is currently used to eliminate thiophenes from petroleum (HDS) due to its great affinity towards sulfur. In addition tungsten together with nickel and cobalt is used for hydrodesulfurization (HDS) in large refineries. The adsorption mechanism of thiophene to transition metals is proposed to occur through the π system, where the organosulfur compound lies almost parallel to the metal surface. Many researchers focus their efforts in optimizing the oxidation state of the transition metals for HDS, like Cu(I) and Ag(II) which together with Pd(0) have proven to be more specific for π bonding with thiophenes of all kinds.

Basis for Odor

Humans and other animals have an exquisitely sensitive sense of smell toward the odor of low-valent organosulfur compounds such as thiols, sulfides, and disulfides. Malodorous volatile thiols are protein-degradation products found in putrid food, so sensitive identification of these compounds is crucial to avoiding intoxication. Low-valent volatile sulfur compounds are also found in areas where oxygen levels in the air are low, posing a risk of suffocation. It has been found that copper is required for the highly sensitive detection of certain volatile thiols and related organosulfur compounds by olfactory receptors in mice. Whether humans, too, require copper for sensitive detection of thiols is not yet known.

Organophosphorus Chemistry

Organophosphorus chemistry presents a groundbreaking resource in this branch of organic chemistry that demonstrates how phosphorus-containing compounds can be manipulated in a variety of organic reactions.

Organophosphorus chemistry is one of numerous organic compounds containing phosphorus that form a broad class. A distinction is made between organophosphorus compounds in which the phosphorus is directly bonded to carbon in the molecules and organophosphorus compounds in which the phosphorus is bonded to the organic part of the molecules by a heteroatom—oxygen, nitrogen, or sulfur (this is mainly the case for esters and other acid derivatives of phosphorus). The latter are widely distributed in nature, primarily as esters of phosphoric, pyrophosphoric, and triphosphoric acids; these include nucleic acids, many important coenzymes, adenosine triphosphate—the energy carrier in living organisms—and a number of vitamins. Organophosphorus compounds containing a phosphorus-carbon bond, for example, β-aminoethylphosphonic acid (ciliatin), were discovered in nature in the 1960's.

Classification: No single classification of organophosphorus compounds exists. Organophosphorus compounds are classified according to various characteristics. For example, according to the number of P—C bonds, they are classified as alkylphospines (RPH_2), dialkylphospines (R_2PH), and trialkylphospines (R_3P) and their derivatives (henceforth R will denote an organic radical).

On the basis of the valence state of phosphorus, researchers distinguish the derivatives of trivalent and pentavalent phosphorus. Also known are compounds of bicoordinated, tetracoordinated, pentacoordinated, and hexacoordinated phosphorus: in compounds such as tetracoordinated phosphorus, the phosphorus atom carries a positive charge, whereas the charge is negative in the hexacoordinated phosphorus.

Organophosporus compounds can also be classified according to the nature of the phosphorus function as phosphines, phos-phine oxides (R_3PO), phosphine sulfides (R_3PS), phosphine imines

(R_3PNR'), methylene phosphines ($R_3P{=}CR'R''$), and phosphonium compounds ($R_4P^+X^-$). Also included are acids containing oxygen, such as phosphonous acid (RPO_2H_2), phosphinous acid (R_2POH), phosphonic acid (RPO_3H_2), and phosphinic acid (R_2PO_2H) and their various sulfur and nitrogen analogs and derivatives, as well as various organic derivatives (esters, amides, anhydrides) of hypophosphorous acid (H_3PO_2), phosphorous acid (H_3PO_3), phosphoric acid (H_3PO_4) and other acids. In addition, there are known organophosphorus compounds with a P—P bond, for example diphosphines, triphos-phines, and tetraphosphines and the corresponding cyclophos-phines and their derivatives.

Preparation: Methods of C—P bond formation are very important in the synthesis of organophosphorus compounds. These include:

(1) The Arbuzov reaction: $(RO)_3P + R'X \rightarrow R'PO(OR)_2 + RX$;

(2) The Michaelis-Becker reaction: $(RO)_2PONa + R'X \rightarrow R'PO(OR)_2 + NaX$;

(3) Synthesis with organometallic compounds, for example, $PCl_3 + 3RMgX \rightarrow R_3P + 3MgXCl$;

(4) Phosphorylation according to a Friedel-Crafts type of reaction:

$$C_6H_6 + PCL_3 \xrightarrow{AlCL_3} C_6H_5PCl_2 + HCl$$

(5) The addition of phosphorus pentachloride to olefins:

$$C_6H_5CH{=}CH_2 + 2PCl_5 \rightarrow C_6H_5CHCl - CH_2PCl_4 \cdot PCl_5;$$

(6) The alkylation of elemental phosphorus, for example,

$$3RCl + 2P \xrightarrow{PCl_3} RPCl_2 + R_2PCl$$

(7) The diene synthesis reaction:

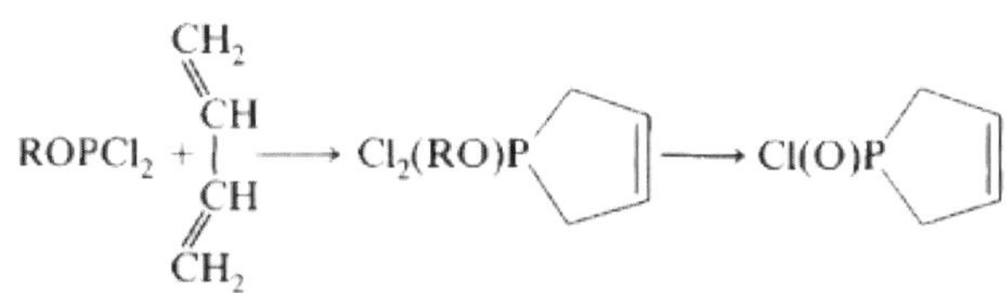

(8) The addition of organophosphorus compounds containing the P—H bond to olefins, carbonyl compounds, and Schiff bases, for example:

$$(RO)_2PHO + NH_3 + CH_2O \rightarrow NH_2CH_2PO(OR)_2.$$

Esters and other acid derivatives of phosphorus are usually obtained by the action of chloranhydrides of these acids on alcohols, often in the presence of bases that bind the released HCl. For example:

$$RPOCl_2 + 2R'OH + 2(C_2H_5)_3N \rightarrow RPO(OR')_2 + 2(C_2H_5)_3N \cdot HCl.$$

Compounds that contain the P=N bond are prepared by the action of azides on trivalent phosphorus compounds: $R_3P + C_6H_5N_3 \rightarrow R_3P = NC_6H_5 + N_2$ or by the phosphase reactions: $RSO_2N_2 + PCl_5 \rightarrow RSO_2N = PCl_3 + 2HCl$. Methylenephosphines are synthesized most often by the action of bases on phosphonium salts:

$$[^+ R_3PCH_2R']Cl^- + NaOR' \rightarrow R_3P = CHR' + NaCl + R'OH$$

Uses: Organophosphorus compounds are used in technology, agriculture, medicine, and scientific research. The manufacture of organophosphorus pesticides has reached large-scale proportions. Although highly effective, such pesticides are toxic for humans and animals, and their application necessitates precautionary measures. One advantage they do possess is that they do not accumulate in the environment.

In medicine, organophosphorus compounds are used mainly in opthalmology. Biologically important phosphates, such as adenosine triphospate, cocarboxylase, and a number of vitamins, are also valuable.

As complex-forming agents, organophosphorus compounds are used in ore concentration and in the production of uranium and other metals. Many organophosphorus compounds are used as additives in lubricants to increase their working properties, in components of plastic and fibers to make them noncombustible (antipyrenes), in solvents, and in hydraulic fluids. Advancements have also been made in the field of organophosphorus complex-ones, which are used for the separation of metals, as well as for other purposes.

Organophosphorus compounds have acquired considerable value in organic synthesis. For example, methylenephosphines are used in the synthesis of olefins from carbonyl compounds (Wittig reaction), and pyrophosphorous acid esters, in peptide synthesis. Various biologically important phosphates are used in biochemical, molecular-biological, and physiological research, and oxides of trialkylphosphines serve as catalysts in carbodi-imide synthesis. Phosphorus-containing polymers, prepared from phosphorus-containing monomers or by the phosphorylation of high-molecular compounds, such as cellulose, polyethylene, and rubber, are also widely used in the preparation of noncombustible materials and ion-exchange resins. Organophosphorus compounds also include such war gases as Sarin, Soman, Tabun, and phosphoryl thiocholines.

References

- Michael A. Urbancic, John R. Shapley (1990). "Pentacarbonylhydridorhenium". Inorganic Syntheses. Inorganic Syntheses. 28: 165–168. doi:10.1002/9780470132593.ch43. ISBN 978-0-470-13259-3
- Organometallic-chemistry: newworldencyclopedia.org, Retrieved 29 March 2018
- W, Uhl and M. R. Halvagar; et al. (2009). "Reducing Ga-H and Ga-C Bonds in Close Proximity to Oxidizing Peroxo Groups: Conflicting Properties in Single Molecules". Chemistry: A European Journal. 15 (42): 11298–11306. doi:10.1002/chem.200900746
- Organosulfur-compound, science: britannica.com, Retrieved 19 May 2018
- Chemistry of aluminium, gallium, indium, and thallium Anthony John Downs (Ed.) ISBN 978-0-7514-0103-5, 1993
- Organophosphorus-Compound: encyclopedia2.thefreedictionary.com, Retrieved 31 July 2018
- $GaCl_3$ in Organic Synthesis Ryo Amemiya and Masahiko Yamaguchi Eur. J. Org. Chem. 2005, 5145–5150 doi:10.1002/ejoc.200500512

8

Acids and Bases

An acid is an ion or a molecule, which donates a hydrogen ion or a proton or alternatively forms a covalent bond with a pair of electrons. Bases are substances that release hydroxide ions in aqueous solution. This chapter discusses in extensive detail the fundamental concepts of acids and bases and includes vital topics such as Lewis acids and bases and the Brønsted–Lowry acid–base theory.

Acids

Acid is any substance that in water solution tastes sour and changes the colour of certain indicators (e.g., reddens blue litmus paper), reacts with some metals (e.g., iron) to liberate hydrogen, reacts with bases to form salts, and promotes certain chemical reactions (acid catalysis). Examples of acids include the inorganic substances known as the mineral acids sulfuric, nitric, hydrochloric, and phosphoric acids and the organic compounds belonging to the carboxylic acid, sulfonic acid, and phenol groups. Such substances contain one or more hydrogen atoms that, in solution, are released as positively charged hydrogen ions.

Broader definitions of an acid, to include substances that exhibit typical acidic behavior as pure compounds or when dissolved in solvents other than water, are given by the Brønsted–Lowry theory and the Lewis theory. Examples of nonaqueous acids are sulfur trioxide, aluminum chloride, and boron trifluoride.

Strong Acids

Strong acids are acids that are completely or nearly 100% ionized in their solutions. Here are some common strong acids: Ionization of a strong acid HA can be represented by:

Strong acids			
Type	Formula		
Hydrogen halides	HCl	HBr	HI
Oxyacids of halogens,	$HClO_3$ $HClO_4$	$HBrO_3$ $HBrO_4$	HIO_3 HIO_4
Sulfuric acid	H_2SO_4		
Nitric acid	HNO_3		

$$HA = H^+ + A^-$$

$$\quad\quad x \quad x$$

where x is the concentration of H^+, $[H^+]$. For a strong acid, $[H^+] = [A^-]$ = concentration of acid (= x), if x is much greater than $1x10^{-7}$ (represented as e^{-7}). For a very dilute strong acid solution with concentration less than 1E-7, the pH is dominated by the autoionization of water,

$$H_2O = H^+ + OH^-, \qquad K_w = 1e - 14 \; at \; 298 \; K.$$

The pH and pOH Scales

The pH and pOH scale at 298 K		
pH	$[H^+]$	pOH
1	0.1	13
2	0.01	12
3	0.001	11
4	1e-4	10
5	1e-5	9
6	1e-6	8

The pH scale is defined as the negative log of the concentration of H^+:

$pH = -log[H^+]$

The pOH scale is defined as the negative log of the concentration of OH^-, $[OH^-]$:

$pOH = -log[OH^-]$

With this scale, calculating the pOH can be done in the same manner as the pH scale.

Example: Calculate the pH of a solution with 1.2345e-4MHCl

Solution,

The solution of a strong acid is completely ionized. Thus, $[H^+]$ = 1.234e-4.

üüüüüüüü(üüüü)

Discussion,

What is the pH for a solution containing 1.234M [HCl]? pH = 0.0913

Example: Calculate the pH of a stock HCl solution that is 32% by mass HCl.

Solution,

The density of such a solution is needed before we can calculate the pH. Since the density is not on the label, we need to find it from the Material Safety Data Sheet, which gives the specific gravity of 1.150. Thus, the amount of acid in 1.0 L is 1150 g.

The amount of HCl = 1000*1.150*0.32

= 368 g (1 mol/36.5 g <- molar mass of HCl)

= 10.08 M

$= [H^+]$

pH = -log(10..08) = -1.003

Discussion,

Yes, pH have negative values if $[H^+] > 1.0$

Check out the information on nitric acid, and calculate the pH of a stock nitric acid solution.

Example: Calculate the pH of a solution containing 1.00e-7 of HCl.

Solution,

$[H^+]$ = 1.0e-7 M from the strong acid, and if x is the amount from the ionization of water, then we have the equilibrium due to the auto ionization of water:

$$\begin{array}{lll} HCl = H^+ + Cl^- & & \\ \quad 1e-7 \quad 1e-7 & <--- & [H^+] \textit{ from the acid} \end{array}$$

$$\begin{array}{llll} H_2O = & H^+ & + \; OH^- & \\ & (1e-7)+x & x & <--- \textit{ we don't know yet} \end{array}$$

Recall that $K_w = [H^+][OH^-]$ = 1e-14, due to the ionization equilibrium of water in the solution:

$\{(1.00e\text{-}7)+x\}\, x = 1e\text{-}14$

$x^2 + 1.00e\text{-}7x - 1.00e\text{-}14 = 0$

Solving this equation for x results in

$$\begin{aligned} x &= \{-1.00e-7 + (1.00e-14 + 4*1.00e-14)^{1/2}\} / 2 \\ &= 0.61e-7 \\ [H^+] &= (1.00 + 0.61)e-7\ M \\ pH &= -log(1.61e-7) = 6.79 \end{aligned}$$

Arrhenius Acids

Svante Arrhenius

The Swedish chemist Svante Arrhenius attributed the properties of acidity to hydrogen ions (H^+) or protons in 1884. An Arrhenius acid is a substance that, when added to water, increases the concentration of H^+ ions in the water. Note that chemists often write $H^+(aq)$ and refer to the hydrogen ion when describing acid-base reactions but the free hydrogen nucleus, a proton, does not exist alone in water, it exists as the hydronium ion, H_3O^+. Thus, an Arrhenius acid can also be described as a substance that increases the concentration of hydronium ions when added to water. Examples include molecular substances such as HCl and acetic acid.

An Arrhenius base, on the other hand, is a substance which increases the concentration of hydroxide (OH^-) ions when dissolved in water. This decreases the concentration of hydronium because the ions react to form H_2O molecules:

$$H_3O^+_{(aq)} + OH^-_{(aq)} \rightleftharpoons H_2O_{(l)} + H_2O_{(l)}$$

Due to this equilibrium, any increase in the concentration of hydronium is accompanied by a decrease in the concentration of hydroxide. Thus, an Arrhenius acid could also be said to be one that decreases hydroxide concentration, while an Arrhenius base increases it.

In an acidic solution, the concentration of hydronium ions is greater than 10^{-7} moles per liter. Since pH is defined as the negative logarithm of the concentration of hydronium ions, acidic solutions thus have a pH of less than 7.

Brønsted–lowry Acids

Acetic acid, a weak acid, donates a proton (hydrogen ion, highlighted in green) to water in an equilibrium reaction to give the acetate ion and the hydronium ion. Red: oxygen, black: carbon, white: hydrogen.

While the Arrhenius concept is useful for describing many reactions, it is also quite limited in its scope. In 1923 chemists Johannes Nicolaus Brønsted and Thomas Martin Lowry independently recognized that acid-base reactions involve the transfer of a proton. A Brønsted-Lowry acid (or simply Brønsted acid) is a species that donates a proton to a Brønsted-Lowry base. Brønsted-Lowry acid-base theory has several advantages over Arrhenius theory. Consider the following reactions of acetic acid (CH_3COOH), the organic acid that gives vinegar its characteristic taste:

$$CH_3COOH + H_2O \rightleftharpoons CH_3COO^- + H_3O^+$$

$$CH_3COOH + NH_3 \rightleftharpoons CH_3COO^- + NH^+_4$$

Both theories easily describe the first reaction: CH_3COOH acts as an Arrhenius acid because it acts as a source of H_3O^+ when dissolved in water, and it acts as a Brønsted acid by donating a proton to water. In the second example CH_3COOH undergoes the same transformation, in this case donating a proton to ammonia (NH_3), but does not relate to the Arrhenius definition of an acid because the reaction does not produce hydronium. Nevertheless, CH_3COOH is both an Arrhenius and a Brønsted-Lowry acid.

Brønsted-Lowry theory can be used to describe reactions of molecular compounds in nonaqueous solution or the gas phase. Hydrogen chloride (HCl) and ammonia combine under several different conditions to form ammonium chloride, NH_4Cl. In aqueous solution HCl behaves as hydrochloric acid and exists as hydronium and chloride ions. The following reactions illustrate the limitations of Arrhenius's definition:

1. $H_3O^+_{(aq)} + Cl^-_{(aq)} + NH_3 \rightarrow Cl^-_{(aq)} + NH^+_{4(aq)} + H_2O$
2. $HCl_{(benzene)} + NH_{3(benzene)} \rightarrow NH_4Cl_{(s)}$
3. $HCl(g) + NH_3(g) \rightarrow NH_4Cl(s)$

As with the acetic acid reactions, both definitions work for the first example, where water is the solvent and hydronium ion is formed by the HCl solute. The next two reactions do not involve the formation of ions but are still proton-transfer reactions. In the second reaction hydrogen chloride and ammonia (dissolved in benzene) react to form solid ammonium chloride in benzene solvent and in the third gaseous HCl and NH_3 combine to form the solid.

Lewis Acids

A third, only marginally related concept was proposed in 1923 by Gilbert N. Lewis, which includes reactions with acid-base characteristics that do not involve a proton transfer. A Lewis acid is a species that accepts a pair of electrons from another species; in other words, it is an electron pair acceptor. Brønsted acid-base reactions are proton transfer reactions while Lewis acid-base reactions are electron pair transfers. Many Lewis acids are not Brønsted-Lowry acids. Contrast how the following reactions are described in terms of acid-base chemistry:

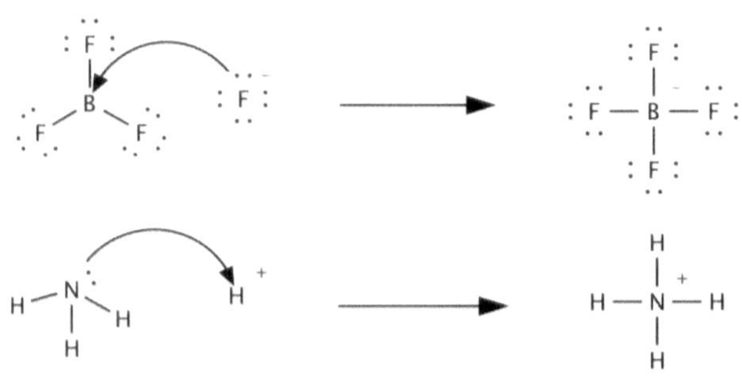

In the first reaction a fluoride ion, F^-, gives up an electron pair to boron trifluoride to form the product tetrafluoroborate. Fluoride "loses" a pair of valence electrons because the electrons shared in the B—F bond are located in the region of space between the two atomic nuclei and are therefore more distant from the fluoride nucleus than they are in the lone fluoride ion. BF_3 is a Lewis acid because it accepts the electron pair from fluoride. This reaction cannot be described in terms of Brønsted theory because there is no proton transfer. The second reaction can be described using either theory. A proton is transferred from an unspecified Brønsted acid to ammonia, a Brønsted base; alternatively, ammonia acts as a Lewis base and transfers a lone pair of electrons to form a bond with a hydrogen ion. The species that gains the electron pair is the Lewis acid; for example, the oxygen atom in H_3O^+ gains a pair of electrons when one of the H—O bonds is broken and the electrons shared in the bond become localized on oxygen. Depending on the context, a Lewis acid may also be described as an oxidizer or an electrophile. Organic Brønsted acids, such as acetic, citric, or oxalic acid, are not Lewis acids. They dissociate in water to produce a Lewis acid, H^+, but at the same time also yield an equal amount of a Lewis base (acetate, citrate, or oxalate, respectively, for the acids mentioned).

Dissociation and Equilibrium

Reactions of acids are often generalized in the form $HA \rightleftharpoons H^+ + A^-$, where HA represents the acid and A^- is the conjugate base. This reaction is referred to as protolysis. The protonated form (HA) of an acid is also sometimes referred to as the free acid.

Acid-base conjugate pairs differ by one proton, and can be interconverted by the addition or removal of a proton (protonation and deprotonation, respectively). Note that the acid can be the charged species and the conjugate base can be neutral in which case the generalized reaction scheme could be written as $HA^+ \rightleftharpoons H^+ + A$. In solution there exists an equilibrium between the acid and its conjugate base. The equilibrium constant K is an expression of the equilibrium concentrations of the molecules or the ions in solution. Brackets indicate concentration, such that $[H_2O]$ means *the concentration of* H_2O. The acid dissociation constant K_a is generally used in the context of acid-base reactions. The numerical value of K_a is equal to the product of the concentrations of the products divided by the concentration of the reactants, where the reactant is the acid (HA) and the products are the conjugate base and H^+.

$$K_a = \frac{[H^+][A^-]}{[HA]}$$

The stronger of two acids will have a higher K_a than the weaker acid; the ratio of hydrogen ions to acid will be higher for the stronger acid as the stronger acid has a greater tendency to lose its proton. Because the range of possible values for K_a spans many orders of magnitude, a more manageable constant, pK_a is more frequently used, where $pK_a = -\log_{10} K_a$. Stronger acids have a smaller pK_a than weaker acids. Experimentally determined pK_a at 25° C in aqueous solution are often quoted in textbooks and reference material.

Nomenclature

In the classical naming system, acids are named according to their anions. That ionic suffix is dropped and replaced with a new suffix (and sometimes prefix), according to the table. For example, HCl has chloride as its anion, so the -ide suffix makes it take the form hydrochloric acid. In the IUPAC naming system, "aqueous" is simply added to the name of the ionic compound. Thus, for hydrogen chloride, the IUPAC name would be aqueous hydrogen chloride. The prefix "hydro-" is added only if the acid is made up of just hydrogen and one other element.

Classical naming system:

Anion prefix	Anion suffix	Acid prefix	Acid suffix	Example
per	ate	per	ic acid	perchloric acid ($HClO_4$)
	ate		ic acid	chloric acid ($HClO_3$)
	ite		ous acid	chlorous acid ($HClO_2$)
hypo	ite	hypo	ous acid	hypochlorous acid (HClO)
	ide	hydro	ic acid	hydrochloric acid (HCl)

Chemical Characteristics

Monoprotic Acids

Monoprotic acids are those acids that are able to donate one proton per molecule during the process of dissociation (sometimes called ionization) as shown below (symbolized by HA):

$$HA_{(aq)} + H_2O_{(l)} \rightleftharpoons H_3O^+_{(aq)} + A^-_{(aq)} K_a$$

Common examples of monoprotic acids in mineral acids include hydrochloric acid (HCl) and nitric acid (HNO_3). On the other hand, for organic acids the term mainly indicates the presence of one carboxylic acid group and sometimes these acids are known as monocarboxylic acid. Examples in organic acids include formic acid (HCOOH), acetic acid (CH_3COOH) and benzoic acid (C_6H_5COOH).

Polyprotic Acids

Polyprotic acids, also known as polybasic acids, are able to donate more than one proton per acid molecule, in contrast to monoprotic acids that only donate one proton per molecule. Specific types of polyprotic acids have more specific names, such as diprotic acid (two potential protons to donate) and triprotic acid (three potential protons to donate).

A diprotic acid (here symbolized by H_2A) can undergo one or two dissociations depending on the pH. Each dissociation has its own dissociation constant, K_{a1} and K_{a2}.

$$H_2A_{(aq)} + H_2O_{(l)} \rightleftharpoons H_3O^+_{(aq)} + HA^-_{(aq)} K_{a1}$$

$$HA^-_{(aq)} + H_2O_{(l)} \rightleftharpoons H_3O^+_{(aq)} + A^{2-}_{(aq)} K_{a2}$$

The first dissociation constant is typically greater than the second; i.e., $K_{a1} > K_{a2}$. For example, sulfuric acid (H_2SO_4) can donate one proton to form the bisulfate anion (HSO_4^-), for which K_{a1} is very large; then it can donate a second proton to form the sulfate anion (SO_4^{2-}), wherein the K_{a2} is intermediate strength. The large K_{a1} for the first dissociation makes sulfuric a strong acid. In a similar manner, the weak unstable carbonic acid (H_2CO_3) can lose one proton to form bicarbonate anion (HCO_3^-) and lose a second to form carbonate anion (CO_3^{2-}). Both K_a values are small, but $K_{a1} > K_{a2}$.

A triprotic acid (H_3A) can undergo one, two, or three dissociations and has three dissociation constants, where $K_{a1} > K_{a2} > K_{a3}$.

$$H_3A_{(aq)} + H_2O_{(l)} \rightleftharpoons H_3O^+_{(aq)} + H_2A^-_{(aq)} K_{a1}$$

$$H_2A^-_{(aq)} + H_2O_{(l)} \rightleftharpoons H_3O^+_{(aq)} + HA^{2-}_{(aq)} K_{a2}$$

$$HA^{2-}_{(aq)} + H_2O_{(l)} \rightleftharpoons H_3O^+_{(aq)} + A^{3-}_{(aq)} K_{a3}$$

An inorganic example of a triprotic acid is orthophosphoric acid (H_3PO_4), usually just called phosphoric acid. All three protons can be successively lost to yield $H_2PO_4^-$, then HPO_4^{2-}, and finally PO_4^{3-}, the orthophosphate ion, usually just called phosphate. Even though the positions of the three protons on the original phosphoric acid molecule are equivalent, the successive K_a values differ since it is energetically less favorable to lose a proton if the conjugate base is more negatively charged.

An organic example of a triprotic acid is citric acid, which can successively lose three protons to finally form the citrate ion.

Although the subsequent loss of each hydrogen ion is less favorable, all of the conjugate bases are present in solution. The fractional concentration, α (alpha), for each species can be calculated. For example, a generic diprotic acid will generate 3 species in solution: H_2A, HA^-, and A^{2-}. The fractional concentrations can be calculated as below when given either the pH (which can be converted to the $[H^+]$) or the concentrations of the acid with all its conjugate bases:

$$\alpha_{H_2A} = \frac{[H^+]^2}{[H^+]^2 + [H^+]K_1 + K_1K_2} = \frac{[H_2A]}{[H_2A]+[HA^-]+[A^{2-}]}$$

$$\alpha_{HA^-} = \frac{[H^+]K_1}{[H^+]^2 + [H^+]K_1 + K_1K_2} = \frac{[HA^-]}{[H_2A]+[HA^-]+[A^{2-}]}$$

$$\alpha_{A^{2-}} = \frac{K_1K_2}{[H^+]^2 + [H^+]K_1 + K_1K_2} = \frac{[A^{2-}]}{[H_2A]+[HA^-]+[A^{2-}]}$$

A plot of these fractional concentrations against pH, for given K_1 and K_2, is known as a Bjerrum plot. A pattern is observed in the above equations and can be expanded to the general n -protic acid that has been deprotonated i -times:

$$\alpha_{H_{n-i}A^{i-}} = \frac{[H^+]^{n-i}\prod_{j=0}^{i} K_j}{\sum_{i=0}^{n}\left[[H^+]^{n-i}\prod_{j=0}^{i} K_j\right]}$$

where, $K_0 = 1$ and the other K-terms are the dissociation constants for the acid.

Neutralization

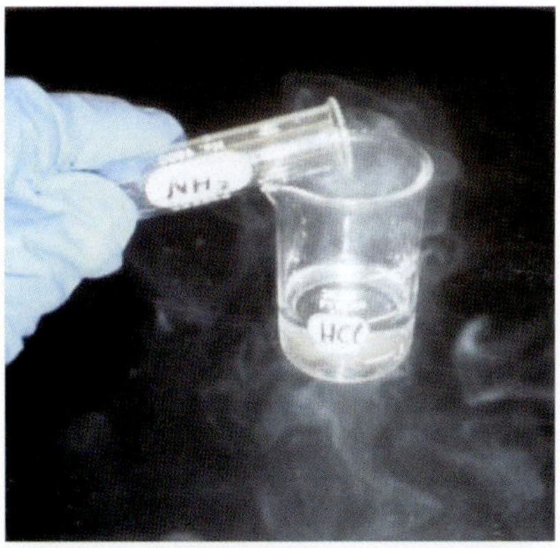

Hydrochloric acid (in beaker) reacting with ammonia fumes to produce ammonium chloride (white smoke).

Neutralization is the reaction between an acid and a base, producing a salt and neutralized base; for example, hydrochloric acid and sodium hydroxide form sodium chloride and water:

$$HCl_{(aq)} + NaOH_{(aq)} \rightarrow H_2O_{(l)} + NaCl_{(aq)}$$

Neutralization is the basis of titration, where a pH indicator shows equivalence point when the equivalent number of moles of a base have been added to an acid. It is often wrongly assumed that neutralization should result in a solution with pH 7.0, which is only the case with similar acid and base strengths during a reaction.

Neutralization with a base weaker than the acid results in a weakly acidic salt. An example is the weakly acidic ammonium chloride, which is produced from the strong acid hydrogen chloride and the weak base ammonia. Conversely, neutralizing a weak acid with a strong base gives a weakly basic salt, e.g. sodium fluoride from hydrogen fluoride and sodium hydroxide.

Weak Acid–weak base Equilibrium

In order for a protonated acid to lose a proton, the pH of the system must rise above the pK_a of the acid. The decreased concentration of H^+ in that basic solution shifts the equilibrium towards the conjugate base form (the deprotonated form of the acid). In lower-pH (more acidic) solutions, there is a high enough H^+ concentration in the solution to cause the acid to remain in its protonated form.

Solutions of weak acids and salts of their conjugate bases form buffer solutions.

Applications of Acids

There are numerous uses for acids. Acids are often used to remove rust and other corrosion from metals in a process known as pickling. They may be used as an electrolyte in a wet cell battery, such as sulfuric acid in a car battery.

Strong acids, sulfuric acid in particular, are widely used in mineral processing. For example, phosphate minerals react with sulfuric acid to produce phosphoric acid for the production of phosphate fertilizers, and zinc is produced by dissolving zinc oxide into sulfuric acid, purifying the solution and electrowinning.

In the chemical industry, acids react in neutralization reactions to produce salts. For example, nitric acid reacts with ammonia to produce ammonium nitrate, a fertilizer. Additionally, carboxylic acids can be esterified with alcohols, to produce esters.

Acids are used as additives to drinks and foods, as they alter their taste and serve as preservatives. Phosphoric acid, for example, is a component of cola drinks. Acetic acid is used in day-to-day life as vinegar. Carbonic acid is an important part of some cola drinks and soda. Citric acid is used as a preservative in sauces and pickles.

Tartaric acid is an important component of some commonly used foods like unripened mangoes and tamarind. Natural fruits and vegetables also contain acids. Citric acid is present in oranges, lemon and other citrus fruits. Oxalic acid is present in tomatoes, spinach, and especially in carambola and rhubarb; rhubarb leaves and unripe carambolas are toxic because of high concentrations of oxalic acid.

Ascorbic acid (Vitamin C) is an essential vitamin for the human body and is present in such foods as amla (Indian gooseberry), lemon, citrus fruits, and guava. Certain acids are used as drugs. Acetylsalicylic acid (Aspirin) is used as a pain killer and for bringing down fevers.

Acids play important roles in the human body. The hydrochloric acid present in the stomach aids digestion by breaking down large and complex food molecules. Amino acids are required for synthesis of proteins required for growth and repair of body tissues. Fatty acids are also required for growth and repair of body tissues. Nucleic acids are important for the manufacturing of DNA and RNA and transmitting of traits to offspring through genes. Carbonic acid is important for maintenance of pH equilibrium in the body.

Acid Catalysis

Acids are used as catalysts in industrial and organic chemistry; for example, sulfuric acid is used in very large quantities in the alkylation process to produce gasoline. Some acids, such as sulfuric, phosphoric, and hydrochloric acids, also effect dehydration and condensation reactions. In biochemistry, many enzymes employ acid catalysis.

Biological Occurrence

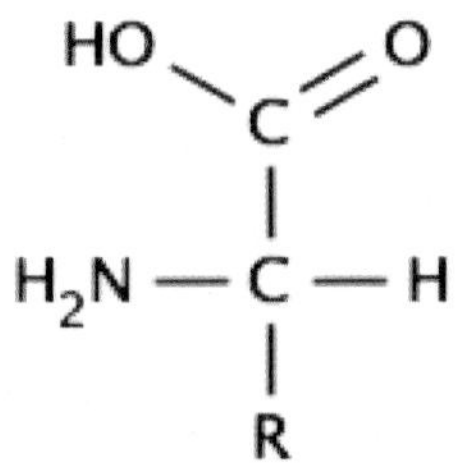

Basic structure of an amino acid.

Many biologically important molecules are acids. Nucleic acids, which contain acidic phosphate groups, include DNA and RNA. Nucleic acids contain the genetic code that determines many of an organism's characteristics, and is passed from parents to offspring. DNA contains the chemical blueprint for the synthesis of proteins which are made up of amino acid subunits. Cell membranes contain fatty acid esters such as phospholipids.

An α-amino acid has a central carbon (the α or *alpha* carbon) which is covalently bonded to a carboxyl group (thus they are carboxylic acids), an amino group, a hydrogen atom and a variable group. The variable group, also called the R group or side chain, determines the identity and many of the properties of a specific amino acid. In glycine, the simplest amino acid, the R group is a hydrogen atom, but in all other amino acids it is contains one or more carbon atoms bonded to hydrogens, and may contain other elements such as sulfur, oxygen or nitrogen. With the exception of glycine, naturally occurring amino acids are chiral and almost invariably occur in the L-configuration. Peptidoglycan, found in some bacterial cell walls contains some D-amino acids. At physiological pH, typically around 7, free amino acids exist in a charged form, where the acidic carboxyl group (-COOH) loses a proton ($-COO^-$) and the basic amine group ($-NH_2$) gains a proton ($-NH_3^+$). The entire molecule has a net neutral charge and is a zwitterion, with the exception of amino acids with basic or acidic side chains. Aspartic acid, for example, possesses one protonated amine and two deprotonated carboxyl groups, for a net charge of −1 at physiological pH.

Fatty acids and fatty acid derivatives are another group of carboxylic acids that play a significant role in biology. These contain long hydrocarbon chains and a carboxylic acid group on one end. The cell membrane of nearly all organisms is primarily made up of a phospholipid bilayer, a micelle of hydrophobic fatty acid esters with polar, hydrophilic phosphate "head" groups. Membranes contain additional components, some of which can participate in acid-base reactions.

In humans and many other animals, hydrochloric acid is a part of the gastric acid secreted within the stomach to help hydrolyze proteins and polysaccharides, as well as converting the inactive pro-enzyme, pepsinogen into the enzyme, pepsin. Some organisms produce acids for defense; for example, ants produce formic acid.

Acid-base equilibrium plays a critical role in regulating mammalian breathing. Oxygen gas (O_2) drives cellular respiration, the process by which animals release the chemical potential energy stored in food, producing carbon dioxide (CO_2) as a byproduct. Oxygen and carbon dioxide are exchanged in the lungs, and the body responds to changing energy demands by adjusting the rate of ventilation. For example, during periods of exertion the body rapidly breaks down stored carbohydrates and fat, releasing CO_2 into the blood stream. In aqueous solutions such as blood CO_2 exists in equilibrium with carbonic acid and bicarbonate ion.

$$CO_2 + H_2O \rightleftharpoons H_2CO_3 \rightleftharpoons H^+ + HCO_3^-$$

It is the decrease in pH that signals the brain to breathe faster and deeper, expelling the excess CO_2 and resupplying the cells with O_2.

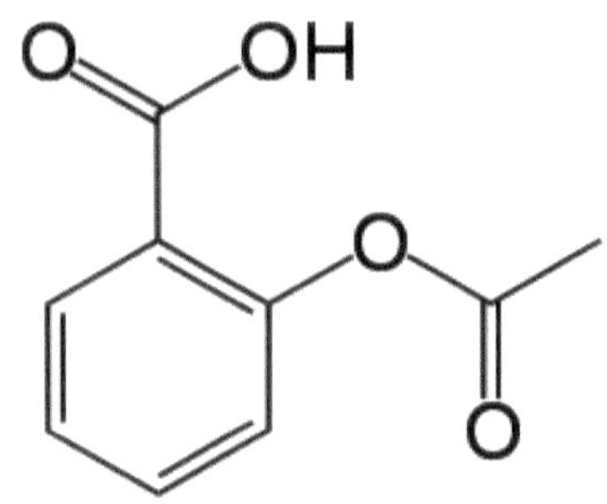

Aspirin (acetylsalicylic acid) is a carboxylic acid.

Cell membranes are generally impermeable to charged or large, polar molecules because of the lipophilic fatty acyl chains comprising their interior. Many biologically important molecules, including a number of pharmaceutical agents, are organic weak acids which can cross the membrane in their protonated, uncharged form but not in their charged form (i.e. as the conjugate base). For this reason the activity of many drugs can be enhanced or inhibited by the use of antacids or acidic foods. The charged form, however, is often more soluble in blood and cytosol, both aqueous environments. When the extracellular environment is more acidic than the neutral pH within the cell, certain acids will exist in their neutral form and will be membrane soluble, allowing them to cross the phospholipid bilayer. Acids that lose a proton at the intracellular pH will exist in their soluble, charged form and are thus able to diffuse through the cytosol to their target. Ibuprofen, aspirin and penicillin are examples of drugs that are weak acids.

Acid Dissociation Constants Ka

1. According to the Arrhenius theory of acids and bases, acids are those which contain hydrogen ion and bases or alkalis are those which contain hydroxyl ion.

2. The chemical activity and the electrical conductivity experiments are the basis of this theory.
3. Thus the proton (H^+) and the hydroxyl ion (OH^-) are responsible for acidic and basic properties of compounds respectively.
4. The reaction product of an acid and base is called a salt. This theory was a distinct advancement in the study of compounds but it failed to explain the similar properties exhibited by some other compounds which are not having either a proton to liberate or a hydroxyl ion.
5. Water is considered as a very good solvent for ionic compounds because of the reason that it can stabilize the ions that are formed due to the dissociation by attracting them strongly.
6. This attraction is strong particularly because of the asymmetry in the shape of water molecule which results in the asymmetrical distribution of the charges.
7. Ions get hydrated by many water molecules that surround them and hence the stability of the ions.
8. A more powerful theory was proposed to explain the general concept of acids and bases and it is known as Lowry-Bronsted definition.
9. According to this theory, an acid is a species having a tendency to donate or lose a proton and a base is a species having the tendency to accept or add a proton.
10. Water can act as an acid or a base. When it hydrates a proton it acts as a base and when it the hydrated water gets dehydrated it acts as an acid.

In acid-base reaction H^+ ion is transferred from acid to a base. Acid is generally represented as HA while corresponding base represented as A^-.

HA(aq) acid + H_2O(l) base → H_3O^+(aq) corresponding acid + A^-(aq) corresponding base

The strength of the acid can be defined from the equilibrium position of the above equation.

$$\text{ka} = \frac{[H_3O^+][A^-]}{[HA]}$$

Or

$$\text{ka} = \frac{[H^+][A^-]}{[HA]}$$

The equilibrium constant Ka, in the case of acid is called dissociation constant of acid or simply strength constant. There is no difference between H_3O^+ and H^+ since the proton in aqueous medium always exists as hydronium ion.Similarly for bases too there is a dissociation constant which is given by Kb. Based on the above equation the by expressing in logarithmic expression:

$$\text{pKa} = -\log\left[\text{Ka}\right]$$

Acid Dissociation Constant of Water

As indicated above water acts as both acid and base. In presence f a stronger acid, it acts as a base by accepting a proton while in presence of a stronger base it donates the proton. This property is often referred to as ampholyte. This may also give information that water reacts in itself and is in equilibrium.

$$H_2O(\text{liquid}) + H_2O(\text{liquid}) \leftrightarrow H_3O+(aq) + OH^-(aq)$$

The equilibrium position in the above equation is far to the left in pure conditions.

Hence the equilibrium constant of water is,

$$Kw = 1.0 \text{ X } 10^{-14} \text{ M}$$

Since the dissociation equilibrium is more towards H_2O, and by the dissociation of water the ratio of H_3O^+ and OH^- ions is 1:1,

$$[H_3O]^+ = [OH]^- = 1.0 \text{ X } 10^{-7} \text{ M}$$

Determination of the Dissociation Constant of a Weak Acid

Weak acids are those which will not get dissociated completely in aqueous solution. An non dissociated amount of acid will still remain at the point of equilibrium It is difficult to find the Dissociation constant for such equilibrium where all the entities are unknowns.

If it is a monobasic acid it is easy to equate the formed H_3O^+ ions and the A^- ions and take it as $[H_3O^+]^2$ so that the two unknowns can be converted to one. But the non dissociated acid still remains an unknown which will leave all unknowns.

In order to solve this predicament, an approximation is required. This is done by taking excess of pure weak acid so that the concentration of the remaining undissociated can be taken as negligible in the whole amount. Since we know the amount and purity of the added acid we can get the approximate value for the un dissociated weak acid. Thus we can find the dissociation constant of the weak acid.

Dissociation Constant of Acetic Acid

Acetic acid is a weak acid. Acetic acid will not dissociate completely in water. Its reaction with water is,

$$CH_3COOH + H_2O \leftrightarrow H_3O^+ + CH_3COO^-$$

The equilibrium constant for the equation is,

$$Ka = \frac{[H_3O]^+[CH_3COO]^-}{[CH_3COOH]}$$

Since the acetic acid is a weak acid the dissociation is not complete and it will be partial. Thus the denominator faction will be considerable. The amount of non dissociated acetic acid in the medium can be found out in several ways of experiments.

The dissociation constant of acetic acid is thus,

$$\mathrm{Ka}(\text{acetic acid}) = 1.8 \text{ X } 10^{-5} \text{ M}$$

Monoprotic Acids

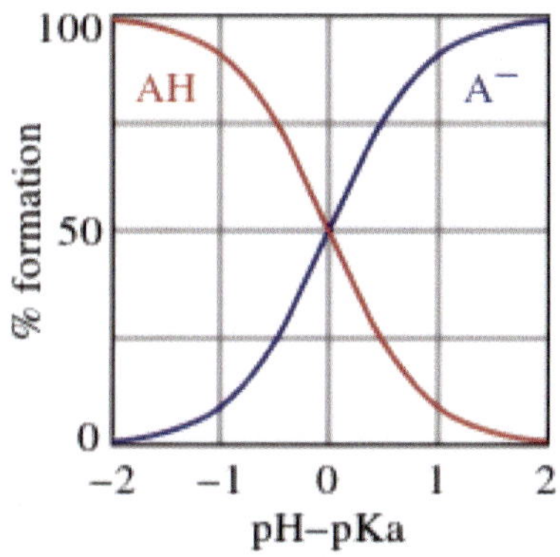

Variation of the % formation of a monoprotic acid, AH, and its conjugate base, A^-, with the difference between the pH and the pK_a of the acid.

After rearranging the expression defining K_a, and putting pH = $-\log_{10}[H^+]$, one obtains

$$\mathrm{pH} = \mathrm{p}K_a + \log\left(\frac{[A^-]}{[HA]}\right)$$

This is the Henderson–Hasselbalch equation, from which the following conclusions can be drawn:

- At half-neutralization the ratio $\frac{[A^-]}{[HA]} = 1$; since log(1) = 0, the pH at half-neutralization is numerically equal to pK_a. Conversely, when pH = pK_a, the concentration of HA is equal to the concentration of A^-.
- The buffer region extends over the approximate range $pK_a \pm 2$. Buffering is weak outside the range $pK_a \pm 1$. At pH $\leq pK_a - 2$ the substance is said to be fully protonated and at pH $\geq pK_a + 2$ it is fully dissociated (deprotonated).
- If the pH is known, the ratio may be calculated. This ratio is independent of the analytical concentration of the acid.

In water, measurable pK_a values range from about −2 for a strong acid to about 12 for a very weak acid (or strong base). A buffer solution of a desired pH can be prepared as a mixture of a weak acid and its conjugate base. In practice the mixture can be created by dissolving the acid in water, and adding the requisite amount of strong acid or base. When the pK_a and analytical concentration of the acid are known, the extent of dissociation and pH of a solution of a monoprotic acid can be easily calculated using an ICE table.

Strong Acids and Bases

A strong acid is effectively completely dissociated in aqueous solution below certain concentration threshold values in concentrated solutions of acids like hydrochloric acid or nitric acid above 2-3 M. A tacit assumption occurs here about hydrogen ion concentration in dilute solutions being equal to analytic concentration of the monobasic acid as a result of total dissociation. All acids with a pK_a value of less than −2 are more than 99% dissociated at pH0 (1M acid). This is known as solvent leveling since such acids are brought to the same level of being *strong acids*, regardless of their pK_a values. Likewise, all bases with a pK_a value larger than the upper limit are more than 99% dissociated at all attainable pH values and are classified as strong bases.

An example of a strong acid is hydrochloric acid, HCl, which has a pK_a value, estimated from thermodynamic quantities, of −9.3 in water. The concentration of undissociated HCl molecules in aqueous solution will be less than 0.01% of the analytical concentration of the acid. Hydrochloric acid is said to be fully dissociated in aqueous solution because the amount of undissociated acid, in equilibrium with the dissociation products, is below the detection limit.

Polyprotic Acids

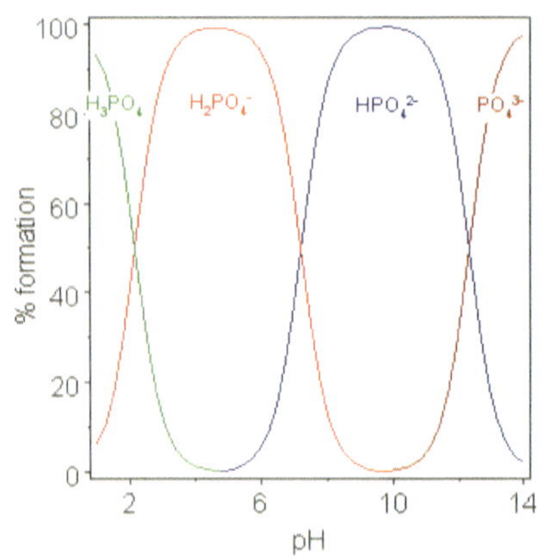

Phosphoric acid speciation

Polyprotic acids are acids that can lose more than one proton. The constant for dissociation of the first proton may be denoted as K_{a1} and the constants for dissociation of successive protons as K_{a2}, etc. Phosphoric acid, H_3PO_4, is an example of a polyprotic acid as it can lose three protons.

Equilibrium	pK_a value
$H_3PO_4 \rightleftharpoons H_2PO_4^- + H^+$	pK_{a1} = 2.14
$H_2PO_4^- \rightleftharpoons HPO_4^{2-} + H^+$	pK_{a2} = 7.20
$HPO_4^{2-} \rightleftharpoons PO_4^{3-} + H^+$	pK_{a3} = 12.37

When the difference between successive pK values is about four or more, as in this example, each species may be considered as an acid in its own right; In fact salts of $H_2PO_4^-$ may be crystallised from solution by adjustment of pH to about 5.5 and salts of HPO_4^{2-} may be crystallised from solution by adjustment of pH to about 10. The species distribution diagram shows that the concentrations of the two ions are maximum at pH 5.5 and 10.

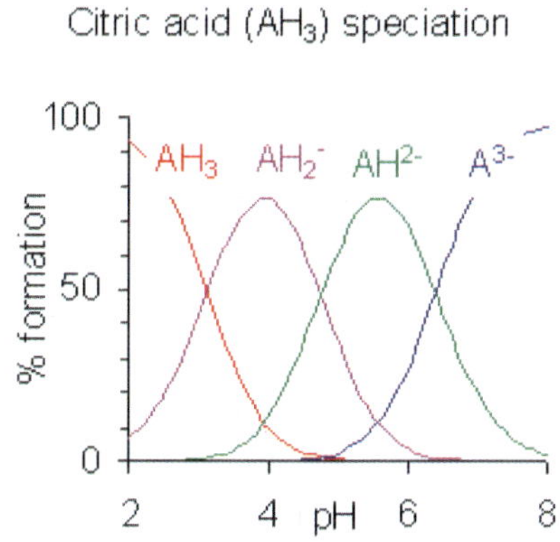

% species formation calculated with the program HySS for a 10 millimolar solution of citric acid. pK_{a1} = 3.13, pK_{a2} = 4.76, pK_{a3} = 6.40.

When the difference between successive p*K* values is less than about four there is overlap between the pH range of existence of the species in equilibrium. The smaller the difference, the more the overlap. The case of citric acid is shown at the right; solutions of citric acid are buffered over the whole range of pH 2.5 to 7.5.

According to Pauling's first rule, successive p*K* values of a given acid increase ($pK_{a2} > pK_{a1}$). For oxyacids with more than one ionizable hydrogen on the same atom, the pK_a values often increase by about 5 units for each proton removed, as in the example of phosphoric acid above.

In the case of a diprotic acid, H_2A, the two equilibria are:

$$H_2A \rightleftharpoons HA^- + H^+$$

$$HA^- \rightleftharpoons A^{2-} + H^+$$

it can be seen that the second proton is removed from a negatively charged species. Since the proton carries a positive charge extra work is needed to remove it; that is the cause of the trend noted above. Phosphoric acid values (above) illustrate this rule, as do the values for vanadic acid, H_3VO_4. When an exception to the rule is found it indicates that a major change in structure is occurring. In the case of VO_2^+ (aq), the vanadium is octahedral, 6-coordinate, whereas vanadic acid is tetrahedral, 4-coordinate. This is the basis for an explanation of why $pK_{a1} > pK_{a2}$ for vanadium(V) oxoacids.

Equilibrium	pK_a
$[VO_2(H_2O)_4]^+ \rightleftharpoons H_3VO_4 + H^+ + 2H_2O$	$pK_{a1} = 4.2$
$H_3VO_4 \rightleftharpoons H_2VO_4^- + H^+$	$pK_{a2} = 2.60$
$H_2VO_4^- \rightleftharpoons HVO_4^{2-} + H^+$	$pK_{a3} = 7.92$
$HVO_4^{2-} \rightleftharpoons VO_4^{3-} + H^+$	$pK_{a4} = 13.27$

Isoelectric Point

For substances in solution the isoelectric point is defined as the pH at which the sum, weighted by charge value, of concentrations of positively charged species is equal to the weighted sum of concentrations of negatively charged species. In the case that there is one species of each type, the isoelectric point can be obtained directly from the p*K* values. Take the example of glycine, defined as AH. There are two dissociation equilibria to consider:

$$AH_2^+ \rightleftharpoons AH + H^+; [AH][H^+] = K_1[AH_2^+]$$
$$AH \rightleftharpoons A^- + H^+; [A^-][H^+] = K_2[AH]$$

Substitute the expression for [AH] into the first equation,

$$[A^-][H^+]^2 = K_1K_2[AH_2^+]$$

At the isoelectric point the concentration of the positively charged species, AH_2^+, is equal to the concentration of the negatively charged species, A^-, so

$$[H^+]^2 = K_1K_2$$

Therefore, taking cologarithms, the pH is given by,

$$pI = \frac{pK_1 + pK_2}{2}$$

p*I* values for amino acids are listed at Proteinogenic amino acid#Chemical properties. When more than two charged species are in equilibrium with each other a full speciation calculation may be needed.

Protonation Constants

The dissociation of a monoprotic acid can also be described as the protonation of the conjugate base of the acid

$$A^- + H^+ \rightleftharpoons AH$$

This leads to the definition of an *association* (protonation) constant, denoted here as $K_{association}$, as

$$K_{association} = \frac{[HA]}{[A^-][H^+]}$$

The *dissociation* (deprotonation) constant definition can be written as

$$K_{dissociation} = \frac{[A^-][H^+]}{[HA]}$$

The definitions show that the values of the two constants are reciprocals of each other and,

$$pK_{dissociation} = \log(K_{association})$$

The situation is a little more complicated with polybasic acids. For example, with phosphoric acid,

$$pK_{a1} = \log(K_{association,3})$$

$$pK_{a2} = \log(K_{association,2})$$

$$pK_{a3} = \log(K_{association,1})$$

Amphoteric Substances

An amphoteric substance is one that can act as an acid or as a base, depending on pH. Water is amphoteric. Another example of an amphoteric molecule is the bicarbonate ion HCO_3^- that is the conjugate base of the carbonic acid molecule H_2CO_3 in the equilibrium,

$$H_2CO_3 + H_2O \rightleftharpoons HCO_3^- + H_3O^+$$

but also the conjugate acid of the carbonate ion CO_3^{2-} in (the reverse of) the equilibrium,

$$HCO_3^- + OH^- \rightleftharpoons CO_3^{2-} + H_2O.$$

Carbonic acid equilibria are important for acid–base homeostasis in the human body.

An amino acid is also amphoteric with the added complication that the neutral molecule is subject to an internal acid–base equilibrium in which the basic amino group attracts and binds the proton from the acidic carboxyl group, forming a zwitterion.

$$NH_2CHRCO_2H \rightleftharpoons NH_3^+CHRCO_2^-$$

At pH less than about 5 both the carboxylate group and the amino group are protonated. As pH increases the acid dissociates according to

$$NH_3^+CHRCO_2H \rightleftharpoons NH_3^+CHRCO_2^- + H^+$$

At high pH a second dissociation may take place.

$$NH_3^+CHRCO_2^- \rightleftharpoons NH_2CHRCO_2^- + H^+$$

Thus the zwitterion, $NH_3^+CHRCO_2^-$, is amphoteric because it may either be protonated or deprotonated.

Bases and Basicity

The equilibrium constant K_b for a base is usually defined as the *association* constant for protonation of the base, B, to form the conjugate acid, HB^+.

$$B + H_2O \rightleftharpoons HB^+ + OH^-$$

Using similar reasoning to that used before

$$K_b = \frac{[HB^+][OH^-]}{[B]}$$

$$pK_b = -\log_{10}(K_b)$$

K_b is related to K_a for the conjugate acid. In water, the concentration of the hydroxide ion, $[OH^-]$, is related to the concentration of the hydrogen ion by $K_w = [H^+][OH^-]$, therefore,

$$[OH^-] = \frac{K_w}{[H^+]}$$

Substitution of the expression for $[OH^-]$ into the expression for K_b gives,

$$K_b = \frac{[HB^+]K_w}{[B][H^+]} = \frac{K_w}{K_a}$$

When K_a, K_b and K_w are determined under the same conditions of temperature and ionic strength, it follows, taking cologarithms, that $pK_b = pK_w - pK_a$. In aqueous solutions at 25 °C, pK_w is 13.9965. So,

$$pK_b \approx 14 - pK_a$$

with sufficient accuracy for most practical purposes. In effect there is no need to define pK_b separately from pK_a, but it is done here as often only pK_b values can be found in the older literature.

For metal hydroxides K_b can also be defined as the *dissociation* constant for loss of a hydroxide ion: $B(OH) \rightleftharpoons B^+ + OH^-$ or $B(OH)_2 \rightleftharpoons B(OH)^+ + OH^-$. This is the reciprocal of a stability constant for formation of the complex.

Because the relationship $pK_b = pK_w - pK_a$ holds only in aqueous solutions (though analogous relationships apply for other solvents), subdisciplines of chemistry like organic chemistry that usually deal with nonaqueous solutions generally do not use pK_b as a measure of basicity. Instead, the pK_a of the conjugate acid, denoted by pK_{aH}, is quoted when basicity needs to be quantified. A higher value for pK_{aH} corresponds to a stronger base. For example, the values $pK_{aH}(C_5H_5N) = 5.25$ and $pK_{aH}((CH_3CH_2)_3N) = 10.75$ indicate that triethylamine is a stronger base than pyridine.

Temperature Dependence

All equilibrium constants vary with temperature according to the van 't Hoff equation

$$\frac{d\ln(K)}{dT} = \frac{\Delta H^\ominus}{RT^2}$$

R is the gas constant and T is the absolute temperature. Thus, for exothermic reactions, (the standard enthalpy change, $\Delta H^\ominus$, is negative) K decreases with temperature, but for endothermic reactions ($\Delta H^\ominus$ is positive) K increases with temperature.

The standard enthalpy change for a reaction is itself a function of temperature, according to Kirchhoff's law of thermochemistry:

$$\left(\frac{\partial \Delta H}{\partial T}\right)_p = \Delta C_p$$

where ΔC_p is the heat capacity change at constant pressure. In practice $\Delta H^{\ominus}$ may be taken to be constant over a small temperature range.

Acidity in Nonaqueous Solutions

A solvent will be more likely to promote ionization of a dissolved acidic molecule in the following circumstances:

1. It is a protic solvent, capable of forming hydrogen bonds.
2. It has a high donor number, making it a strong Lewis base.
3. It has a high dielectric constant (relative permittivity), making it a good solvent for ionic species.

pK_a values of organic compounds are often obtained using the aprotic solvents dimethyl sulfoxide (DMSO) and acetonitrile (ACN).

Solvent properties at 25 °C		
Solvent	Donor number	Dielectric constant
Acetonitrile	14	37
Dimethylsulfoxide	30	47
Water	18	78

DMSO is widely used as an alternative to water because it has a lower dielectric constant than water, and is less polar and so dissolves non-polar, hydrophobic substances more easily. It has a measurable pK_a range of about 1 to 30. Acetonitrile is less basic than DMSO, and, so, in general, acids are weaker and bases are stronger in this solvent. Some pK_a values at 25 °C for acetonitrile (ACN) and dimethyl sulfoxide (DMSO) are shown in the tables. Values for water are included for comparison.

pK_a values of acids			
$HA \rightleftharpoons A^- + H^+$	ACN	DMSO	Water
p-Toluenesulfonic acid	8.5	0.9	strong
2, 4-Dinitrophenol	16.66	5.1	3.9
Benzoic acid	21.51	11.1	4.2
Acetic acid	23.51	12.6	4.756
Phenol	29.14	18.0	9.99
$BH^+ \rightleftharpoons B + H^+$	ACN	DMSO	Water
Pyrrolidine	19.56	10.8	11.4
Triethylamine	18.82	9.0	10.72
Proton sponge	18.62	7.5	12.1
Pyridine	12.53	3.4	5.2
Aniline	10.62	3.6	4.6

Ionization of acids is less in an acidic solvent than in water. For example, hydrogen chloride is a weak acid when dissolved in acetic acid. This is because acetic acid is a much weaker base than water.

$$HCl + CH_3CO_2H \rightleftharpoons Cl^- + CH_3C(OH)_2^+$$

acid + base ⇌ conjugate base + conjugate acid

Compare this reaction with what happens when acetic acid is dissolved in the more acidic solvent pure sulfuric acid

$$H_2SO_4 + CH_3CO_2H \rightleftharpoons HSO_4^- + CH_3C(OH)_2^+$$

The unlikely geminal diol species $CH_3C(OH)_2^+$ is stable in these environments. For aqueous solutions the pH scale is the most convenient acidity function. Other acidity functions have been proposed for non-aqueous media, the most notable being the Hammett acidity function, H_o, for superacid media and its modified version H_- for superbasic media.

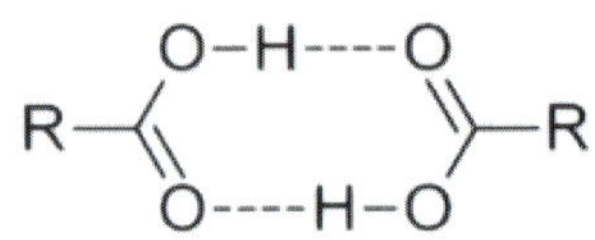

Dimerization of a carboxylic acid.

In aprotic solvents, oligomers, such as the well-known acetic acid dimer, may be formed by hydrogen bonding. An acid may also form hydrogen bonds to its conjugate base. This process, known as homoconjugation, has the effect of enhancing the acidity of acids, lowering their effective pK_a values, by stabilizing the conjugate base. Homoconjugation enhances the proton-donating power of toluenesulfonic acid in acetonitrile solution by a factor of nearly 800. In aqueous solutions, homoconjugation does not occur, because water forms stronger hydrogen bonds to the conjugate base than does the acid.

Mixed Solvents

When a compound has limited solubility in water it is common practice (in the pharmaceutical industry, for example) to determine pK_a values in a solvent mixture such as water/dioxane or water/methanol, in which the compound is more soluble. In the example shown at the right, the pK_a value rises steeply with increasing percentage of dioxane as the dielectric constant of the mixture is decreasing.

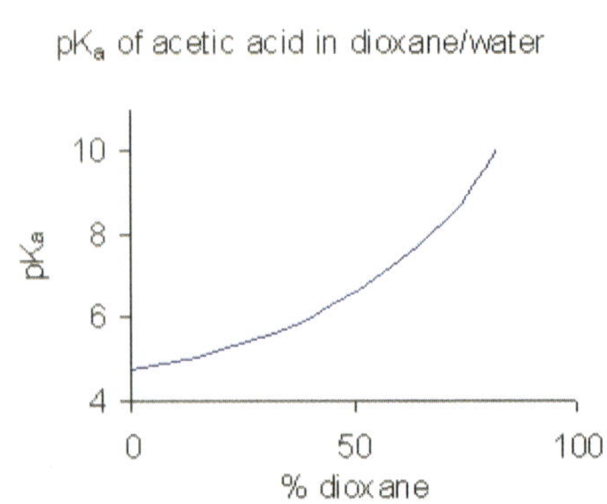

pK_a of acetic acid in dioxane/water mixtures. Data at 25 °C

A pK_a value obtained in a mixed solvent cannot be used directly for aqueous solutions. The reason for this is that when the solvent is in its standard state its activity is *defined* as one. For example, the standard state of water:dioxane mixture with 9:1 mixing ratio is precisely that solvent mixture, with no added solutes. To obtain the pK_a value for use with aqueous solutions it has to be extrapolated to zero co-solvent concentration from values obtained from various co-solvent mixtures.

These facts are obscured by the omission of the solvent from the expression that is normally used to define pK_a, but pK_a values obtained in a *given* mixed solvent can be compared to each other, giving relative acid strengths. The same is true of pK_a values obtained in a particular non-aqueous solvent such a DMSO.

As of 2008, a universal, solvent-independent, scale for acid dissociation constants has not been developed, since there is no known way to compare the standard states of two different solvents.

Factors that Affect pKa Values

Pauling's second rule is that the value of the first pK_a for acids of the formula $XO_m(OH)_n$ depends primarily on the number of oxo groups m, and is approximately independent of the number of hydroxy groups n, and also of the central atom X. Approximate values of pK_a are 8 for $m = 0$, 2 for $m = 1$, −3 for $m = 2$ and < −10 for $m = 3$. Alternatively, various numerical formulas have been proposed including p$K_a = 8 - 5m$ (known as Bell's rule), p$K_a = 7 - 5m$, or p$K_a = 9 - 7m$. The dependence on m correlates with the oxidation state of the central atom, X: the higher the oxidation state the stronger the oxyacid. For example, pK_a for HClO is 7.2, for $HClO_2$ is 2.0, for $HClO_3$ is −1 and $HClO_4$ is a strong acid (p$K_a \ll 0$). The increased acidity on adding an oxo group is due to stabilization of the conjugate base by delocalization of its negative charge over an additional oxygen atom. This rule can help assign molecular structure: for example phosphorous acid (H_3PO_3) has a pK_a near 2 suggested that the structure is $HPO(OH)_2$, as later confirmed by NMR spectroscopy, and not $P(OH)_3$ which would be expected to have a pK_a near 8.

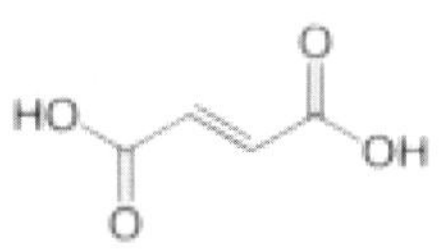

Fumaric acid

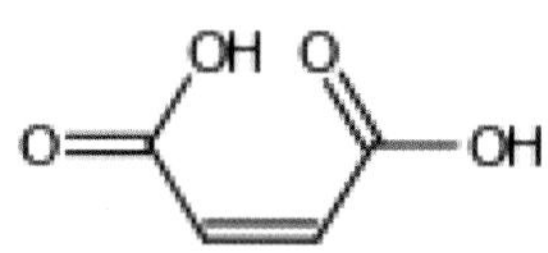

Maleic acid

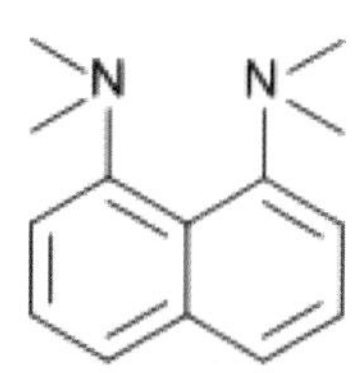

Proton sponge

With organic acids inductive effects and mesomeric effects affect the pK_a values. A simple example is provided by the effect of replacing the hydrogen atoms in acetic acid by the more electronegative chlorine atom. The electron-withdrawing effect of the substituent makes ionisation easier, so successive pK_a values decrease in the series 4.7, 2.8, 1.4, and 0.7 when 0, 1, 2, or 3 chlorine atoms are present. The Hammett equation, provides a general expression for the effect of substituents.

$$\log(K_a) = \log(K_a^0) + \rho\sigma.$$

K_a is the dissociation constant of a substituted compound, K_a^o is the dissociation constant when the substituent is hydrogen, ρ is a property of the unsubstituted compound and σ has a particular value for each substituent. A plot of $\log(K_a)$ against σ is a straight line with intercept $\log(K_a^o)$

and slope ρ. This is an example of a linear free energy relationship as $\log(K_a)$ is proportional to the standard free energy change. Hammett originally formulated the relationship with data from benzoic acid with different substiuents in the *ortho-* and *para-* positions: some numerical values are in Hammett equation. This and other studies allowed substituents to be ordered according to their electron-withdrawing or electron-releasing power, and to distinguish between inductive and mesomeric effects.

Alcohols do not normally behave as acids in water, but the presence of a double bond adjacent to the OH group can substantially decrease the pK_a by the mechanism of keto–enol tautomerism. Ascorbic acid is an example of this effect. The diketone 2, 4-pentanedione (acetylacetone) is also a weak acid because of the keto–enol equilibrium. In aromatic compounds, such as phenol, which have an OH substituent, conjugation with the aromatic ring as a whole greatly increases the stability of the deprotonated form.

Structural effects can also be important. The difference between fumaric acid and maleic acid is a classic example. Fumaric acid is (E)-1,4-but-2-enedioic acid, a *trans* isomer, whereas maleic acid is the corresponding *cis* isomer, i.e. (Z)-1,4-but-2-enedioic acid. Fumaric acid has pK_a values of approximately 3.0 and 4.5. By contrast, maleic acid has pK_a values of approximately 1.5 and 6.5. The reason for this large difference is that when one proton is removed from the *cis* isomer (maleic acid) a strong intramolecular hydrogen bond is formed with the nearby remaining carboxyl group. This favors the formation of the maleate H^+, and it opposes the removal of the second proton from that species. In the *trans* isomer, the two carboxyl groups are always far apart, so hydrogen bonding is not observed.

Proton sponge, 1,8-bis(dimethylamino)naphthalene, has a pK_a value of 12.1. It is one of the strongest amine bases known. The high basicity is attributed to the relief of strain upon protonation and strong internal hydrogen bonding.

Effects of the solvent and solvation should be mentioned also in this topic. It turns out, these influences are more subtle than that of a dielectric medium mentioned above. For example, the expected (by electronic effects of methyl substituents) and observed in gas phase order of basicity of methylamines, $Me_3N > Me_2NH > MeNH_2 > NH_3$, is changed by water to $Me_2NH > MeNH_2 > Me_3N > NH_3$. Neutral methylamine molecules are hydrogen-bonded to water molecules mainly through one acceptor, N–HOH, interaction and only occasionally just one more donor bond, $NH–OH_2$. Hence, methylamines are stabilized to about the same extent by hydration, regardless of the number of methyl groups. In stark contrast, corresponding methylammonium cations always utilize all the available protons for donor $NH–OH_2$ bonding. Relative stabilization of methylammonium ions thus decreases with the number of methyl groups explaining the order of water basicity of methylamines.

Thermodynamics

An equilibrium constant is related to the standard Gibbs energy change for the reaction, so for an acid dissociation constant

$$\Delta G^{\ominus} = -RT \ln K_a \approx 2.303RT \, pK_a .$$

R is the gas constant and T is the absolute temperature. Note that pK_a = $-\log(K_a)$ and 2.303 ≈

ln(10). At 25 °C, $\Delta G^\ominus$ in kJ·mol⁻¹ ≈ 5.708 pK_a (1 kJ·mol⁻¹ = 1000 joules per mole). Free energy is made up of an enthalpy term and an entropy term.

$$\Delta G^\ominus = \Delta H^\ominus - T\Delta S^\ominus$$

The standard enthalpy change can be determined by calorimetry or by using the van 't Hoff equation, though the calorimetric method is preferable. When both the standard enthalpy change and acid dissociation constant have been determined, the standard entropy change is easily calculated from the equation above. In the following table, the entropy terms are calculated from the experimental values of pK_a and $\Delta H^\ominus$. The data were critically selected and refer to 25° C and zero ionic strength, in water.

Acids					
Compound	Equilibrium	pK_a	$\Delta G^\ominus$ (kJ·mol⁻¹)	$\Delta H^\ominus$ (kJ·mol⁻¹)	$-T\Delta S^\ominus$ (kJ·mol⁻¹)
HA = Acetic acid	$HA \rightleftharpoons H^+ + A^-$	4.756	27.147	−0.41	27.56
H_2A^+ = GlycineH$^+$	$H_2A^+ \rightleftharpoons HA + H^+$	2.351	13.420	4.00	9.419
	$HA \rightleftharpoons H^+ + A^-$	9.78	55.825	44.20	11.6
H_2A = Maleic acid	$H_2A \rightleftharpoons HA^- + H^+$	1.92	10.76	1.10	9.85
	$HA^- \rightleftharpoons H^+ + A^{2-}$	6.27	35.79	−3.60	39.4
H_3A = Citric acid	$H_3A \rightleftharpoons H_2A^- + H^+$	3.128	17.855	4.07	13.78
	$H_2A^- \rightleftharpoons HA^{2-} + H^+$	4.76	27.176	2.23	24.9
	$HA^{2-} \rightleftharpoons A^{3-} + H^+$	6.40	36.509	−3.38	39.9
H_3A = Boric acid	$H_3A \rightleftharpoons H_2A^- + H^+$	9.237	52.725	13.80	38.92
H_3A = Phosphoric acid	$H_3A \rightleftharpoons H_2A^- + H^+$	2.148	12.261	−8.00	20.26
	$H_2A^- \rightleftharpoons HA^{2-} + H^+$	7.20	41.087	3.60	37.5
	$HA^{2-} \rightleftharpoons A^{3-} + H^+$	12.35	80.49	16.00	54.49
HA^- = Hydrogen sulfate	$HA^- \rightleftharpoons A^{2-} + H^+$	1.99	11.36	−22.40	33.74
H_2A = Oxalic acid	$H_2A \rightleftharpoons HA^- + H^+$	1.27	7.27	−3.90	11.15
	$HA^- \rightleftharpoons A^{2-} + H^+$	4.266	24.351	−7.00	31.35

1. $\Delta G^\ominus \approx 2.303RTpK_a$
2. Computed here, from ΔH and ΔG values supplied in the citation, using $-T\Delta S^\ominus = \Delta G^\ominus - \Delta H^\ominus$

Conjugate acids of bases				
Compound	Equilibrium	pK_a	$\Delta H^\ominus$ (kJ·mol⁻¹)	$-T\Delta S^\ominus$ (kJ·mol⁻¹)
B = Ammonia	$HB^+ \rightleftharpoons B + H^+$	9.245	51.95	0.8205
B = Methylamine	$HB^+ \rightleftharpoons B + H^+$	10.645	55.34	5.422
B = Triethylamine	$HB^+ \rightleftharpoons B + H^+$	10.72	43.13	18.06

The first point to note is that, when pK_a is positive, the standard free energy change for the dissociation reaction is also positive. Second, some reactions are exothermic and some are endothermic, but, when $\Delta H^\ominus$ is negative $T\Delta S^\ominus$ is the dominant factor, which determines that $\Delta G^\ominus$ is positive. Last, the entropy contribution is always unfavourable ($\Delta S^\ominus < 0$) in these reactions. Ions in aqueous solution tend to orient the surrounding water molecules, which orders the solution and decreases the entropy. The contribution of an ion to the entropy is the partial molar entropy which is often negative, especially for small or highly charged ions. The ionization of a neutral acid involves formation of two ions so that the entropy decreases ($\Delta S^\ominus < 0$). On the second ionization of the same acid, there are now three ions and the anion has a charge, so the entropy again decreases.

Note that the *standard* free energy change for the reaction is for the changes *from* the reactants in their standard states *to* the products in their standard states. The free energy change *at* equilibrium is zero since the chemical potentials of reactants and products are equal at equilibrium.

Experimental Determination

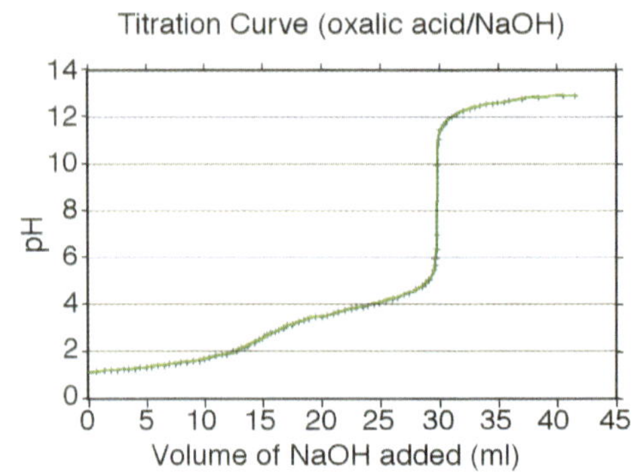

A calculated titration curve of oxalic acid titrated with a solution of sodium hydroxide

The experimental determination of pK_a values is commonly performed by means of titrations, in a medium of high ionic strength and at constant temperature. A solution of the compound in the medium is acidified with a strong acid to the point where the compound is fully protonated. The solution is then titrated with a strong base until all the protons have been removed. At each point in the titration pH is measured using a glass electrode and a pH meter. The equilibrium constants are found by fitting calculated pH values to the observed values, using the method of least squares.

The total volume of added strong base should be small compared to the initial volume of titrand solution in order to keep the ionic strength nearly constant. This will ensure that pK_a remains invariant during the titration.

A calculated titration curve for oxalic acid is shown at the right. Oxalic acid has pK_a values of 1.27 and 4.27. Therefore, the buffer regions will be centered at about pH1.3 and pH4.3. The buffer regions carry the information necessary to get the pK_a values as the concentrations of acid and conjugate base change along a buffer region.

Between the two buffer regions there is an end-point, or equivalence point, at about pH3. This end-point is not sharp and is typical of a diprotic acid whose buffer regions overlap by a small amount: $pK_{a2}-pK_{a1}$ is about three in this example. (If the difference in pK values were about two or less, the end-point would not be noticeable.) The second end-point begins at about pH6.3 and is sharp. This indicates that all the protons have been removed. When this is so, the solution is not buffered and the pH rises steeply on addition of a small amount of strong base. However, the pH does not continue to rise indefinitely. A new buffer region begins at about pH 11 (pK_w-3), which is where self-ionization of water becomes important.

It is very difficult to measure pH values of less than two in aqueous solution with a glass electrode, because the Nernst equation breaks down at such low pH values. To determine pK values of less than about 2 or more than about 11 spectrophotometricor NMR measurements may be used instead of, or combined with, pH measurements.

When the glass electrode cannot be employed, as with non-aqueous solutions, spectrophotometric methods are frequently used. These may involve absorbance or fluorescence measurements. In both cases the measured quantity is assumed to be proportional to the sum of contributions from each photo-active species; with absorbance measurements the Beer-Lambert law is assumed to apply.

Aqueous solutions with normal water cannot be used for ^{1}H NMR measurements but heavy water, D_2O, must be used instead. ^{13}C NMR data, however, can be used with normal water and ^{1}H NMR spectra can be used with non-aqueous media. The quantities measured with NMR are time-averaged chemical shifts, as proton exchange is fast on the NMR time-scale. Other chemical shifts, such as those of ^{31}P can be measured.

Micro-constants

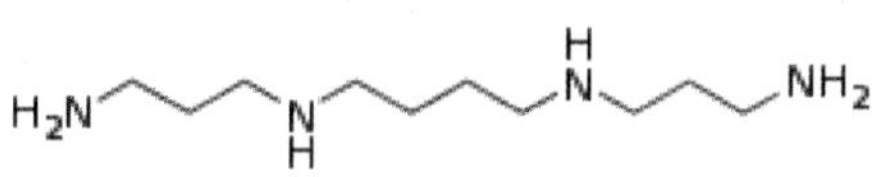

Spermine

A base such as spermine has a few different sites where protonation can occur. In this example the first proton can go on the terminal $-NH_2$ group, or either of the internal –NH– groups. The pK_a values for dissociation of spermine protonated at one or other of the sites are examples of micro-constants. They cannot be determined directly by means of pH, absorbance, fluorescence or NMR measurements. Nevertheless, the site of protonation is very important for biological function, so mathematical methods have been developed for the determination of micro-constants.

Acid Dissociation Constant Table

Acid	Formula	Ka_1	Ka_2	Ka_3
Acetic	CH_3COOH	1.75×10^{-5}		
Ammonium Ion	NH_4^+	5.70×10^{-10}		
Arsenic	H_3AsO_4	5.8×10^{-3}	1.1×10^{-7}	3.2×10^{-12}
Benzoic	C_6H_5COOH	6.28×10^{-5}		
Boric	H_3BO_3	5.81×10^{-10}		
Carbonic	H_2CO_3	4.45×10^{-7}	4.69×10^{-11}	
Citric	$HOOC(OH)C(CH_2COOH)_2$	7.45×10^{-4}	1.73×10^{-5}	4.02×10^{-7}
Formic	HCOOH	1.80×10^{-4}		

Glycolic	$HOCH_2COOH$	1.47 x 10^{-4}		
Hydrazinium Ion	$H_2NNH_3^+$	1.05 x 10^{-8}		
Hydrogen Peroxide	H_2O_2	2.2 x 10^{-12}		
Hydrogen Sulfide	H_2S	9.6 x 10^{-8}	1.3 x 10^{-14}	
Hydroxyl Ammonium Ion	$HONH_3^+$	1.10 x 10^{-6}		
Tartaric acid	$HOOC(CHOH)_2COOH$	9.20 x 10^{-4}	4.31 x 10^{-5}	
Succinic	$HOOCCH_2CH_2COOH$	6.21 x 10^{-5}	2.31 x 10^{-6}	
Iodic acid	HIO_3	1.7 x 10^{-1}		
Phosphoric acid	H_3PO_4	7.11 x 10^{-3}	6.32 x 10^{-8}	4.5 x 10^{-13}
Sulfamic	H_2NSO_3H	1.03 x 10^{-1}		
Lactic acid	$CH_3CHOHCOOH$	1.38 x 10^{-4}		
Maleic acid	cis-HOOCCH:CHCOOH	1.3 x 10-2	5.9 x 10^{-7}	
Nitrous acid	HNO_2	7.1 x 10^{-4}		
Hypochlorous	HOCl	3.0 x 10^{-8}		

Bases

In chemistry, a base is thought of as a substance which can accept protons or any chemical compound that yields hydroxide ions (OH^-) in solution. It is also commonly referred to as any substance that can react with an acid to decrease or neutralize its acidic properties, change the color of indicators (e.g. turn red litmus paper blue), feel slippery to the touch when in solution, taste bitter, react with acids to form salts, and promote certain chemical reactions (e.g. base catalysis). Example of simple bases are sodium hydroxide and ammonia. Sodium hydroxide (NaOH), also known as caustic soda or lye, dissociates in water to form hydroxide ions (OH^-) and sodium ions (Na^+).

Bases have many practical uses, and several of them are commonly found in the home. Household ammonia is a familiar cleaning agent. Lye is used for cleaning clogs and sink drains. Potassium hydroxide, also called caustic potash, is used to make soft soap that dissolves in water with ease. Magnesium hydroxide in water (also called milk of magnesia) is used as an antacid or laxative.

In a basic solution, phenolphthalein has a pink or red color.

Alkali and Base

An alkali was initially thought of as the antithesis of an acid. The formation of salts from the acid and alkali reaction led to the view that salts can be derived from two constituents of opposite natures.

Yet, not all non-acidic constituents possessed alkaline properties. Examples are oxides and hydroxides of heavy metals. Hence, the concept of "base" was born. This concept was first introduced by the French chemist Guillaume François Rouelle in 1754. He noted that acids, which in those days were mostly volatile liquids, such as acetic acid turned into solid salts only when combined with specific substances. These substances formed a concrete *base* for the salt, and hence the name.

Confusion between Base and Alkali

The terms "base" and "alkali" are often used interchangeably, because most common bases are alkalis. It is common to speak of "measuring the alkalinity of soil" when what is actually meant is the measurement of the pH (base property). Similarly, bases which are not alkalis, such as ammonia, are sometimes erroneously referred to as alkaline.

Note that not all or even most salts formed by alkali metals are alkaline; this designation applies only to those salts which are basic.

While most electropositive metal oxides are basic, only the soluble alkali metal and alkaline earth metal oxides can be correctly called alkalis.

This definition of an alkali as a basic salt of an alkali metal or alkaline earth metal does appear to be the most common, based on dictionary definitions, however, conflicting definitions of the term alkali do exist. These include:

- Any base that is water solubleAlkali, Farlex. This is more accurately called an Arrhenius base.
- The solution of a base in water.

Acids and Bases

Acids and bases form complementary pairs, so their definitions need to be considered together. There are three common groups of definitons- the Arrhenius, Brønsted-Lowry, and Lewis definitions, in order of increasing generality:

- Arrhenius: According to this definition, an acid is a substance that increases the concentration of hydronium ion (H_3O^+) when dissolved in water, while bases are substances that increase the concentration of hydroxide ions (OH^-). This definition limits acids and bases to substances that can dissolve in water. Around 1800, many French chemists, including Antoine Lavoisier, incorrectly believed that all acids contained oxygen. Indeed the modern German word for Oxygen is Sauerstoff. Englishchemists, including Sir Humphry Davy at the same time believed all acids contained hydrogen. The Swedish chemist Svante Arrhenius used this belief to develop this definition of acid.
- Brønsted-Lowry: According to this definition, an acid is a proton (hydrogen nucleus) donor

and a base is a proton (hydrogen nucleus) acceptor. The acid is said to be dissociated after the proton is donated. An acid and the corresponding base are referred to as conjugate acid-base pairs. Brønsted and Lowry formulated this definition, which includes water-insoluble substances not in the Arrhenius definition.

- Lewis: According to this definition, an acid is an electron-pair acceptor and a base is an electron-pair donor. (These are frequently referred to as "Lewis acids" and "Lewis bases," and are electrophiles and nucleophiles, respectively, in organic chemistry; Lewis bases are also ligands in coordination chemistry.) Lewis acids include substances with no transferable protons (i.e. H^+hydrogen ions), such as iron(III) chloride, and hence the Lewis definition of an acid has wider application than the Brønsted-Lowry definition. The Lewis definition can also be explained with molecular orbital theory. In general, an acid can receive an electron pair in its lowest unoccupied orbital (LUMO) from the highest occupied orbital (HOMO) of a base. That is, the HOMO from the base and the LUMO from the acid combine to a bonding molecular orbital.

General Properties

Some general properties of bases include:

- Taste: Bitter taste (opposed to sour taste of acids and sweetness of aldehydes and ketones);
- Touch: Slimy or soapy feel on fingers;
- Reactivity:Caustic on organic matter, react violently with acidic or reducible substances;
- Electric conductivity: Aqueous solutions or molten bases dissociate in ions and conduct electricity;
- Litmus test: Bases turn red litmus paper blue.

Chemical Characteristics

Bases Ionization Constant and pH

A general equation can be written for the acceptance of H^+ ions from water by a molecular base, B, to form its conjugate acid, BH^+.

$$B(aq) + H_2O(l) \rightleftharpoons BH^+(aq) + OH^-(aq)$$

$$K_b = \frac{[ConjugateAcid]\cdot[OH^-]}{[ConjugateBase]}$$

Then, $K_b = \dfrac{[BH^+]\cdot[OH^-]}{[BH]}$

The equilibrium constant Kb is also called the Base Ionization Constant. It refers to the reaction in which a base forms its conjugate acid by removing an H^+ ion from water.

The pH of (impure) water is a measure of its acidity. In pure water, about one in ten million molecules dissociate into hydronium ions (H_3O^+) and hydroxide ions (OH^-), according to the following equation:

$$2H_2O(l) \rightleftharpoons H_3O^+(aq) + OH^-(aq)$$

A base accepts (removes) hydronium ions (H_3O^+) from the solution, or donates hydroxide ions (OH^-) to the solution. Both actions will lower the concentration of hydronium ions, and thus raise pH. By contrast, an acid donates H_3O^+ ions to the solution or accepts OH^-, thus lowering pH.

For example, if 1 mole of sodium hydroxide (40 g) is dissolved in 1 litre of water, the concentration of hydroxide ions becomes $[OH^-] = 1$ mol/L. Therefore $[H^+] = 10^{-14}$ mol/L, and $pH = -\log 10^{-14} = 14$.

The basicity constant or pK_b is a measure of basicity and related to the pKa by the simple relationship $pK_a + pK_b = 14$.

Base Strength

A "Strong Base" is one which hydrolyzes completely, deprotonating acids in an acid-base reaction, hence, raising the pH of the solution towards 14. Compounds with a pH of more than about 13 are called strong bases. Strong bases, like strong acids, attack living tissue and cause serious burns. They react differently to skin than acids do so while strong acids are corrosive, we say that strong bases are caustic. Common examples of strong bases are the hydroxides of alkali metals and alkaline earth metals like NaOH and $Ca(OH)_2$. Very strong bases are even able to deprotonate very weakly acidic C-H groups in the absence of water. Superbases are a class of especially basic compounds and harpoon bases are a special class of strong bases with poor nucleophilicity.

Examples of Strong Bases (Hydroxide compounds) in descending strength:

- Potassium hydroxide (KOH);
- Barium hydroxide ($Ba(OH)_2$);
- Cesium hydroxide (CsOH);
- Sodium hydroxide (NaOH);
- Strontium hydroxide ($Sr(OH)_2$);
- Calcium hydroxide ($Ca(OH)_2$);
- Lithium hydroxide (LiOH);
- Rubidium hydroxide (RbOH).

Even stronger bases are:

- Sodium hydride (NaH);
- Lithium diisopropylamide (LDA) ($C_6H_{14}LiN$);
- Sodium amide ($NaNH_2$).

A "Weak Base" is one that does not fully ionize in solution. When a base ionizes, it takes up a hydrogen ion from the water around it, leaving an OH- ion behind. Weak bases have a higher H^+

concentration than strong bases. Weak bases exist in chemical equilibrium in the same way weak acids do. The Base Ionization Constant K_b indicates the strength of the base. Large K_bs belong to stronger bases. The pH of a base is greater than 7, normally up to 14. Common example of a weak base is ammonia, which is used for cleaning.

Examples of Weak Bases:

- Alanine ($C_3H_5O_2NH_2$);
- Ammonia (water) (NH_3 (NH_4OH));
- Dimethylamine ($(CH_3)_2NH$);
- Ethylamine ($C_2H_5NH_2$);
- Glycine ($C_2H_3O_2NH_2$);
- Hydrazine (N_2H_4);
- Methylamine (CH_3NH_2);
- Trimethylamine ($(CH_3)_3N$).

Acid - Base Neutralization

Bases can be thought of as the chemical opposite of acids. A reaction between an acid and base is called neutralization. Bases and acids are seen as opposites because the effect of an acid is to increase the hydronium ion (H_3O^+) concentration in water, where as bases reduce this concentration. Bases react with acids to produce salts and water.

A salts positive ion comes from the base and its negative ion comes from the acid. Considering a metal hydroxide as a base the general reaction is:

$$\underset{\text{acid}}{HX(aq)} + \underset{\text{base}}{MOH(aq)} \rightarrow \underset{\text{salt}}{MX(aq)} + \underset{\text{water}}{HOH(l)}$$

Salts of strong bases and strong acids

A strong acid HCl (hydrochloric acid) reacts with a strong base NaOH (sodium hydroxide) to form NaCl (salt = sodium chloride) and water. If the amounts of the acid and the base are in the correct stoichiometric ratio, then the reaction will undergo complete neutralization where the acid and the base both will lose their respective properties.

$$\underset{\text{strong acid}}{HCL(aq)} + \underset{\text{strong base}}{NaOH(aq)} \rightarrow \underset{\text{salt}}{NaCl(aq)} + \underset{\text{water}}{H_2O(l)}$$

Salts of strong bases and weak acids

A strong base NaOH (sodium hydroxide) added to a weak acid CH_3COOH (acetic acid) in 1L of solution, forming $NaCH_3COO$ (sodium acetate) and water.

$CH_3COOH (aq) + NaOH(aq) \rightarrow NaCH_3COO (aq) + H_2O(l)$

weak acid	weak base	salt	water

Salts of Weak Bases and Strong Acids

Weak bases react with strong acids to form acidic salt solutions. The conjugate acid of the weak base determines its pH. For example, NH_3 (ammonia) is added to HCl (hydrochloric acid) to form NH_4Cl (ammonium chloride).

$NH_3(aq) + HCl(aq) \rightarrow NH_4Cl(aq)$

weak base	strong acid	salt

Salts of Weak Bases and Weak Acids

Salt solutions containing acidic cations and basic anions such as NH_4F (ammonium fluoride) have two possible reactions:

$NH_4^+(aq) + H_2O(l) \leftrightarrow H_3O^+(aq) + NH_3(aq)$ $\quad K_a(NH_4^+) = 5.6 \times 10\text{-}10$

$F^-(aq) + H_2O(l) \leftrightarrow HF(aq) + OH^-(aq)$ $\quad K_b(F^-) = 1.4 \times 10\text{-}11$

Since $K_a(NH_4^+) > K_b(F^-)$, the reaction of ammonia with water is more favorable. Therefore, the resulting solution is slightly acidic.

Alkalis

Alkali Salts

Most basic salts are alkali salts, of which common examples are:

- Sodium hydroxide (often called "caustic soda");
- Potassium hydroxide (commonly called "potash");
- Lye (generic term, for either of the previous two, or even for a mixture);
- Calcium carbonate (sometimes called "free lime");
- Magnesium hydroxide is an example of an atypical alkali: it is a weak base (cannot be detected by phenolphthalein) and it has low solubility in water.

Alkaline Soil

Soil with a pH value higher than 7.4 is normally referred to as alkaline. This soil property can occur naturally, due to the presence of alkali salts. Although some plants do prefer slightly basic soil (including vegetables like cabbage and fodder like buffalo grass), most plants prefer a mildly acidic soil (pH between 6.0 and 6.8), and alkaline soils can cause problems.

Alkali Lakes

In alkali lakes (a type of salt lake), evaporation concentrates the naturally occurring alkali salts, often forming a crust of mildly basic salt across a large area.

Examples of alkali lakes:

- Red berry Lake, Saskatchewan, Canada.
- Tramping Lake, Saskatchewan, Canada.

Alkalinity of Non-hydroxides

Both sodium carbonate and ammonia are bases, although neither of these substances contains OH^- groups. That is because both compounds accept H^+ when dissolved in water:

$$Na_2CO_3 + H_2O \rightarrow 2\,Na^+ + HCO_3^- + OH^-$$

$$NH_3 + H_2O \rightarrow NH_4^+ + OH^-$$

Bases as Heterogeneous Catalysts

Basic substances can be used as insoluble heterogeneous catalysts for chemical reactions. Examples are metal oxides such as magnesium oxide, calcium oxide, and barium oxide as well as potassium fluoride on alumina and some zeolites. A great deal of transition metals make good catalysts, many of which form basic substances. Basic catalysts have been used for hydrogenations, the migration of double bonds, in the Meerwein-Ponndorf-Verlay reduction, the Michael reaction, and many other reactions.

Practical Base Chemistry

In addition to their uses in industries, bases have many applications around the home. Antacids are used to neutralize stomach acidity; gardeners use bases such as Lime (CaO) to make soil more basic. Mild bases are used to clean everything from dishes and clothes to vehicles and the family dog.

Neutralizing Stomach Acidity

An antacid is a base that is used to neutralize excess stomach acid. The recommended dose is the amount of base required to neutralize *some*, but not all, of the stomach acid.

The Acid - Base Chemistry of some Antacids:

Compound	Chemical Formula	Chemical Reaction
Aluminum hydroxide	$Al(OH)_3$	$Al(OH)_3(s) + 3HCl(aq)$ -----> $AlCl_3(aq) + 3H_2O(l)$
Calcium carbonate	$CaCO_3$	$CaCO_3(s) + 2HCl(aq)$ -----> $CaCl_2(aq) + H_2O(l) + CO_2(g)$
Magnesium carbonate	$MgCO_3$	$MgCO_3(s) + 2HCl(aq)$ -----> $MgCl_2(aq) + H_2O(l) + CO_2(g)$

Magnesium hydroxide	$Mg(OH)_2$	$Mg(OH)_2(s) + 2HCl(aq)$ -----> $MgCl_2(aq) + 2H_2O(l)$
Sodium bicarbonate	$NaHCO_3$	$NaHCO_3(aq) + HCl(aq)$ -----> $NaCl(aq) + H_2O(l) + CO_2(g)$

Household Cleaners

Most cleaning compounds such as dishwashing detergents, scouring powders, laundry detergents, and oven cleaners are basic. Many decades ago homemade lye soap was used to clean clothes as well as people's skin. The closest thing to lye soap we see today is dishwashing detergent. Really tough cleaning jobs around the home require chemically aggressive cleaners. Highly basic cleaners are used to get rid of dirt, grease, or stains. Drain and oven cleaners are at the other end of the pH spectrum, having pHs of 12 or higher. They usually contain a strong base such as NaOH that reacts with fats and grease to form a soluble soap. All strongly basic solutions, both in the lab as well as in homes are hazardous and should be handled with caution at all times.

Brønsted–Lowry Acid–base Theory

The Brønsted-Lowry theory describes acid-base interactions in terms of proton transfer between chemical species. A Brønsted-Lowry acid is any species that can donate a proton, H^+, start superscript, plus, end superscript, and a base is any species that can accept a proton. In terms of chemical structure, this means that any Brønsted-Lowry acid must contain a hydrogen that can dissociate as H^+, start superscript, plus, end superscript. In order to accept a proton, a Brønsted-Lowry base must have at least one lone pair of electrons to form a new bond with a proton.

Using the Brønsted-Lowry definition, an acid-base reaction is any reaction in which a proton is transferred from an acid to a base. We can use the Brønsted-Lowry definitions to discuss acid-base reactions in any solvent, as well as those that occur in the gas phase. For example, consider the reaction of ammonia gas, NH_3, (g), with hydrogen chloride gas, $HCl(g)$, to form solid ammonium chloride, $NH_4Cl(s)$:

$$NH_3(g) + HCl(g) \rightarrow NH_4Cl(s)$$

This reaction can also be represented using the Lewis structures of the reactants and products, as seen below:

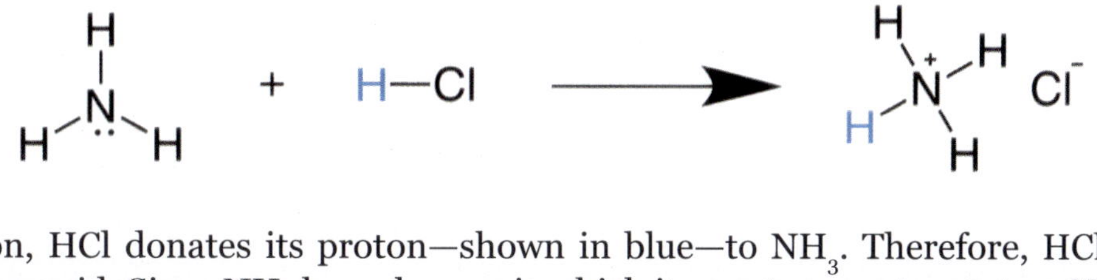

In this reaction, HCl donates its proton—shown in blue—to NH_3. Therefore, HCl is acting as a Brønsted-Lowry acid. Since NH_3 has a lone pair which it uses to accept a proton, NH_3 is a Brønsted-Lowry base.

Note that according to the Arrhenius theory, the above reaction would *not* be an acid-base reaction because neither species is forming H^+ or OH^- in water. However, the chemistry involved -a proton transfer from HCl to NH_3 to form NH_4Cl^- is very similar to what would occur in the aqueous phase.

Identifying Brønsted-lowry Acids and Bases

In the reaction between nitric acid and water, nitric acid, HNO_3 donates a proton—shown in blue—to water, thereby acting as a Brønsted-Lowry acid.

$$HNO_3(aq) + H_2O(l) \rightarrow H_3O^+(aq) + NO_3^-(aq)$$

Since water accepts the proton from nitric acid to form H_3O^+,water acts as a Brønsted-Lowry base. This reaction highly favors the formation of products, so the reaction arrow is drawn only to the right.

Let's now look at a reaction involving ammonia, NH_3, in water:

$$NH_3(aq) + H_2O(l) \rightleftharpoons NH_4^+(aq) + OH^-(aq)$$

In this reaction, water is donating one of its protons to ammonia. After losing a proton, water becomes hydroxide, OH^-. Since water is a proton donor in this reaction, it is acting as a Brønsted-Lowry acid. Ammonia *accepts* a proton from water to form an ammonium ion, NH_4^+. Therefore, ammonia is acting as a Brønsted-Lowry base.

Aqueous Solutions

Acetic acid, a weak acid, donates a proton (hydrogen ion, highlighted in green) to water in an equilibrium reaction to give the acetate ion and the hydronium ion. Red: oxygen, black: carbon, white: hydrogen.

Consider the following acid–base reaction:

$$CH_3COOH + H_2O \rightleftharpoons CH_3COO^- + H_3O^+$$

Acetic acid, CH_3COOH, is an acid because it donates a proton to water (H_2O) and becomes its conjugate base, the acetate ion (CH_3COO^-). H_2O is a base because it accepts a proton from CH_3COOH and becomes its conjugate acid, the hydronium ion, (H_3O^+).

The reverse of an acid-base reaction is also an acid-base reaction, between the conjugate acid of the base in the first reaction and the conjugate base of the acid. In the above example, acetate is the base of the reverse reaction and hydronium ion is the acid.

$$H_3O^+ + CH_3COO^- \rightleftharpoons CH_3COOH + H_2O$$

The power of the Brønsted–Lowry theory is that, in contrast to Arrhenius theory, it does not require an acid to dissociate.

Amphoteric Substances

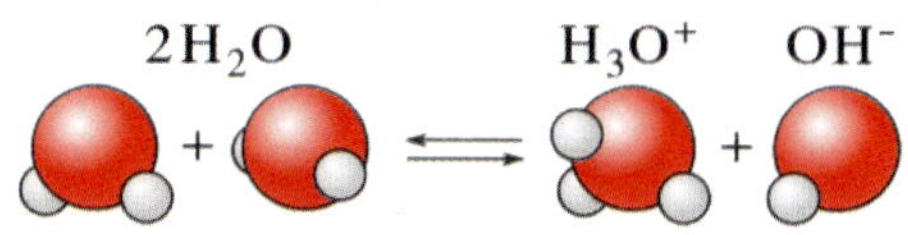

The amphoteric nature of water

The essence of Brønsted–Lowry theory is that an acid only exists as such in relation to a base, and *vice versa*. Water is amphoteric as it can act as an acid or as a base. In the image shown at the right one molecule of H_2O acts as a base and gains H^+ to become H_3O^+ while the other acts as an acid and loses H^+ to become OH^-.

Another example is furnished by substances like aluminium hydroxide, $Al(OH)_3$.

$$Al(OH)_3 + OH^- \rightleftharpoons Al(OH)_4^-\text{, acting as an acid}$$

$$3H^+ + Al(OH)_3 \rightleftharpoons 3H_2O + Al^{3+}(aq)\text{, acting as a base}$$

Non-aqueous Solutions

The hydrogen ion, or hydronium ion, is a Brønsted–Lowry acid in aqueous solutions, and the hydroxide ion is a base, by virtue of the self-dissociation reaction

$$H_2O + H_2O \rightleftharpoons H_3O^+ + OH^-$$

An analogous reaction occurs in liquid ammonia

$$NH_3 + NH_3 \rightleftharpoons NH_4^+ + NH_2^-$$

Thus, the ammonium ion, NH_4^+, plays the same role in liquid ammonia as does the hydronium ion in water and the amide ion, NH , is analogous to the hydroxide ion. Ammonium salts behave as acids, and amides behave as bases.

Some non-aqueous solvents can behave as bases, that is, proton acceptors, in relation to Brønsted–Lowry acids.

$$HA + S \rightleftharpoons A^- + SH^+$$

where S stands for a solvent molecule. The most important such solvents are dimethylsulfoxide, DMSO, and acetonitrile, CH_3CN, as these solvents has been widely used to measure the acid dissociation constants of organic molecules. Because DMSO is a stronger proton acceptor than H_2O the acid becomes a stronger acid in this solvent than in water. Indeed, many molecules behave as acids in non-aqueous solution that do not do so in aqueous solution. An extreme case occurs with carbon acids, where a proton is extracted from a C-H bond.

Some non-aqueous solvents can behave as acids. An acidic solvent will increase basicity of substances dissolved in it. For example, the compound CH_3COOH is known as acetic acid because of its acidic behaviour in water. However it behaves as a base in liquid hydrogen chloride, a much more acidic solvent.

$$HCl + CH_3COOH \rightleftharpoons Cl^- + CH_3C(OH)_2^+$$

Comparison with Lewis Acid–base Theory

In the same year that Brønsted and Lowry published their theory, G. N. Lewis proposed an alternative theory of acid–base reactions. The Lewis theory is based on electronic structure. A Lewis base is defined as a compound that can donate an electron pair to a Lewis acid, a compound that can accept an electron pair. Lewis's proposal gives an explanation to the Brønsted–Lowry classification in terms of electronic structure.

$$HA + B: \rightleftharpoons A:^- + BH^+$$

In this representation both the base B, and the conjugate base A^-, are shown carrying a lone pair of electrons and the proton, which is a Lewis acid, is transferred between them.

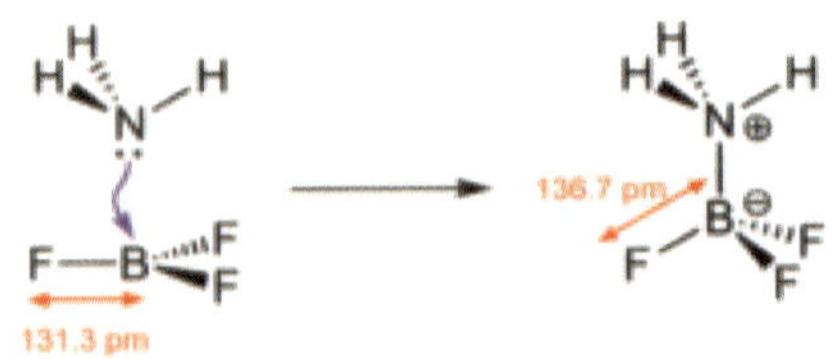

Adduct of ammonia and boron trifluoride

Lewis later wrote in "To restrict the group of acids to those substances that contain hydrogen interferes as seriously with the systematic understanding of chemistry as would the restriction of the term oxidizing agent to substances containing oxygen." In Lewis theory an acid, A, and a base, B:, form an adduct, AB, in which the electron pair is used to form a dative covalent bond between A and B. This is illustrated with the formation of the adduct $H_3N–BF_3$ from ammonia and boron trifluoride, a reaction that cannot occur in aqueous solution because boron trifluoride reacts violently with water in a hydrolysis reaction.

$$BF_3 + 3H_2O \rightarrow B(OH)_3 + 3HF$$

$$HF \rightleftharpoons H^+ + F^-$$

These reactions illustrate that BF_3 is an acid in both Lewis and Brønsted–Lowry classifications and emphasizes the consistency between both theories.

Boric acid is recognized as a Lewis acid by virtue of the reaction

$$B(OH)_3 + H_2O \rightleftharpoons B(OH)_4^- + H^+$$

In this case the acid does not dissociate, it is the base, H_2O that dissociates. A solution of $B(OH)_3$ is acidic because hydrogen ions are liberated in this reaction.

There is strong evidence that dilute aqueous solutions of ammonia contain negligible amounts of the ammonium ion

$$H_2O + NH_3 \rightarrow OH^- + NH_4^+$$

and that, when dissolved in water, ammonia functions as a Lewis base.

Comparison with the Lux–Flood Theory

The reactions between certain oxides in non-aqueous media cannot be explained on the basis of Brønsted–Lowry theory. For example, the reaction

$$2MgO + SiO_2 \rightarrow Mg_2SiO_4$$

does not fall within the scope of the Brønsted–Lowry definition of acids and bases. On the other hand, MgO is basic and SiO_2 is acidic in the Brønsted–Lowry sense, referring to mixtures in water.

$$2H^+ + MgO(s) \rightarrow Mg^{2+}(aq) + 2H_2O$$

$$SiO_2(s) + 2H_2O \rightarrow SiO_4^{4-} + 4H^+ (\equiv Si(OH)_4(aq))$$

Lux–Flood theory also classifies magnesium oxide as a base in non-aqueous circumstances. This classification is important in geochemistry. Minerals such as olivine, (Mg,Fe)SiO_4 are classed as ultramafic; olivine is a compound of a very basic oxide, MgO, with an acidic oxide, silica, SiO_2.

Lewis Acids and Bases

In 1923, G. N. Lewis proposed a generalized definition of acid-base behavior in which acids and bases are identified by their ability to accept or to donate a pair of electrons and form a coordinate covalent bond.

A coordinate covalent bond (or dative bond) occurs when one of the atoms in the bond provides both bonding electrons. For example, a coordinate covalent bond occurs when a water molecule combines with a hydrogen ion to form a hydronium ion. A coordinate covalent bond also results when an ammonia molecule combines with a hydrogen ion to form an ammonium ion. Both of these equations are shown here:

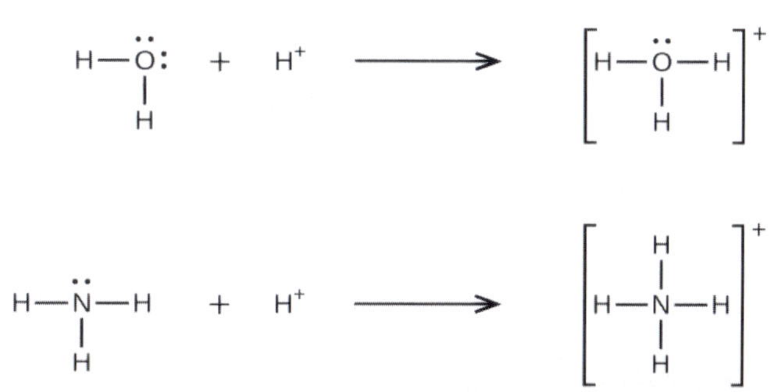

A Lewis acid is any species (molecule or ion) that can accept a pair of electrons, and a Lewis base is any species (molecule or ion) that can donate a pair of electrons.

A Lewis acid-base reaction occurs when a base donates a pair of electrons to an acid. A Lewis acid-base adduct, a compound that contains a coordinate covalent bond between the Lewis acid and the Lewis base, is formed. The following equations illustrate the general application of the Lewis concept.

The boron atom in boron trifluoride, BF_3, has only six electrons in its valence shell. Being short of the preferred octet, BF_3 is a very good Lewis acid and reacts with many Lewis bases; a fluoride ion is the Lewis base in this reaction, donating one of its lone pairs:

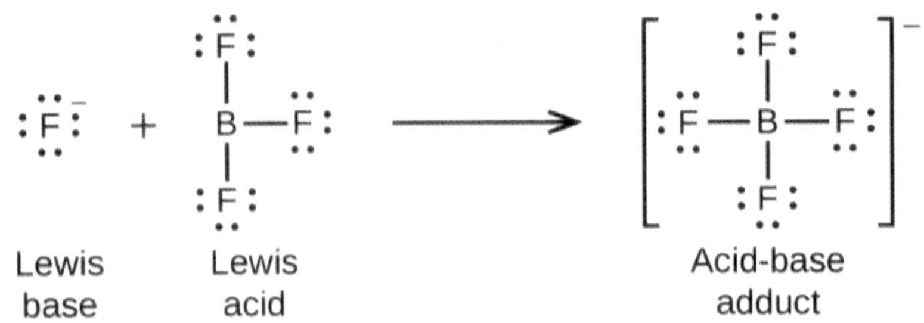

In reactions, each of two ammonia molecules, Lewis bases, donates a pair of electrons to a silver ion, the Lewis acid:

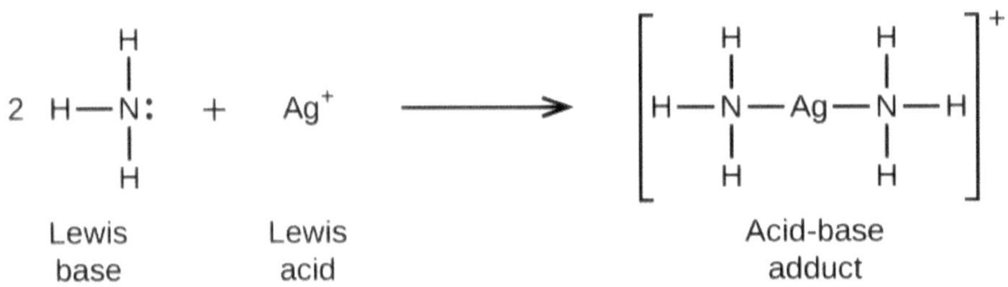

Nonmetal oxides act as Lewis acids and react with oxide ions, Lewis bases, to form oxyanions:

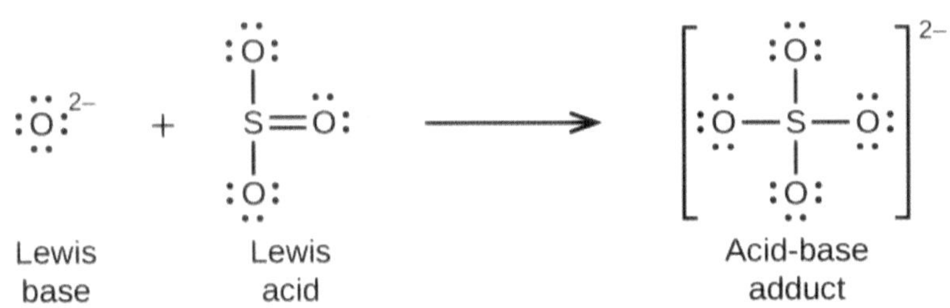

Many Lewis acid-base reactions are displacement reactions in which one Lewis base displaces another Lewis base from an acid-base adduct, or in which one Lewis acid displaces another Lewis acid:

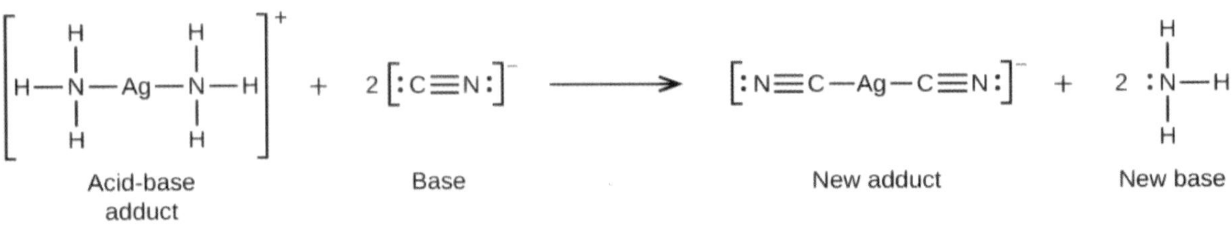

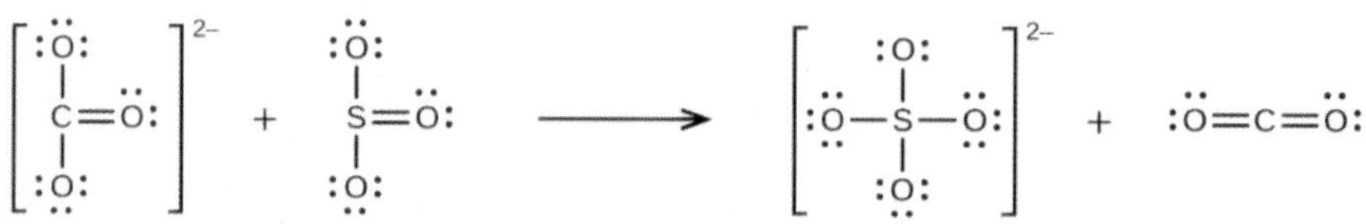

Acid-base adduct | Acid | New adduct | New acid

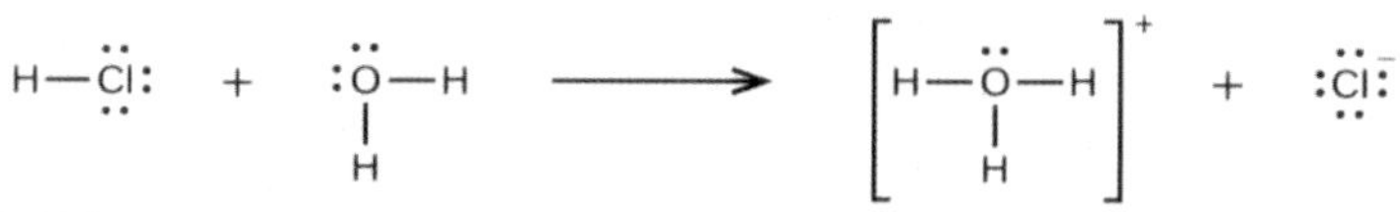

Acid-base adduct | Base | New adduct | New base

The last displacement reaction shows how the reaction of a Brønsted-Lowry acid with a base fits into the Lewis concept. A Brønsted-Lowry acid such as HCl is an acid-base adduct according to the Lewis concept, and proton transfer occurs because a more stable acid-base adduct is formed. Thus, although the definitions of acids and bases in the two theories are quite different, the theories overlap considerably.

Many slightly soluble ionic solids dissolve when the concentration of the metal ion in solution is decreased through the formation of complex (polyatomic) ions in a Lewis acid-base reaction. For example, silver chloride dissolves in a solution of ammonia because the silver ion reacts with ammonia to form the complex ion $Ag(NH_3)_2^+$. The Lewis structure of the $Ag(NH_3)_2^+$ ion is:

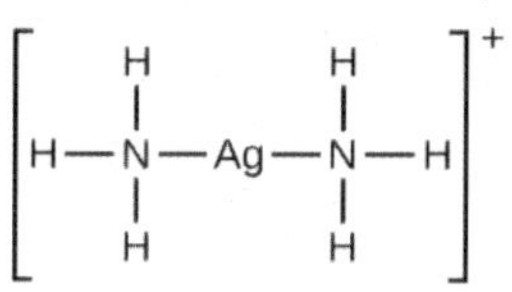

The equations for the dissolution of AgCl in a solution of NH_3 are:

$$AgCl(s) \rightarrow Ag^+(aq) + Cl^-(aq)$$
$$Ag^+(aq) + 2NH_3(aq) \rightarrow Ag(NH_3)_2^+(aq)$$
$$\text{Net}: AgCl(s) + 2NH_3(aq) \rightarrow Ag(NH_3)_2^+(aq) + Cl^-(aq)$$

Aluminum hydroxide dissolves in a solution of sodium hydroxide or another strong base because of the formation of the complex ion $Al(OH)_4^-$. The Lewis structure of the $Al(OH)_4^-$ ion is:

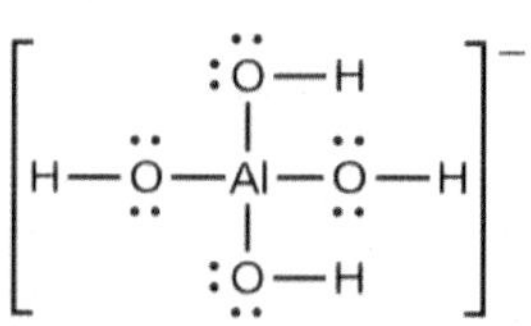

The equations for the dissolution are:

$$Al(OH)_3(s) \rightarrow Al^{3+}(aq) + 3OH^-(aq)$$
$$Al^{3+}(aq) + 4OH^-(aq) \rightarrow Al(OH)_4^-(aq)$$
$$\text{Net}: Al(OH)_3(s) + OH^-(aq) \rightarrow Al(OH)_4^-(aq)$$

Mercury(II) sulfide dissolves in a solution of sodium sulfide because HgS reacts with the S^{2-} ion:

$$HgS(s) \rightarrow Hg^{2+}(aq) + S^{2-}(aq)$$
$$Hg^{2+}(aq) + 2S^{2-}(aq) \rightarrow HgS_2^{2-}(aq)$$
$$\text{Net}: HgS(s) + S^{2-}(aq) \rightarrow HgS_2^{2-}(aq)$$

A complex ion consists of a central atom, typically a transition metal cation, surrounded by ions, or molecules called ligands. These ligands can be neutral molecules like H_2O or NH_3, or ions such as CN^- or OH^-. Often, the ligands act as Lewis bases, donating a pair of electrons to the central atom. The ligands aggregate themselves around the central atom, creating a new ion with a charge equal to the sum of the charges and, most often, a transitional metal ion. This more complex arrangement is why the resulting ion is called a *complex ion*. The complex ion formed in these reactions cannot be predicted; it must be determined experimentally. The types of bonds formed in complex ions are called coordinate covalent bonds, as electrons from the ligands are being shared with the central atom. Because of this, complex ions are sometimes referred to as coordination complexes. This will be studied further in upcoming chapters.

The equilibrium constant for the reaction of the components of a complex ion to form the complex ion in solution is called a formation constant (K_f) (sometimes called a stability constant). For example, the complex ion $Cu(CN)_2^-$ is shown here:

$$:N \equiv C - Cu - C \equiv N:$$

It forms by the reaction:

$$Cu^+(aq) + 2CN^-(aq) \rightleftharpoons Cu(CN)_2^-(aq)$$

At equilibrium:

$$K_f = Q = \frac{[Cu(CN)_2^-]}{[Cu^+][CN^-]^2}$$

The inverse of the formation constant is the dissociation constant (K_d), the equilibrium constant for the *decomposition* of a complex ion into its components in solution.

In general, the larger the formation constant, the more stable the complex; however, as in the case of K_{sp} values, the stoichiometry of the compound must be considered.

Substance	K_f at 25 °C
$[Cd(CN)\ ]^-$	3×10^{18}
$Ag(NH_3)_2^+$	1.7×10^{7}
$[AlF_6]^{3-}$	7×10^{19}

Table: Common Complex Ions by Decreasing Formulation Constants

As an example of dissolution by complex ion formation, let us consider what happens when we add aqueous ammonia to a mixture of silver chloride and water. Silver chloride dissolves slightly in water, giving a small concentration of Ag^+ ($[Ag^+] = 1.3 \times 10^{-5} M$):

$$AgCl(s) \rightleftharpoons Ag^+(aq) + Cl^-(aq)$$

However, if NH_3 is present in the water, the complex ion, $Ag(NH_3)_2^+$, can form according to the equation:

$$Ag^+(aq) + 2NH_3(aq) \rightleftharpoons Ag(NH_3)_2^+(aq)$$

with

$$K_f = \frac{[Ag(NH_3)_2^+]}{[Ag^+][NH_3]^2} = 1.7 \times 10^7$$

The large size of this formation constant indicates that most of the free silver ions produced by the dissolution of AgCl combine with NH_3 to form $Ag(NH_3)_2^+$. As a consequence, the concentration of silver ions, $[Ag^+]$, is reduced, and the reaction quotient for the dissolution of silver chloride, $[Ag^+][Cl^-]$, falls below the solubility product of AgCl:

$$Q = [Ag^+][Cl^-] < K_{sp}$$

More silver chloride then dissolves. If the concentration of ammonia is great enough, all of the silver chloride dissolves.

Example: Dissociation of a Complex Ion

Calculate the concentration of the silver ion in a solution that initially is 0.10 M with respect to $Ag(NH_3)_2^+$.

Solution,

We use the familiar path to solve this problem:

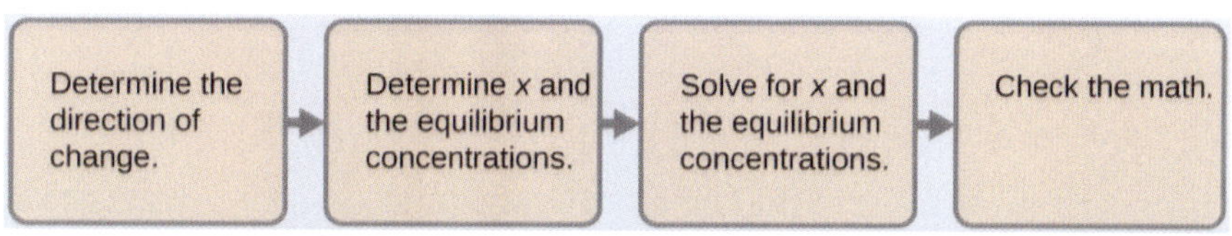

Determine the direction of change. The complex ion $Ag(NH_3)_2^+$ is in equilibrium with its components, as represented by the equation:

$$Ag^+(aq) + 2NH_3(aq) \rightleftharpoons Ag(NH_3)_2^+(aq)$$

1. We write the equilibrium as a formation reaction because Appendix K lists formation constants for complex ions. Before equilibrium, the reaction quotient is larger than the equilibrium constant [$K_f = 1.7 \times 10^7$, and $Q = \frac{0.10}{0 \times 0}$, it is infinitely large], so the reaction shifts to the left to reach equilibrium.

2. *Determine* x *and equilibrium concentrations.* We let the change in concentration of Ag^+ be x. Dissociation of 1 mol of $Ag(NH_3)_2^+$ gives 1 mol of Ag^+ and 2 mol of NH_3, so the change in $[NH_3]$ is $2x$ and that of $Ag(NH_3)_2^+$ is $-x$.

3. *Solve for x and the equilibrium concentrations,* at equilibrium:

$$K_f = \frac{[Ag(NH_3)_2^+]}{[Ag^+][NH_3]^2}$$

$$1.7 \times 10^7 = \frac{0.10 - x}{(x)(2x)^2}$$

Both Q and K_f are much larger than 1, so let us assume that the changes in concentrations needed to reach equilibrium are small. Thus $0.10 - x$ is approximated as 0.10:

$$1.7 \times 10^7 = \frac{0.10 - x}{(x)(2x)^2}$$

$$x^3 = \frac{0.10}{4(1.7 \times 10^7)} = 1.5 \times 10^{-9}$$

$$x = \sqrt[3]{1.5 \times 10^{-9}} = 1.1 \times 10^{-3}$$

Because only 1.1% of the $Ag(NH_3)_2^+$ dissociates into Ag^+ and NH_3, the assumption that x is small is justified.

Now we determine the equilibrium concentrations:

$$[Ag^+] = 0 + x = 1.1 \times 10^{-3}\ M$$

$$[NH_3] = 0 + 2x = 2.2 \times 10^{-3}\ M$$

$$[Ag(NH_3)_2^+] = 0.10 - x = 0.10 - 0.0011 = 0.099$$

The concentration of free silver ion in the solution is 0.0011 M.

4. *Check the work.* The value of Q calculated using the equilibrium concentrations is equal to K_f within the error associated with the significant figures in the calculation.

References

- Kotz, John C.; Treichel, Paul M.; Townsend, John; Treichel, David (2014-01-24). Chemistry & Chemical Reactivity. Cengage Learning. ISBN 9781305176461
- Fraczkiewicz R (2013). "In Silico Prediction of Ionization". In Reedijk J. Reference Module in Chemistry, Molecular Sciences and Chemical Engineering [Online]. (secondary). vol. 5. Amsterdam, The Netherlands: Elsevier. doi:10.1016/B978-0-12-409547-2.02610-X
- Acid, science: britannica.com, Retrieved 12 July 2018
- "Project: Ionic Strength Corrections for Stability Constants". International Union of Pure and Applied Chemistry. Archived from the original on 29 October 2008. Retrieved 2008-11-23

- Hall, Norris F. (March 1940). "Systems of Acids and Bases". Journal of Chemical Education. 17 (3): 124–128. Bibcode:1940JChEd..17..124H. doi:10.1021/ed017p124
- Acid-dissociation-constant, inorganic-chemistry: tutorvista.com, Retrieved 19 May 2018
- Lehninger, Albert L.; Nelson, David L.; Cox, Michael M. (2005-01-01). Lehninger Principles of Biochemistry. Macmillan. ISBN 9780716743392
- Mehta, Akul. "Henderson–Hasselbalch Equation: Derivation of pK_a and pK_b". PharmaXChange. Retrieved 16 November 2014
- Chemistry, Lewis-acids-and-bases-15-2: opentextbc.ca, Retrieved 31 March 2018
- Lowry, T. M. (1923). "The uniqueness of hydrogen". Journal of the Society of Chemical Industry. 42 (3): 43–47. doi:10.1002/jctb.5000420302

Permissions

All chapters in this book are published with permission under the Creative Commons Attribution Share Alike License or equivalent. Every chapter published in this book has been scrutinized by our experts. Their significance has been extensively debated. The topics covered herein carry significant information for a comprehensive understanding. They may even be implemented as practical applications or may be referred to as a beginning point for further studies.

We would like to thank the editorial team for lending their expertise to make the book truly unique. They have played a crucial role in the development of this book. Without their invaluable contributions this book wouldn't have been possible. They have made vital efforts to compile up to date information on the varied aspects of this subject to make this book a valuable addition to the collection of many professionals and students.

This book was conceptualized with the vision of imparting up-to-date and integrated information in this field. To ensure the same, a matchless editorial board was set up. Every individual on the board went through rigorous rounds of assessment to prove their worth. After which they invested a large part of their time researching and compiling the most relevant data for our readers.

The editorial board has been involved in producing this book since its inception. They have spent rigorous hours researching and exploring the diverse topics which have resulted in the successful publishing of this book. They have passed on their knowledge of decades through this book. To expedite this challenging task, the publisher supported the team at every step. A small team of assistant editors was also appointed to further simplify the editing procedure and attain best results for the readers.

Apart from the editorial board, the designing team has also invested a significant amount of their time in understanding the subject and creating the most relevant covers. They scrutinized every image to scout for the most suitable representation of the subject and create an appropriate cover for the book.

The publishing team has been an ardent support to the editorial, designing and production team. Their endless efforts to recruit the best for this project, has resulted in the accomplishment of this book. They are a veteran in the field of academics and their pool of knowledge is as vast as their experience in printing. Their expertise and guidance has proved useful at every step. Their uncompromising quality standards have made this book an exceptional effort. Their encouragement from time to time has been an inspiration for everyone.

The publisher and the editorial board hope that this book will prove to be a valuable piece of knowledge for students, practitioners and scholars across the globe.

Index